LUNAR
IMPACT

LUNAR IMPACT

The  History of PROJECT RANGER

R. CARGILL HALL

INTRODUCTION TO THE DOVER EDITION BY
PAUL DICKSON

DOVER PUBLICATIONS, INC.
MINEOLA, NEW YORK

To Tigger, Melanie, and Brad

Bibliographical Note

This Dover edition, first published in 2010, is an unabridged republication of
the work originally published in the NASA History Series (NASA SP-4210),
Washington, D.C., in 1977. A new Introduction by Paul Dickson has been specially prepared for this edition.

International Standard Book Number

ISBN-13: 978-0-486-47757-2
ISBN-10: 0-486-47757-6

Manufactured in the United States by Courier Corporation
47757601
www.doverpublications.com

INTRODUCTION TO THE DOVER EDITION

In the early days of the Space Age, Ranger was the name given to a series of probes designed to gather data about the moon. NASA initiated Project Ranger, then unnamed, in December 1959, when it asked the Jet Propulsion Laboratory (JPL) to study spacecraft design and a mission to "acquire and transmit a number of images of the lunar surface." This was to be accomplished by probes that would televise images of the moon back to earth in the moments before the probes slammed into the lunar surface.

In February 1960, JPL Director Dr. William H. Pickering recommended that NASA Headquarters approve the name JPL was using for the project, Ranger. The name had been introduced by the JPL program director, Clifford D. Cummings, who had noticed while on a camping trip that his Ford pickup truck was called Ranger. Cummings liked the name and, because it referred to "land exploration activities," suggested it as a name for the lunar impact probe. By May 1960 it was in common use.

Giving a name to the probes had one drawback in that the name Ranger became a synonym for failure, as the first six probes were unsuccessful. Some were total duds and there was at least one spectacular near miss—Ranger 6 crashed out of sight of the earth on the far side of the moon. But then came Ranger 7 which took the first pictures of the moon by a U.S. spacecraft on July 31, 1964 about 17 minutes before impacting the lunar surface during which time it sent back many more clear images. Then came Rangers 8 and 9, which returned even more high-quality images of the moon. The United States had made the first step in getting a close view of the moon with an eye to human exploration. Ranger found at least two spots on the lunar surface which were level enough and seemed appropriate for a landing by a crewed spacecraft but televised images, no matter how clear, could not determine if the surface would support such a effort—the question of whether the surface was hard enough would have to be answered by later soft-landers.

In *Lunar Impact: The NASA History of Project Ranger,* R. Cargill Hall demonstrates why he is regarded as one of the very best of America's aerospace historians. Among other positions he served as a historian at the headquarters of the Strategic Air Command and as the NASA historian at Caltech's Jet Propulsion Laboratory where the Ranger program was managed. It is to the everlasting credit of Hall and NASA that this story is told so frankly and unflinchingly because much of it is not pretty. As Ranger falters careers are derailed, bitter disputes erupt, Congress intervenes with budget cuts and morale plummets. Murphy's Law is in full effect for the first chapters of this story of serial failure.

The project spanned the years 1959 to 1965 and cost $267 million. There were those who questioned whether a project that did little more than yield close-up television pictures of the moon was worth it in terms of the time, money, and careers claimed. Hall's narrative leads the reader to the conclusion that it was an eventual success and deserves major claim to progress in terms of early lunar reconnaissance. He also concluded that the Ranger elements of spacecraft, communications network, and managerial techniques combined with those first images made it an essential prototype for the nation's future instrumented exploration of the deep reaches of outer space. He also builds a case for the Ranger project accelerating the pace of planetary science by making remote visual imaging a basic exploratory tool of that science. Ranger became "an accepted antecedent" to the planning of further experiments in outer space.

The value of *Lunar Impact* is that it gives a deep insight into the realities of a large project guided by the process of trial and error at a time we could not imagine we would ever be routinely sending highly instrumented robotic rovers to distant places in the solar system.

PAUL DICKSON
February 2010

FOREWORD

R. Cargill Hall has written a history. Readers not familiar with the state of writing about twentieth century technology and science may not realize his achievement. Accounts—so-called histories—of recent technology and science are often little more than simplistic narratives focusing almost entirely upon sequences in hardware development or upon scientific idea explication. In commendable contrast, Hall organized a coherent narrative and analysis of complex institutions, people, ideas, and machines changing in character and in relationship one to another over time.

His history of the Ranger Project is also critical and mature. He avoided neither complexity and contradiction nor reasoned analysis and judgments about episodes and people. He allowed for accident, unintended consequences, shifting priorities, budgetary adjustments, and over-determined events. This is evidenced by a frank account of six superficially ignominious Ranger failures, an analysis of the effects of NASA management by committee, an appraisal of the impact of high-priority Project Apollo upon Ranger, and a consideration of the consequences of Ranger's being done at the Jet Propulsion Laboratory, a university rather than an industrial laboratory. He comprehended how these and other factors generally influenced the project and shaped the automatic machine, the exploring spacecraft, at its hard core.

Lunar Impact transforms the records of a technological project into history by applying the canons of historical scholarship. The techniques and modes of interpretation are the general historian's. Because of this, Ranger emerges from the study not simply as a machine designed and operated by technical specialists but, more complexly and convincingly, as the focus of activities resulting from the conflicting interests, the power struggles, and the contrasting objectives of individuals, groups, and institutions. Viewed in this way, the writing of the history of the Ranger project becomes a challenge similar to the writing of the history of political campaigns and business enterprises.

The reader will recognize elements common to many kinds of history; he or she may also note the development of themes often encountered in large-scale technological and scientific projects. Most obvious is the tension between the values and goals of science and of engineering. Throughout Hall's history one encounters scientists striving to shape Ranger so that it could perform a number of complex scientific experiments; the reader also meets engineers endeavoring to design a machine realistically contrived to perform one or two priority tasks like photographing the surface of the moon. Another significant theme concerns the tension permeating the Jet Propulsion Laboratory between the academic spirit of free enquiry and loose disciplinary structure, and the industrial laboratory style of project-oriented organization and highly directed problem-solving. These tensions were severe, sometimes constructive and at other times frustrating. Hall succeeds in seeing the situations in the perspectives of the various principals and principles.

Hall's book is also unusual and interesting because it reckons, as noted, with failure, the frequently ignominious inadequacy of the early machines, launches, and operations. Hall absorbed enough of the wisdom of experienced managers and engineers not to be startled or shocked into rash pronouncements and value judgments about failure. He learned that early Ranger was a high-risk endeavor, venturing into treacherous waters in which a number of careers foundered. In writing of failure, Hall had an advantage over his counterparts doing political histories: he did not have to contend with the host of simplistic myths that accumulate about the sites of political disasters. As yet, technology and science affairs are not easily enough comprehended to attract the sensation seeker and the simplifier.

On the other hand, Hall, unlike the political historian, did have the formidable problem of making complex technology comprehensible to the generally-informed reader. Wisely and considerately, he wrote about technical and scientific matters with his professed audience in mind—historians, interested laymen, and managers; he resisted the temptation to write for the highly sensitive, deeply involved, and specialized readers of the NASA "comment cycle." Also his technical and scientific information is related by him to general themes. The personalities of salient scientists, engineers, and managers are delineated insofar as these influenced the character of the technological artifacts and the scientific ideas.

Finally, the agency sponsoring this study should be commended. A large publicly-funded administrative enterprise such as NASA understandably gravitates toward sustaining public relations and strives for favorable public opinion. Hall's book emerged from the NASA matrix surprisingly free of the constraining influences. The engineers, scientists, and managers at NASA and at the Jet Propulsion Laboratory do not appear in this book as cutouts; they do emerge as believable three-dimensional men involved in an extremely interesting and significant episode in recent history.

Thomas P. Hughes
University of Pennsylvania

CONTENTS

INTRODUCTION TO THE DOVER EDITION iii

Part I. THE ORIGINAL RANGER

1. THE ORIGINS OF RANGER 3
   A Moon Flight Proposal 3
   The First Lunar Flights 6
   Sky and Planetary Science 10
   From Interest to Resolve 14

2. ORGANIZING THE CAMPAIGN 25
   Prospects and Suggestions 25
   Assigning Tasks to People 26
   New Mechanisms for Management 31
   Organizing for Ranger: Huntsville and the Air Force 32
   Organizing for Ranger: JPL and Headquarters 34
   Launch Vehicles: Management Problems Materialize 39
   Reorganizing for Ranger 43

3. SPACE SCIENCE AND THE RANGER MACHINE 46
   A Planetary Machine for Space Science 46
   The Vega-Ranger: Where Planet and Sky Science Meet 49
   Creating the Ranger Machine 54

4. RANGER'S LUNAR OBJECTIVES IN DOUBT 63
   A Difference in Weights and Measures 63
   Preparing for the Test Flights 67
   When and How to Sterilize Spacecraft 71
   Space Science and the Original Ranger Missions 73

5. SUPPORTING THE FLIGHT OPERATIONS 81
   The Deep Space Network 81
   Space Flight Operations 88
   Launch Operations 91

6. TEST FLIGHTS AND DISAPPOINTMENTS 94
   Planning the Ascent 94
   Preparing for Launch 97
   A Learning Experience 99
   The Second Lesson 105
   The Aftermath 109

7. A NEW NATIONAL GOAL 112
    Man on the Moon 112
    Rangers for Apollo 114
    The New Order 120

8. THE QUESTION OF SCIENCE AND RANGER 124
    Lunar Spacecraft Development: Sterilization in Practice 124
    After the Apollo Decision: What Science and Where? 128
    Science Reasserted in Project Ranger 130
    Modifying Ranger for More Space Science 134

9. LUNAR EXPLORATION BEGUN 138
    Preparing to Go 138
    A First Chance at the Moon 143
    Reflections on a Near Miss 147
    Another Chance 150

10. WHICH WAY RANGER? 156
    Family Relations 156
    NASA's Lunar Objectives Reconsidered 160
    One More Time 163

11. IN THE COLD LIGHT OF DAWN 171
    The Ranger Inquiry at JPL 171
    The Kelley Board Investigation 173
    New Management and New Objectives 176
    Space Science Against the Wall 181

Part II. THE NEW RANGER

12. HOMESTRETCH ENGINEERING 185
    Redesigning for Improved Reliability 185
    Launch Vehicles Revisited 187
    Requalifying Ranger: Progress and Problems 191

13. SPACE SCIENCE: A NEW ERA AND HARD TIMES 199
    Guaranteeing Support to Project Apollo 199
    Planning in the Face of Change 202
    Making a Case for More Rangers 205
    Lunar Orbiter and Congress Intervene 209

14. MORE MISSIONS FOR SCIENCE? 212
    Block V Underway 212
    Nonvisual Science: All or Nothing at All 218

15. SIX TO THE MOON 223

Organizing Visual Science for Ranger Block III.............................223
Ready to Go...229
The Flight of Ranger 6...235

16. THE WORST OF TIMES...240
Ranger 6 Inquiry at JPL..241
The NASA Inquiry: Disparate Findings.......................................243
A Public Accounting..246
Action and Reaction..249
Congress Investigates Ranger...252

17. RANGER 7: A CRASHING SUCCESS..256
Reworking Ranger 7...256
On the Trail of Ranger 6...258
Preparing to Go Again..261
The Flight of Ranger 7...264

18. KUDOS AND QUESTIONS...271
Jubilant Days..271
New Interest in a New Era..280
Hard Questions for Space Science...284

19. THE RANGER LEGACY...289
One More for Apollo..289
The Last One for Science...296
Ranger: An Analysis..306
The Ranger Legacy..308

SOURCES...312

APPENDIXES

A. LUNAR THEORY BEFORE 1964...315

B. LUNAR MISSIONS 1958 THROUGH 1965..323

C. SPACECRAFT TECHNICAL DETAILS..327

D. RANGER EXPERIMENTS..333

E. BLOCK III VISUAL SCIENCE: MEMORANDUM OF AGREEMENT.......................339

F. RANGER SCHEDULE HISTORY...345

G. RANGER FINANCIAL HISTORY..349

H. RANGER PERFORMANCE HISTORY..353

I. A BIBLIOGRAPHY OF SCIENTIFIC FINDINGS...................................357

NOTES...367

INDEX..445

ILLUSTRATIONS

1. JPL Director William Pickering in 1954...5
2. STL Pioneer Lunar Probe (Courtesy TRW)...7
3. JPL Pioneer Lunar Probe...9
4. The Able Lunar Orbiter (Courtesy TRW)...14
5. Atlas-Vega Launch Vehicle...16
6. Luna 1..19
7. NASA Administrator Keith Glennan, Deputy Administrator Hugh
   Dryden, and Associate Administrator Richard Horner.....................21
8. Luna 3..22
9. NASA Space Flight Programs Director Abe Silverstein.....................27
10. JPL Lunar Program Director Clifford Cummings................................29
11. JPL Ranger Project Manager James Burke...30
12. Marshall Agena Systems Manager Friedrich Duerr.............................33
13. Air Force Ranger Manager John Albert with James Burke.................35
14. JPL Ranger Project Assistant Manager Gordon Kautz.......................37
15. NASA Lunar and Planetary Programs Chief Edgar Cortright..............38
16. 1960 Ranger Project Organization..40
17. Lockheed Agena Manager Harold Luskin..41
18. Spacecraft Attitude Stabilization in Three Axes................................48
19. Vega Spacecraft Model...50
20. NASA Space Flight Programs Assistant Director for Space Sciences
    Homer Newell...51
21. Ranger Spacecraft Preliminary Design..55
22. Spacecraft Packaging..55
23. Typical Assembly Package..56
24. Ranger 1 and 2 Spacecraft Design...58
25. Ranger 3, 4, and 5 Superstructure Design..62
26. Venera 1...66
27. Ranger Block I Proof Test Model and Nose Shroud Mockup...........68
28. Ranger 1 in Systems Test Complex...69
29. NASA Lunar Flight Systems Chief Oran Nicks....................................71
30. Scientific Experiments on the Block I Spacecraft..............................77
31. Scientific Experiments on the Block II Spacecraft.............................80
32. JPL Deep Space Instrumentation Facility Director Eberhardt Rechtin....83
33. Station Locations in the Deep Space Network....................................85
34. Twenty-Six-Meter Radio Antenna at Goldstone................................86
35. The Deep Space Network–Spacecraft Link...87
36. JPL Space Flight Control Center, 1961..90
37. Space Trajectory Selected for Rangers 1 and 2..................................95

38. Agena B Satellite Configuration 96
39. Ranger Ascent Sequence 98
40. Three-Meter Antenna at Johannesburg 100
41. Launch of Ranger 1 102
42. Deep Space Tracking Coverage as a Function of Spacecraft Altitude ... 104
43. Ranger 2 Countdown Progresses Beneath a Full Moon 107
44. JPL Ranger Spacecraft Systems Manager Allen Wolfe 117
45. Ranger Block III Television Camera Sequencing 118
46. Model of Ranger Block III Preliminary Design 119
47. NASA Administrator James Webb 121
48. NASA Ranger Program Chief William Cunningham 123
49. Sterile Assembly of a Seismometer Capsule at Aeronutronic ... 126
50. Placement of Sky Science Experiments on Ranger Block III
    Spacecraft, Showing Solar Panel Requirements 136
51. Lunar Launch Constraints 140
52. Technicians Make Final Adjustments to Ranger 3 at
    Cape Canaveral 141
53. Launch of Ranger 3 144
54. Ranger Block II Midcourse Maneuver Sequence 146
55. Ranger Block II Terminal Maneuver Sequence 148
56. Technicians Prepare Ranger 4 for Launch at Cape Canaveral 151
57. Red, White, and Blue Cross (Copyright, *Los Angeles Times;*
    Reprinted with Permission) 155
58. The Mariner R Spacecraft and Launch Vehicle 161
59. JPL Space Flight Control Center Readied for Ranger 5 165
60. Ranger 5 Ignition 168
61. Officials Assembled for the Ranger 5 Postlaunch Press
    Conference at Cape Canaveral 169
62. JPL Lunar and Planetary Program Director Robert Parks 178
63. JPL Ranger Project Manager Harris Schurmeier 179
64. The Ranger Block III Spacecraft as Viewed from Above 187
65. The Ranger Block III Spacecraft as Viewed from Below 188
66. Assembly of Ranger 6: Installation of the Midcourse Motor 194
67. The Television Subsystem Is Readied for Tests at JPL 196
68. Northrop Manager of Project Ranger William Howard 213
69. JPL Ranger Block V Project Manager Geoffrey Robillard 216
70. JPL Ranger Project Scientist Thomas Vrebalovich 224
71. Block III Principal Investigator Gerard Kuiper 227
72. Block III Coexperimenter Eugene Shoemaker 228
73. Block III Coexperimenter Harold Urey 229
74. Block III Coexperimenter Raymond Heacock 230
75. Block III Coexperimenter Ewen Whitaker 231
76. At JPL Technicians Make Final Adjustments on Ranger 6 232

77. The Audience in Von Karman Auditorium Hears Downhower
    Describe the Final Moments in the Flight of Ranger 6 238
78. Homer Newell, William Pickering, and Harris Schurmeier Answer
    Newsmen's Questions at Ranger 6 Postflight Press Conference...............239
79. "You're Shy—I'm Shy 28 Million Bucks" (Courtesy Fort Wayne
    [Indiana] *News-Sentinel*...............241
80. NASA Deputy Associate Administrator for Industry Affairs Earl
    Hilburn...............245
81. Dick Tracy Investigates, With Note From Burke to Schurmeier
    (Created by Chester Gould © Chicago Tribune-New York News
    Syndicate)...............258
82. Exposed Pins in the Agena Umbilical Connector...............259
83. Net Control Area in the New Space Flight Operations Facility...............263
84. Launch of Ranger 7...............265
85. Newell, Pickering, and Cortright Confer at the Flight Control
    Center Early on July 31, 1964...............267
86. Cheering...............268
87. And Weeping...............269
88. Newell and Pickering Shake Hands Before the Ranger 7
    Postflight Press Conference...............272
89. Experimenters Heacock, Kuiper, and Whitaker Examine Ranger 7
    Pictures at the Flight Control Center...............274
90. Ranger 7 Closeup Pictures of the Sea of Clouds...............275
91. "Howdy" (Courtesy Tom Little in *The* [Nashville] *Tennessean*)...............278
92. Newell and Pickering Brief President Johnson...............279
93. Shoemaker and Kuiper Answer Newsmen's Questions at the
    Interim Scientific Results Conference...............286
94. Ranger Block III Terminal Maneuver...............291
95. Kuiper, Heacock, and Whitaker Examine Ranger 8 Pictures
    at the Flight Control Center...............292
96. Ranger 8 Pictures of the Sea of Tranquility...............293
97. At Headquarters, NASA Administrator James Webb Explains
    Lunar Surface Model for President Johnson and Vice-President
    Humphrey...............297
98. "Successful Launch" (Courtesy Gene Basset, Scripps-Howard
    Newspapers)...............298
99. Urey, Whitaker, and Shoemaker Watch Ranger 9 Pictures
    "Live" at the Flight Control Center...............301
100. Urey, Kuiper, and Shoemaker Confer Before Ranger 9
    Experimenters' Press Conference...............302
101. Ranger 9 Pictures of the Crater Alphonsus...............303
102. "Ranger 9 Touch" (Courtesy Tom Little in *The* [Nashville]
    Tennessean)...............305

103. White House Awards Ceremony 306
104. X-Ray of Human Skull and X-Ray Enhanced by Computer
Processing..311

PREFACE

Ranger was the first successful American project of lunar exploration. It was an enterprise sponsored by the National Aeronautics and Space Administration (NASA) and executed by the Jet Propulsion Laboratory (JPL). Including authorization, design, development, and flights, the project spanned the years 1959–1965 and culminated in closeup television pictures of the moon. This project produced most of the basic management techniques, flight operating procedures, and technology for NASA's later unmanned lunar and planetary missions. It also established methods for selecting scientific experiments and integrating them with the spacecraft. The history of Ranger is thus essential to understanding the evolution and operational form of NASA's continuing program of unmanned exploration of deep space.

I have prepared this history with three particular audiences in mind: scholars concerned with the history of American science and technology; practicing managers involved in large government research and development projects; and scientists, engineers, or laymen interested in the nation's space program. I have arranged the story topically within chronological sections, and paid considerable attention in the analysis to the bureaucratic problems of science and technology as they were expressed and resolved by individuals. I have attempted further to capture the adventure of probing the unknown and the frequent drama in the interplay of the expectations of American scientists, popular American interests in space, the satisfaction of engineering requirements, restricted budgets, human frustration, congressional impatience, and institutional tension.

In the writing of this history, William Pickering, the Director of the Jet Propulsion Laboratory, granted me full access to the JPL files and records, and allowed me complete freedom to interpret the material as I wished. The NASA History Office made available the pertinent NASA records at Headquarters in Washington, D. C. I also made extensive use of interviews, which were taped and transcribed, with virtually all of the key participants in Project Ranger. Although allowance had to be made for subjective proclivities and faded memories, sometimes in conflict with the written record, the interviews proved valuable in recapturing the excitement of tense moments. They also furnished many of the private thoughts and motives of the participants, providing color and substance otherwise missing in the written record. Finally, I asked some of these same individuals to read and comment on portions or all of the draft manuscript. Though employed at Caltech/JPL, I served functionally as a member of the NASA history program, where this "comment cycle" is typical. I found the comment cycle helpful rather than inhibiting. At no time was my freedom of interpretation infringed upon.

From the beginning I had the encouragement and unstinting assistance of the NASA History Office: NASA Historian Eugene M. Emme, Director of the History Program Monte D. Wright, Publications Manager Frank W. Anderson, and Archivist Lee D. Saegesser. At JPL, Laboratory Director Pickering and Assistant Laboratory Director Walter H. Padgham supplied the resources and patience needed to see the work completed. I am grateful to Richard K. Smith, who criticized the early chapters, and to Rodman W. Paul of the Caltech Division of Humanities and Social Sciences, who helped out in a variety of important ways. I owe a special debt to Daniel J. Kevles, also of the Caltech Humanities Division, who counseled with me throughout the work, criticized every chapter, and edited the penultimate draft of the manuscript.

In this brief space, I can only acknowledge the generous additional support I received from many other individuals and groups. Reference librarians at the JPL and Caltech Millikan Libraries patiently answered innumerable queries, and Caltech students past and present—P. Thomas Carroll, Lee W. Vibber, and Hallie A. Poore—assisted in the research and footnoting. The following, in particular, helped with interviews, critiques of the manuscript, or both:

John G. Albert	Alvin R. Luedecke
James R. Arnold	Homer E. Newell
James D. Burke	Oran W. Nicks
Edgar M. Cortright	Eberhardt Rechtin
Clifford I. Cummings	Geoffrey Robillard
N. William Cunningham	Harris M. Schurmeier
Lee A. DuBridge	Robert C. Seamans, Jr.
Friedrich Duerr	Eugene M. Shoemaker
T. Keith Glennan	Homer J. Stewart
Walter E. Jakobowski	Thomas Vrebalovich
Gordon P. Kautz	James E. Webb
Clayton R. Koppes	Ewen A. Whitaker
Donald R. Latham	James H. Wilson

The artist Robert Shepard rendered the print of Ranger used for the front cover of this book, and it appears courtesy of the *TRW Space Log*. JPL artist Arthur Beeman furnished the print of the deep space radio antenna that appears on the back cover. All illustrations not credited otherwise are NASA/JPL figures.

Finally, heartfelt thanks are due Helena Rosen, who, over the past four years, made the manuscript a reality, typing and retyping "adjusted" chapters more times than either she or I care to remember. Needless to say, the final responsibility for this history, including its interpretations and whatever inadequacies or weaknesses it may contain, is mine alone.

R. Cargill Hall

Pasadena, California
October, 1976

LUNAR
IMPACT

Part I

THE ORIGINAL RANGER

Chapter One

THE ORIGINS OF RANGER

"FROM what they say they have put one small ball in the air," President Dwight D. Eisenhower declared at his news conference on October 9, 1957, adding, "at this moment you [don't] have to fear the intelligence aspects of this."[1] But despite the Presidential assurance, the Soviet satellite, Sputnik 1, launched a few days earlier as part of the world-wide scientific program of the International Geophysical Year,[2] had shattered confidence in American technical preeminence. In the wake of the Soviet triumph, many Americans concluded that the United States must undertake a vigorous space program of its own.

A MOON FLIGHT PROPOSAL

Shortly after the launch of Sputnik 1, William H. Pickering,[3] the Director of the Jet Propulsion Laboratory in Pasadena, California, advanced a novel idea. The Soviet Union had merely hurled a satellite into orbit around the earth. Pickering, along with a number of his staff at the Laboratory, wanted the United States to meet the Russian space challenge by sending a spacecraft to the moon.

The Jet Propulsion Laboratory, or JPL, had been pursuing exotic projects in its field for years. Begun in 1936 under the auspices of the Guggenheim Aeronautical Laboratory of the California Institute of Technology, it had originated as a student rocket research project when the scientific community generally regarded rockets as an indulgence best left to students. In 1940 the Caltech rocket experimenters acquired an Army Air Corps contract and built facilities in northwestern Pasadena, at the foot of the San Gabriel Mountains in the Arroyo Seco wash. There they developed the first solid- and liquid-propellant rocket motors for jet-assisted takeoff of military aircraft. The enterprise was reorganized and named the Jet Propulsion Laboratory when, in 1944, after the advent of the German V-2 rocket, U.S. Army Ordnance awarded Caltech a contract to develop tactical ballistic missiles.[4]

Continuing to work for the Army into the 1950s, JPL engineers and scientists designed and developed the liquid-propellant WAC Corporal sounding rocket, the Corporal tactical missile, and the solid-propellant Sergeant tactical missile system. The Laboratory also pioneered in the development of radio telemetry and of various radio and inertial guidance systems for the Army's Redstone rocket arsenal in Huntsville, Alabama, where the director of research was Wernher von Braun. All the while, JPL, whose facilities were owned by the government, remained an Army establishment under the contract management of Caltech. Its posture and atmosphere were free-wheeling, academic, and innovative. By 1957 Director Pickering, a professor on the Institute faculty, presided over a considerable laboratory complex nestled in the Arroyo Seco and populated by some 2,000 employees.

Like the Laboratory, Pickering, too, had come a long way from his beginnings—the small fishing village of Havelock, New Zealand, where he had attended the same primary school as the famed pioneer in nuclear physics, Ernest Rutherford. Displaying an aptitude for mathematics and science, Pickering was sent to high school in Wellington, the capital of New Zealand, where he built and operated wireless sets and performed extracurricular chemistry experiments in a classmate's cellar. Lured to Caltech in 1929 by an uncle in Los Angeles, he embarked on a career in electrical engineering, but by 1936 emerged with a Ph.D. in physics and an appointment to the Caltech faculty. Applying his capabilities in electrical engineering to one of the central research subjects in the physics of the day, he joined Robert A. Millikan and H. Victor Neher in research on the absorption properties of primary cosmic rays using instrumented balloon sondes.

During World War II Pickering organized and taught electronics courses at Caltech for military personnel, which brought him into contact with the Radiation Laboratory at MIT, including its director, the physicist Lee A. DuBridge. In 1944 he went out to JPL to design and develop telemetering and instrumentation equipment for the long-range missiles. When DuBridge was named the new president of Caltech in 1946, Pickering was busy perfecting the telemetry system to be used in the Laboratory's rocket research vehicles.[5] He preferred to work at the forefront of applied engineering research and development. Appointed the Director of JPL in 1954, he began turning that preference into Laboratory policy (Figure 1). By the time of Sputnik, JPL was equipped to contribute to the nation's first response to the Soviet space challenge: the orbiting of the Explorer 1 satellite. Pickering's laboratory supplied the solid-propellant upper stages of the launch rocket, furnished the space-to-ground communications equipment and instrumentation for the satellite, and helped integrate into it the radiation monitoring experiment of James Van Allen.

But Pickering, spare, intense, reserved, and in a quiet way implacable, was determined to mount a JPL program of lunar flights. To his mind, and to DuBridge's, such flights were an appropriate entrant in the emerging Soviet-American space race. Like rockets a generation before, lunar flights might once have been a subject fit only for science fiction, but now they were on the reachable

Fig. 1. JPL Director William Pickering in 1954

frontier of engineering science, exactly the frontier where Pickering wanted JPL to be. Using the technology available, the United States could launch a simple, spin-stabilized vehicle, similar to the Explorer satellite in design, on reasonably short notice, possibly as early as June 1958. Three weeks after the launch of Sputnik 1, Pickering, with DuBridge's support, had ready a JPL moon flight proposal. Designated "Project Red Socks," the proposal declared it "imperative" for the nation to "regain its stature in the eyes of the world by producing a significant technological advance over the Soviet Union" in rocketry and space flight. Pickering wanted the Department of Defense to approve JPL's embarking immediately on a series of nine rocket flights to the moon.[6]

Pickering and DuBridge got nowhere with their Red Socks lunar proposal in the Defense Department[7] until early 1958, when it came under the consideration of the new Advanced Research Projects Agency, or ARPA, whose responsibilities temporarily included the direction of all U.S. space projects. The new ARPA director and former General Electric Company executive, Roy Johnson, was eager "to surpass the Soviet Union in any way possible,"[8] and he could choose to do it from a host of unsolicited flight proposals. In fact, "after we had been in business a short time," his deputy Rear Admiral John E. Clark recalled, "it seemed to me that everybody in the country had come in with a proposal except Fanny Farmer Candy, and I expected them at any minute."[9] Because the Soviet Union had not yet launched a rocket to the moon, an unmanned lunar program appeared to be the most promising approach to "beat the Russians" in space.

With the President's approval, on March 27, 1958, Secretary of Defense Neil McElroy announced that ARPA's space program would advance space flight technology and "determine our capability of exploring space in the vicinity of the moon, to obtain useful data concerning the moon, and to provide a close look at the moon."[10] Conducted as part of the United States contribution to the International Geophysical Year, the lunar project would consist of three Air Force launches using modified Thor ballistic missiles with liquid-propellant Vanguard upper stages, followed by two Army launches using modified Jupiter-C missiles and JPL solid-propellant upper stages. JPL was to design the Army's lunar probe and arrange for the necessary instrumentation and tracking. ARPA directed the Air Force to launch its lunar probes "as soon as possible consistent with the requirement that a minimal amount of useful data concerning the moon be obtained."[11]

THE FIRST LUNAR FLIGHTS

The ARPA lunar program approved in March 1958, generally known as the "Pioneer program," offered five flight opportunities, three for the Air Force and two for the Army. Space Technology Laboratories, the West Coast-based contract manager of the Air Force ballistic missile program, which was assigned responsibility for the technical direction of the Air Force lunar missions, also furnished the spacecraft. Shaped like two truncated cones back-to-back (Figure 2), the fiberglass lunar probe, 74 centimeters (29 inches) in diameter and 46 centimeters (18 inches) long, carried 17.5 kilograms (39 pounds) of scientific instruments, battery power, transmitter and antenna, and a retrorocket system designed to slow the vehicle into lunar orbit.

In keeping with the original ARPA requirements, this spacecraft also supported a small facsimile television system. But engineers at the Space Technology Laboratories had barely completed the spacecraft design in June 1958 when the discovery of the first Van Allen radiation belt stimulated scientists to issue urgent requests for more, improved experiments to measure charged particles in near-earth space. Though retaining the television camera, the firm's researchers directed the rest of the scientific instrumentation toward fields and particles in space: a magnetometer to measure the magnetic fields of the earth and moon, and a micrometeoroid impact counter to survey the flux and energy of micrometeoroids

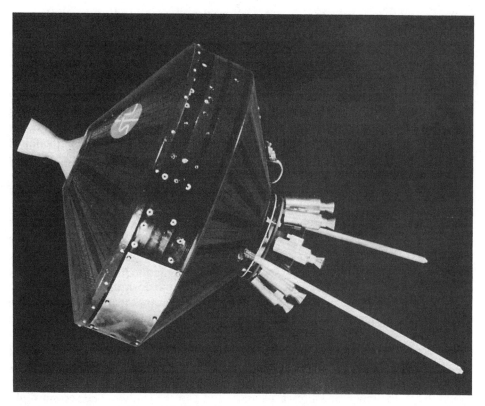

Fig. 2. STL Pioneer Lunar Probe (Courtesy TRW)

between these two bodies. To obtain more information on the distribution of radiation in space, they installed a Van Allen-supplied ion chamber on flight two, and augmented it with a proportional counter from the University of Chicago on the third flight.[12] The final report summarized the extent of this thoroughgoing change: "To the maximum extent possible, within the weight and power restrictions, experiments were designed to obtain scientific measurement of the environment in cislunar space."[13]

The first lunar flight of the Air Force Thor–Able launch vehicle rose from Cape Canaveral on August 17 and ended 77 seconds later in a pyrotechnic display above the beach. This "catastrophic failure," the investigative report declared, was caused when a turbo pump bearing seized in the main-stage rocket engine.[14] The embarrassing flight went officially unnamed by the Air Force, though informally it became known as "Pioneer O." After corrective measures had been taken with the main-stage engine, the Air Force launched the second vehicle on October 11, 1958. This time a guidance system error caused an early shutdown of the second-stage engine. Upon completion of burning of the third-stage engine, the velocity attained was less than that required to escape the earth's gravity. The spacecraft separated properly from its third-stage rocket and continued to ascend to an altitude of 115,000 kilometers (71,700 miles), about one-third the distance to the moon, before falling back to be incinerated in the earth's upper atmosphere.

This second flight, promptly christened Pioneer 1, though of course precluding photography of the moon, did yield good scientific data from the magnetometer and micrometeoroid detector. The ionization chamber measuring radiation intensity developed a leak; much of its radiation information, at first unintelligible, was subsequently unscrambled and recovered.[15] The last flight in the Air Force series, Pioneer 2, followed on November 8. The third-stage engine failed to ignite, and the vehicle rose only 1,550 kilometers (963 miles) before falling back to earth. It returned no significant experimental data.[16]

While the Air Force lunar flights were underway, JPL completed design of the Army's lunar probe, which would also separate from its fourth-stage rocket, and of the necessary instrumentation and tracking facilities. The JPL design called for a cone-shaped, fiberglass instrument package, 51 centimeters (20 inches) long and 25.5 centimeters (10 inches) in diameter at its base (Figure 3).[17] The scientific experiment consisted of a small camera weighing 1.5 kilograms (3.3 pounds), capable of photographing the moon. The lunar image on 35-millimeter film was to be developed by a wet process, scanned by optical means, transmitted from the spacecraft via telemetry code, and reconstructed on earth by facsimile methods at a ground receiving station. Snapped at closest approach, 24,000 kilometers (15,000 miles) from the moon's surface, the picture would provide a resolution of 32 kilometers (20 miles).[18] Flight plans called for the first of the Army–NASA lunar probes to carry and test a special shutter-trigger mechanism: photoelectric cells would "see" the moon at a preset distance, and trip the shutter. The second flight would then carry the complete camera on a looping trajectory around the moon, with the aim of returning one good photograph of the far side. On the first of these two probes, two Geiger–Muller tubes furnished by Van Allen to measure charged radiation particles in space were added in place of the camera.

Data returned by United States IGY satellites Explorers 1, 3, and 4, meantime, revealed more details of the high-intensity radiation surrounding the earth. In the absence of heavy shielding, such radiation could fog the film in the photographic experiment planned for the Army lunar probe. Consequently, the Army canceled the camera experiment and, in August 1958, JPL began to develop a small, lightweight, slow-scan television camera and magnetic-tape recording and transmission system, all of which were to be functionally insensitive to radiation in space, in time for flight in early 1959.[19]

On December 6, 1958, the Army launched the first of its lunar probes, Pioneer 3, on a trajectory that was supposed to carry it past the moon into solar orbit. This probe, like its predecessor Pioneer 1, did not attain escape velocity because the first-stage propulsion system ceased ignition prematurely. Aloft for 38 hours, Pioneer 3 ascended 101,000 kilometers (63,500 miles) before falling back to earth. Nevertheless, sky scientists acquired valuable information from Van Allen's two Geiger–Muller tubes. These data revealed the existence of two primary bands,

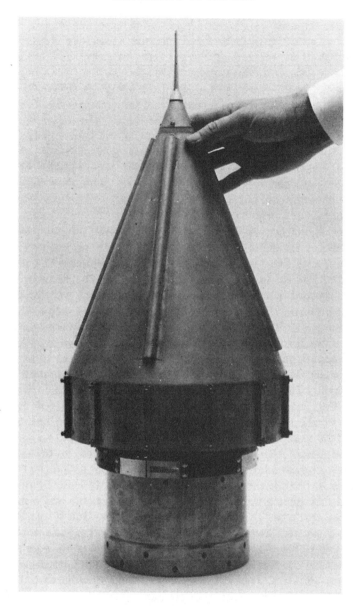

Fig. 3. JPL Pioneer Lunar Probe

or shells, of high-intensity radiation about the earth at approximately 4,800 and 16,000 kilometers (3,000 and 10,000 miles) altitude, with radiation intensity near the earth progressively diminishing between 48,000 and 96,000 kilometers (30,000 and 60,000 miles).[20]

The data returned by Pioneer 3's radiation experiments, and the discovery of the second radiation shell, heightened scientific interest in charged particles in near-earth space all the more. A few weeks later, in early 1959, James Van Allen and his associate George Ludwig of the State University of Iowa urged that their radiation package be flown again in place of the television system on Pioneer 4 to obtain more radiation data and to further refine information already secured. "We happened to have," a JPL official later explained, "a bonanza in the original Explorer by carrying Van Allen's experiment along and obtaining so much information. You just couldn't go wrong by proposing to gain more information of that type for the next several years."[21] Space officials approved the change. Pioneer 4 would also be launched on a lunar flyby trajectory rather than on a circumlunar trajectory for photographic purposes, so as to measure radiation between the earth and the moon.[22]

Pioneer 4, the last of the ARPA-initiated lunar probes, rose from Cape Canaveral without incident on March 3, 1959, carrying scientific instruments virtually identical to those of Pioneer 3. The only distinctive new feature was a small amount of lead shielding, added to one of the two Geiger tubes to screen out low-intensity charged particles. In this flight, the launch vehicle provided the spacecraft earth-escape velocity, and the craft passed by the moon at a distance of 60,000 kilometers (35,500 miles).[23] The shielded Geiger counter showed a lower level of radiation in the low-altitude shell than that detected by Pioneer 3, and almost no radiation in the high-altitude shell. The second monitor, acting as a counter and scaler, also detected both bands, but the peak radiation was slightly broader in extent. These scientific data added further support to the hypothesis that the earth's magnetic field acted as a trap for charged particles that accumulated, slowly dispersed, and then built up again as a function of activity on the surface of the sun.[24]

All the same, ARPA's lunar program had failed to reassert American superiority in technical affairs. Instead, the Soviet Luna 1, launched a few weeks before Pioneer 4, became the first unmanned craft to fly close by the moon. The distinction of first photographing the far side of the moon was also claimed by another Soviet machine, Luna 3, a short time later. But if the ARPA flight objectives had shifted away from the moon to other scientific targets of opportunity, the ARPA effort had after all been intended as no more than a quick, flexible response to the challenge of Sputnik. Indeed, ARPA itself had been given only temporary authority over space activities, lunar or otherwise. By mid-1958 responsibility for a coherent program of civilian space research had been vested in the new National Aeronautics and Space Administration, familiarly known as NASA.

SKY AND PLANETARY SCIENCE

The new NASA brought together a number of the government's disparate facilities for space and aeronautical research, including JPL. Leaving all defense and military matters to the armed services, the new agency was awarded responsibility for all civilian aeronautical and space activities, including, in the language of its organic act, the "expansion of human knowledge of phenomena in the atmosphere and space." A simple enough phrase, yet, as was evident from the shifting objectives of the pioneer lunar flights, space science encompassed diverse, sometimes conflicting subjects, generally grouped into two categories—planetary science and sky science.*

Sky scientists sought to understand the mysteries of the upper atmosphere and the fields and charged particles surrounding the earth. They emphasized meteoritics, solar and cosmic ray physics, plasma dynamics, and the interaction of the solar and terrestrial electromagnetic fields.[25] Planetary scientists, on the other hand, were occupied with the origin, composition, and evolution of planetary bodies in our solar system, including the moon. They wondered whether the earth and its moon had condensed out of the same blob of stellar gas and dust, whether the moon had spun off from the forming earth, or whether it had once been a separate planet captured by the earth's gravitational field. They puzzled over its surface features: whether its craters were caused by internal plutonic forces or by the impact of planetesimals and meteors. They were eager to know the distribution of the moon's mass, its magnetic field, the composition of its atmosphere and crust, and the seismic properties of its interior.[26]

By the late 1950s, sky scientists had a well-developed array of instruments for pursuing their subject. They used Geiger counters and ion chambers, plasma probes, magnetometers, and electrostatic analyzers to trap and observe charged particles or to measure electromagnetic fields. For decades they had been lofting their instruments into space by balloons; since World War II, by rockets. Over the centuries planetary scientists had of course explored their subject with telescopes and, after World War II, with radar. They also talked about visual imaging instruments that would radio back pictures of the moon, discussed surface sampling penetrometers, drills, and apparatus for materials analysis, along with seismometers, magnetometers, and gamma-ray spectrometers. And finally, they dreamed of actually delivering these instruments to the moon and planets with automatic orbiters, soft landers, and possibly even soft landers with the ability to return to earth.[27] In the late 1950s, unlike sky science, planetary science depended upon a technology and, more important, complex interplanetary flight capability, yet to be created.

When the space program got underway, the planetary scientists most active in theorizing about the origins and evolution of the moon included the chemists Harold Urey, a Nobel Laureate. James Arnold, and Harrison Brown; the

---

* Life science and astronomy, not treated here, would shortly become the third and fourth branches of the space science tree. Life science evolved to include experiments with man and living organisms in space, and the quest for extraterrestrial life forms. With space vehicles above the atmosphere, astronomy, specifically that portion concerned with celestial phenomena at galactic distances, would begin to move from an observational to an experimental science.

astronomers Gerard Kuiper, Dinsmore Alter, and Thomas Gold; and the geologists Frank Press, Maurice Ewing, and Eugene Shoemaker. Most were associated with universities, observatories, and other nongovernmental research institutions. Their scientific experience derived from geology, geochemistry, surveying and geodesy, or planetary astronomy.[28] In early 1958, planetary scientists had begun to gather informally with engineers in Lunar and Planetary Colloquia on the West Coast.[29] Several of them, members of the National Academy of Sciences, were appointed to the Academy's Space Science Board, which was formed in June 1958 to advise NASA on its space science research program.[30] By the end of 1958, lunar exploration had been dubbed a meritorious scientific objective for NASA by the Board as well as by the President's Science Advisory Committee.[31] But planetary science was not nearly so well organized, well established, or, most important, well placed in the new NASA as sky science.

Among the notable American scientists active in sky research were the physicists James Van Allen, Hugh Odishaw, Lloyd Berkner, Homer Newell, Joseph Kaplan, Marcus O'Day, John Townsend, Charles Sonett, and S. Fred Singer; the astronomers Fred Whipple and Lyman Spitzer; the meteorologists Harry Wexler, William Kellogg, and Verner Suomi, and the meteorologist–oceanographer Athelstan Spilhaus. These men represented prestigious universities, observatories, and research institutions. Moreover, a significant fraction were under contract to or directly affiliated with government agencies that provided funds for ongoing upper air research using rocket and balloon ascents. They had been largely responsible for inaugurating the International Geophysical Year (IGY).[32] When NASA began operating in late 1958, sky scientists were a cohesive group, well organized and reasonably well publicized. Active in government-sponsored upper air and satellite research projects for over a decade, they sat together on the various official and semiofficial panels that counseled federal agencies on prospective projects, judged experiment proposals, and allocated space for the experiments on rocket sondes and IGY satellites.[33] "Many of these meetings," a participant remarked, "were held consecutively with practically the only changes in the group being the presiding officer and the secretary."[34] Sky scientists, together with engineers, largely staffed the U.S. National Committee for the International Geophysical Year, located in the National Academy of Sciences, a subpanel of which selected the experiments to be carried aloft by the Vanguard, Explorer, and Pioneer spacecraft.[35]

It was largely individuals with such sky science backgrounds, linked organizationally, familiar with each other, and experienced in sounding rocket research, who first came to occupy space science positions in NASA. It was largely a sky science program, both in practice and in a final recommendatory report, that NASA inherited from its predecessor agency, the National Advisory Committee for Aeronautics.[36] Of course, planetary science held a certain standing in the new NASA from the endorsement of the President's Science Advisory Board and the Academy's Space Sciences Board. But, all things considered, sky science held the upper hand in the space agency. And it was thus no surprise that planetary science fared poorly in the ARPA-sponsored Pioneer missions, or in the first lunar project authorized by NASA: Atlas-Able V.

Conceived at the Space Technology Laboratories and approved by NASA Headquarters, the Atlas-Able was to be managed by the Space Technology Laboratories and launched by the Air Force. Originally, plans called for two missions to Venus, followed by two lunar orbiting missions; however, in the spring of 1959, prompted by the Russian success with Luna 1, NASA reprogrammed the Venus flights as lunar orbiters, too.[37] The Atlas-Able vehicle was to consist of a liquid-propellant Atlas intercontinental ballistic missile modified to support the same Vanguard upper stages employed on the Thor-Able Pioneers, together with a correspondingly larger and more versatile spacecraft. This spacecraft, like its Pioneer predecessors, featured the technology of the day—a technology well suited to the purposes of sky science.[38] The specifications submitted to NASA in January 1959 detailed a spin-stabilized 122-kilogram (273-pound), aluminum-alloy spheroid. Almost a meter (3 feet) in diameter, the machine was to incorporate vernier rockets at each pole to decelerate it into lunar orbit. Four "paddle wheel" solar arrays coupled with batteries would provide electrical power. Paint patterns on the highly polished reflective surface and an arrangement of novel cruciform temperature control vanes would dissipate the heat generated by the sun outside and instruments within the spacecraft (Figure 4).[39]

The planned lunar orbits of the four flights could be expected to yield a measurement of the mean moment of inertia of the moon. The small Pioneer television camera designed by the Space Technology Laboratories "to get a crude outline of the moon's surface" were to complement such a planetary measurement on the first two flights. But because of launch vehicle failures, none of the Atlas-Able V probes launched during 1959 and 1960 even left the earth's atmosphere. Vagrant behavior of the very large liquid-propellant rockets—the transportation system needed for planetary research—would remain a serious problem for all American space flight projects well into the 1960s. But even had the Atlas-Able rockets reached interplanetary space, lunar science would scarcely have benefited. In short order, nonvisual sky science experiments had supplanted the television cameras. In fact, of the nine scientific experiments carried on the last two flights, only one, a magnetometer, was directed toward investigating the moon. One science planner at NASA Headquarters aptly explained the latter missions: The "payload instrumentation itself has no value for the lunar program, although it can make an important contribution to the study of solar–terrestrial relationships, serving as an anchored space probe, placed far enough away to be unaffected by our own atmosphere."[40] If in 1959 and 1960 obtaining a sufficiently powerful and reliable space launch vehicle occupied NASA as a whole, acquiring effective leverage in the agency's space research program was a major challenge for the nation's planetary, especially lunar, scientists.

Fig. 4. The Able Lunar Orbiter (Courtesy TRW)

FROM INTEREST TO RESOLVE

The space program might at first have been directed to reassert American supremacy in technical affairs, and sky scientists might have controlled the important scientific posts, but within NASA at the end of 1958, planetary scientists were already on the march. At Headquarters, the charge of all NASA flight projects was vested in the Office of Space Flight Development, directed by Abe Silverstein. Silverstein's Assistant Director of Space Sciences, Homer Newell, established a sciences division staffed by part of the Naval Research Laboratory upper air research group that had moved to NASA, and organized it to satisfy the interests of sky science.[41] In November 1958 Newell created a companion "theoretical division" to devote attention to basic research in cosmology, astronomy, and planetary sciences. To head this new division, Newell selected Robert Jastrow, a physicist and sky science colleague who had come to NASA with the Naval Research Laboratory upper air group.

Late in November, Jastrow, desirous of learning all he could about his new assignment, "traveled across the United States to the Laboratory at La Jolla, California, to visit a man who, I had been told, would be able to give me some advice." Harold Urey, the Nobel Laureate chemist, opened his book on the planets to the chapter on the moon, and explained the "unique importance" of the moon for understanding the origin of the earth and the other planets. "I was fascinated by his story, which had never been told to me before in fourteen years of study and research in physics," Jastrow recalled.[42] He at once became a convert and champion of lunar exploration. A week later in December, Jastrow brought Urey and his new-found enthusiasm to a meeting with Newell at Headquarters. The visitors pointed out that NASA did not have a firm program for lunar exploration beyond the projects inherited from ARPA; they convinced Newell that such a program should be undertaken. "The Ranger Project," Newell reflected some years later, "was in effect born on [that] day..."[43] Whatever the case, certainly born on that day was the resolve that NASA should have a serious program of lunar exploration directed toward the goals espoused by planetary scientists.

Even more important, perhaps, was the subsequent introduction of planetary scientists into the planning structure of NASA. In January 1959, Newell formed an ad hoc Working Group on Lunar Exploration. Its members included Harold Urey, James Arnold, Frank Press, and Harrison Brown. Chaired by Jastrow, the new lunar working group was to operate as a forum for the exchange of views between scientists at NASA and in the academic world—an important function intended by Newell—and it had charge of evaluating and recommending to NASA the experiments to be placed in orbit about the moon or landed on its surface.[44] From this time forward lunar enthusiasts had a voice at Headquarters—and Headquarters soon expected to have a rocket capable of carrying the instruments of planetary science reliably to the moon.

The vehicle was the Atlas-Vega space launch rocket (Figure 5). A liquid-propellant Atlas was to be modified to accommodate upper-stage rockets; the second stage would be powered by a General Electric Vanguard first-stage rocket engine modified for high-altitude operation. For those missions requiring a high-velocity increment, such as deep space missions, there would be a third stage which was already under development at the Jet Propulsion Laboratory. Meanwhile, along with developing the third stage of the general-purpose Vega launch vehicle, JPL was mandated by NASA to conduct unmanned "deep space" exploration—research at lunar distances and beyond. By the end of 1959, NASA would specifically direct JPL to undertake a series of unmanned lunar missions. In the

Fig. 5. Atlas-Vega Launch Vehicle: Atlas First Stage (in two parts), G.E. Second Stage, JPL Third Stage, Spacecraft, and Nose Fairing

vanguard of the budding program of lunar exploration—even then embracing tentative plans for manned landings on the moon[45]—would be Project Ranger.*

Actually, despite Pickering's post-Sputnik bid for lunar flights, many JPL engineers and scientists tended to favor investigating the planets and space medium ahead of the moon. Flights past other planets more distant than the moon offered the kind of technical and experimental challenge that appealed to them. In Pasadena, prevailing opinion held that moon missions, whose launch opportunities occurred every lunar month, could and should be deferred in order to capitalize on the more infrequent planetary opportunities.[46] The inner planets, Mars and Venus, appeared particularly attractive objects of inquiry because they approached the earth more closely than did the other planets in the solar system, permitting a maximum number of scientific instruments to be carried with less powerful launch vehicles. The celestial period of closest planetary approach occurred approximately once every 25 months for Mars, and every 18 months for Venus. The next launch opportunity for Mars would fall in October 1960; for Venus, in December 1960–January 1961.

On April 30, 1959, JPL issued to NASA a five-year plan for deep space exploration. This ambitious prospectus emphasized planetary investigation in an alternating series of flights. Individual projects were not identified, but the prospectus outlined a series of progressively more sophisticated missions, together with recommended scientific and engineering features. The proposed flight schedule, qualified as "consistent with scientific potentialities and astronomical dates," hewed closely to contemporary Vega planning, and was predicated upon the rapid and concurrent development of launch vehicles, spacecraft, and scientific instruments (Table I).[47]

JPL officials on the West Coast could emphasize planetary exploration, but the NASA ad hoc Working Group on Lunar Exploration in Washington, which confined its attention exclusively to the moon,[48] had finished evaluating the scientific instruments it preferred for unmanned lunar missions. At its first meetings in February, the lunar working group divided prospective missions among uncontrolled impact, rough-landing—where experiments survived the crash,

---

* The conceptual distinction between program and project evolved at NASA Headquarters during 1959 and 1960. Basically, a *program* was accepted as a related series of undertakings to accomplish a broad scientific or technical goal over a prolonged period. Attainment of the goal would be secured through implementing specific *projects* with a scheduled beginning and ending. By the end of 1959 NASA could describe its "lunar and planetary exploration program" as including "lunar and planetary probes, orbiters, rough landings, soft landings, and mobile vehicles for unmanned exploration" (NASA, *Long Range Plan,* December 16, 1959, p. 33). These specific program components began to receive project names in the early months of 1960 (e.g., lunar rough landings became Project Ranger). This definition of program and project was formalized in NASA General Management Instruction 4-1-1, *Planning and Implementation of NASA Projects,* January 19, 1961. By the mid-1960s, however, with Apollo predominating in NASA affairs, and with the divorce in organization between manned and unmanned flight projects in shared program areas, the original distinction became blurred. In a sense, the program came to support a project.

Table I. JPL Plan for Deep Space Exploration, April 1959

Flight No.	Goal	Date	Vehicle
1	Lunar miss (eng. test)	August 1960	Vega
2[a]	Mars flyby	October 1960	Vega
3[a]	Venus flyby	January 1961	Vega
4	Lunar rough landing	June 1961	Vega
5	Lunar Orbiter	September 1961	Vega
6	Venus Orbiter	August 1962	Vega
7	Venus entry	August 1962	Vega
8	Mars Orbiter	November 1962	Saturn 1
9	Mars entry	November 1962	Vega
10	Lunar Orbiter and return	February 1963	Saturn 1
11	Lunar soft landing	June 1963	Saturn 1
12	Venus soft landing	March 1964	Saturn 1

[a]Flights 2 and 3 to Mars and Venus had already been incorporated in Vega launch schedules issued by NASA.[49]

soft-landing, and orbit. Members favored an early rough-landing mission. Although technically more demanding than lunar impact, in which the craft and all scientific experiments would be destroyed on hitting the moon, they judged rough-landing more useful to science because it would deposit operating instruments on the lunar surface.

Members of the group advised Homer Newell to support development of a seismometer, batteries and communications system, and a protective rough-landing capsule. They also recommended other nonvisual planetary science instruments that could be operated during approach to the moon: a gamma-ray spectrometer to detect and measure radioactive minerals during descent, a magnetometer to measure the magnetic field in one direction during descent, and an X-ray fluorescent spectroscope to assay surface material after landing.[50] Newell's chief, Silverstein, approved these scientific recommendations, and directed that they be budgeted and scheduled for flight. Thereafter, members of the lunar working group and their academic colleagues familiar with its deliberations submitted proposals for the design and development of a number of these instruments directly to Newell's shop.[51] Despite the research preferences evident at JPL, NASA's leadership favored an expanded lunar program, especially after the Russians launched Luna 1, which bore instruments to measure the moon's magnetic field, radioactivity, and fields and particles in interplanetary space and carried metal pennants stamped with the Soviet coat of arms (Figure 6).[52]

Indeed, the January 1959 flight of Luna 1 provoked wide public discussion of space flight advances that might reasonably be expected in the near future, including photographing the hidden or far side of the moon and manned lunar landings. It also helped secure Congressional approval of the Atlas-Vega launch vehicle[53] and to focus attention at NASA on the deliberations of its lunar working

Fig. 6. Luna 1

group. The chairman of the lunar working group, Robert Jastrow, told Newell: "the national space program will be open to strong criticism if a very early and vigorous effort is not made in the program of lunar exploration...The criticism will be especially strong if it turns out that a slow-paced U.S. lunar program must be contrasted with early Soviet achievements in this field."[54] At Headquarters, calculations showed that Atlas-Vega could be used for both rough landers and lunar orbiting vehicles in the near future, and these moon missions were described to Congress during the NASA authorization hearings.[55]

In a meeting at Headquarters on May 25, 1959, Silverstein and Newell determined to reprogram two Atlas-Vega flights as lunar orbiters, and, following one Vega lunar rough-landing mission, to employ Atlas-Centaurs for two lunar soft-landing missions.[56] In June, Silverstein ordered JPL to cancel the Mars flight scheduled for October 1960 and to design its Vega planetary spacecraft for the lunar orbiting mission.[57] Newell, meantime, formed a new Lunar and Planetary Program Office to direct these missions,[58] and reconstituted the ad hoc Working Group on Lunar Exploration as a standing committee known as the Lunar Science Group.[59] As the summer began and Vega schedules commenced to slip—threatening also to void the first launch opportunity to Venus—NASA Administrator T. Keith Glennan endorsed an immediate and even more extensive program of unmanned lunar exploration. Glennan, for the preceding 10 years President of the Case Institute of Technology in Cleveland, Ohio, had been selected by President Eisenhower as the first Administrator of NASA; the Senate had consented to his appointment, and to the appointment of Hugh L. Dryden, since 1947 the Director of the National Advisory Committee for Aeronautics (NACA) and the postwar architect of American aeronautical research, as NASA Deputy Administrator (Figure 7).

On July 23, Glennan, Dryden, and NASA Associate Administrator Richard E. Horner joined key administration officials at the "White House Annex," just around the corner from NASA Headquarters in the Dolley Madison House on Lafayette Square, to consider the status and goals of the United States space program. Meeting with them were George B. Kistiakowsky, who had succeeded James Killian as Special Assistant to the President for Science and Technology, Gordon Gray, Special Assistant to the President for National Security Affairs, Karl G. Harr, Jr., Special Assistant to the President for Security Operations Coordination, Charles Sullivan of the Department of State, and Foster Collins of the Central Intelligence Agency. Glennan recommended that the nation concentrate its deep space efforts on lunar flights to achieve the short-term objectives called for in a policy paper recently prepared by the National Security Council. His proposal was approved. A few days later, Silverstein instructed JPL to cancel the mission to Venus scheduled for January 1961, and prepare a new Vega flight schedule containing only lunar and earth satellite missions.[60]

Then, on September 13, the Soviet Luna 2 crashed on the surface of the moon, and in early October Luna 3 took its photographs of the hidden side of the moon and radioed them to earth (Figure 8). In the face of an active and very

Fig. 7. NASA Administrator Keith Glennan (center), Deputy Administrator Hugh Dryden (left), and Associate Administrator Richard Horner (right)

Fig. 8. Luna 3

successful Soviet program of lunar exploration, NASA revised its lunar program plans further. In May 1959 the space agency had become aware that the Air Force was developing the Atlas-Agena B launch vehicle; the second-stage Agena B rocket was capable of restarting its engine in orbit. By October 1959 the two-stage Atlas-Agena B had been evaluated as capable of carrying more usable weight into earth orbit than the two-stage Vega, and to be almost equal to the three-stage Vega for all but the highest-energy deep space flights. Early in November, while the NASA Administrator mulled the situation of duplicated space launch vehicles, representatives of NASA and JPL decided to replace all Vega earth satellite vehicles with Air Force-furnished Atlas-Agena Bs. The six remaining Vegas were to be used exclusively for lunar exploration, the group of flights to be divided between lunar photography from orbiting spacecraft and the depositing of instruments on the moon in rough-landed capsules as recommended by NASA's lunar science group.[61] Directed to lunar research, these missions would "contribute to the understanding of the moon's origin and evolution, and provide data on surface structure and environment."[62] But four weeks later, on December 11, coincident with an agency-wide reorganization, NASA Administrator Glennan cancelled the entire Atlas-Vega project. In the interests of economy and improved reliability, he had decided that the Air Force Atlas-Agena B would replace Atlas-Vega as the interim launcher for *all* of NASA's initial space missions.[63]

On December 21, 1959, Silverstein assigned JPL seven new flights in place of the cancelled Vega missions. The first five, to be launched by Atlas-Agena B vehicles, would reconnoiter the moon during 1961 and 1962. Among the prospective planetary experiments, NASA now judged visual imaging most urgent: to obtain high-resolution pictures of the lunar surface "in the period immediately preceding impact" for use in an integrated and continuing lunar exploration program. Photography from lunar orbiters, determined to be too complex for the present state of technology, had been eliminated. But, responding to all of the scientific preferences urged at Headquarters, Silverstein requested that JPL examine the feasibility of carrying a basic group of sky science experiments on the spacecraft for use in measuring fields and particles on the way to the moon—and to reexamine proposals for depositing an instrument package on the moon that would "survive impact and then transmit significant data."[64]

Silverstein's directive called for completion of this project within 36 months, including development of the new launch vehicle, attitude-stabilized spacecraft, scientific experiments, and the communications and operational systems formulated in Project Vega. This unmanned lunar project was acknowledged to be a high-risk undertaking on short-term schedules, geared "to seize the initiative in space exploration from the Soviets as well as obtain important scientific information about the moon."[65] It would also meet another need, publicly expressed by JPL Director Pickering: to demonstrate the superiority of the "American way" to uncommitted states in the international community.[66] At NASA Headquarters, where such sentiment did not find its way into press releases, Homer Newell privately confided in a memo to the file: "In the matter of Russian competition, it is clearly understood that whether it be stated openly or not, the United States is in competition with Russia, and the stakes are very high indeed..."[67]

At JPL, Clifford Cummings, the former Vega Project Director, suggested a name for the lunar missions—Ranger. The name caught on rapidly at the Laboratory and at NASA.[68] But naming the project marked only the beginning. NASA had yet to forge a management structure capable of directing the efforts of the diverse organizations now involved, JPL had to translate esoteric concepts into functioning hardware, and planetary scientists had to validate claims to the experiments that would be carried to the moon in Project Ranger. It was a tall order to squeeze into thirty-six months.

Chapter Two

ORGANIZING THE CAMPAIGN

THE decision to adopt the Atlas-Agena B as NASA's interim launch vehicle raised thorny questions. Which of the planned Air Force Agena B versions was best suited for the Ranger lunar project? What were the technical prospects for these Agena-based missions? And how would the Air Force effort for NASA be organized and directed? To consider these important questions, in the closing days of 1959 the NASA Director of Launch Vehicle Programs appointed an Agena Survey Team.[1]

PROSPECTS AND SUGGESTIONS

The Survey Team had little difficulty concluding that the Discoverer version of the Agena B could be adapted to the lunar mission, and that Ranger should proceed using this vehicle. It was mandatory, the members emphasized, that the civilian space agency acquire early experience with the next generation of American spacecraft for deep space missions—vehicles attitude stabilized on three axes and guided by means of midcourse and terminal (lunar or planetary approach) maneuvers—before trying to develop still larger spacecraft. Moreover, with Vega canceled, other available launch vehicles would not support the payload weights required for planetary research; finally, and more persuasive, if NASA elected not to use the Agena B for Ranger, then it could not hope to challenge seriously the Soviet program of lunar exploration until the more advanced Atlas-Centaur rocket became available in 1962—a two-year delay.

To the Survey Team, such a delay was absolutely unacceptable. It appeared reasonable to expect Atlas-Agena B vehicles to perform useful spacecraft-development missions during 1961 and 1962, and the members asserted that Agena flights were "justified on this basis alone." The estimates of success for the proposed lunar missions, nevertheless, were reserved. The two NASA lunar objectives specified closeup reconnaissance of the moon and the depositing of an operating scientific instrument on its surface. The Survey Team found "a smaller but still finite expectation that during five firings either or both [objectives]...may be met."[2] But the reason for this caveat was as much the untested state of the spacecraft and ground control systems as of the launch vehicles.

Emphasizing this reservation and the importance of developing the next-generation planetary spacecraft and their guidance systems, the Team judged a lunar effort of five Agena flights to be a "minimum program." The first two of these missions were not even to be directed to the moon, but rather launched for the purpose of engineering development without the important features of midcourse or terminal guidance. NASA could attempt a lunar impact only on the third and following missions. The Survey Team's technical prognosis flashed a clear caution signal to NASA: road under construction, proceed with extreme caution, watch for falling rocks.

Management factors compounded the technical risk. Agena B meant more project participants. And in the conclusion of its interim report issued on January 15, 1960, the Team members found the overall prospects for Ranger uncertain at best. There simply was no assurance that additional agencies could or would work in harmony and for a common purpose in building NASA's lunar program. They informed Headquarters that the success or failure of the entire project would likely depend not so much upon technical complications as upon the "difficult managerial circumstances" anticipated.[3] These circumstances might be surmounted if responsibilities were clearly defined and if the major participating agencies could agree on the political importance of these missions. To ensure the timely accomplishment of the lunar missions, and to tip probabilities onto the side of success, the Team advised NASA to establish a tight management structure for this effort—a management structure "with sufficient authority," in the words of the report, to "assure rapid and effective action."[4]

It was one thing to call for the creation of a suitable management structure for the program; but, given the old rivalries and new allegiances that swirled within and around the infant NASA, it was quite another to establish an effective one.

ASSIGNING TASKS TO PEOPLE

The Agena Survey Team's managerial apprehension in early 1960 stemmed from the complex organizational and working relationships that it anticipated. Project Ranger would incorporate four major groups, each with a separate institutional identity, each confident in its own expertise, and each with its own strongly held liking for bureaucratic independence. Each would also have a hand on some part of the project's tiller.

First, and most recently formed, was NASA Headquarters, which expected to direct the enterprise.

Second was Caltech's JPL, the space agency's contract manager for Project Ranger.

Third was the Army's von Braun missile team with its German World War II experience in rocketry, still smarting as a result of government decisions that favored Air Force control of intermediate-range missiles.

Fourth was the Air Force, which, with its launch vehicle contractors, controlled the lion's share of military space and missile developments, but which still nursed a deep resentment over a civilian space agency's preempting a field it called its own.

The management framework for Ranger forged among these agencies was indeed complicated; in fact, it was at first so complex that the overlapping skeins of authority, running up and down as well as across the organizational charts, almost defied understanding.

Titular responsibility for guiding the entire effort naturally rested with NASA Headquarters. Reorganized at the end of 1959, the space agency's most powerful office in terms of program content, scope, and dollars was the Office of Space Flight Programs. Its director, Abe Silverstein, a dynamic, frequently outspoken shirt-sleeves engineer, had served as Assistant Director of NACA's Lewis Aeronautical Laboratory in Cleveland before coming to Washington with NASA (Figure 9). He possessed an enormous capacity for work, and he had

Fig. 9. NASA Space Flight Programs Director Abe Silverstein

gathered about himself a staff of talented young engineers. Until he returned to Lewis as its Director in 1961, Silverstein's influence was pervasive. Not overly concerned with the niceties of formal organization, he preferred to make or concur in all office decisions. Indeed, Silverstein so tightly centralized decision-making in his hands that even his superiors often felt they were left without any real choice of alternatives.

Silverstein's proclivity for centralized decision-making had provoked opposition among the staff at JPL. By the time Vega gave way to Ranger at the end of 1959, JPL Director Pickering already had made clear his opposition to Silverstein's managerial practices, insisting that decisions on questions of a technical nature should be made by those at the field installation having mission or task responsibility—this had been NACA and Army practice—and not in Washington.[5]

To meet Ranger and other deep space assignments for NASA and to bolster JPL's technical responsiveness, Pickering had reorganized the Laboratory at the end of 1959 and in early 1960. Lunar and Planetary Program offices appeared for the first time on organization charts, superimposed upon JPL's functionally arrayed and historic technical divisions. Pickering selected Clifford I. Cummings and James D. Burke, Vega Program Director and Deputy Director, respectively, for equivalent posts in the new Lunar Program Office.[6]

A Caltech graduate who had come to JPL after receiving his B.S. degree in physics in 1944, Cummings worked directly under Pickering on the Corporal missile project before his appointment to head the Vega Program in 1959. Deeply religious and forthright, he was a man of unquestionable integrity with a firm belief in operating projects in a tightly structured, hierarchical fashion (Figure 10).

During a fifteen-year career in missile research and development for the Army, Cummings had learned to depend and insist upon an explicit chain of command. To Pickering at JPL and Silverstein at NASA, he urged the importance of establishing a clear definition of authority and responsibility for all of the organizations involved in NASA's lunar program.[7]

In the spring of 1960, Cummings devoted increasing attention to the Centaur-launched Surveyor soft-landing project; he turned Ranger affairs over to his deputy and long-time colleague, James Burke. By the fall of the year, still serving as Cummings' deputy on the Lunar Program, Burke became the Ranger Spacecraft Project Manager at JPL as well.[8]

Burke had graduated in mechanical engineering from Caltech a class or two behind Cummings in 1945. After a stint as a naval aviator he returned to the Institute for an M.S. degree, joining the Laboratory at graduation in 1949. With a bent for the theoretical, Burke combined a swift grasp of complex systems and supportive detail and an unusual ability to devise and integrate the most promising mechanical and electrical features to achieve technical objectives. With two associates, he had solved the major guidance problem, velocity control,

Fig. 10. JPL Lunar Program Director Clifford Cummings

associated with solid-propellant ballistic missiles.[9] Soon recognized as one of the Laboratory's most perceptive research engineers, he had rapidly advanced to become deputy to Cummings on the Vega Program. Looking younger than his 35 years, often wearing an old naval aviator's jacket astride his bright green motorcycle, Burke could easily be mistaken for a Caltech student instead of JPL's manager of Project Ranger. Articulate, with a quick smile and hearty laugh, he galvanized those who worked with or for him by force of logic, persuasion, and sheer enthusiasm (Figure 11).

Burke the individual and JPL the institution remained responsible to Silverstein at Headquarters for three out of four of Ranger's system components: the new spacecraft, the deep space tracking and control network, and space flight operations and data reduction.[10] As the contract field center holding Project Ranger as part of its deep space mission assignment for the space agency, JPL was also accountable for the fourth component, launch vehicles. But authority in that area rested at Headquarters in the Office of Launch Vehicle Programs, which was directed by Major General Don R. Ostrander. Detailed to NASA from the Air Force, Ostrander was to facilitate the agency's relations in the crucial matter of launch vehicles with his fellow service officers who ran Cape Canaveral on the East Coast and procured Thor or Atlas-Agena vehicles on the West Coast. On his part, Ostrander had delegated the actual direction of NASA procurement of Agena

Fig. 11. JPL Ranger Project Manager James Burke

and Centaur launch vehicles, including those for Ranger, to Wernher von Braun's Army missile team at the newly-formed George C. Marshall Space Flight Center in Huntsville, Alabama.

Thus, whatever JPL's accountability for all of Ranger, it actually shared authority in the important area of launch vehicles with groups in Washington and in the field. This functional alignment of tasks apportioned responsibility between two field centers (Pickering's JPL and von Braun's Marshall) and between two Headquarters offices (Ostrander's Launch Vehicles and Silverstein's Flight Programs), thereby separating the spacecraft and launch vehicle components that together made up individual flight projects. Burke appreciated the management difficulties that might attend this division of responsibilities. So did the Agena Survey Team—of which Burke had been a member—whose interim report recommended to NASA a strong management structure for Agena flight projects.

And so did an increasing number of NASA officials who were bothered by the emergence of general management difficulties throughout the agency.

NEW MECHANISMS FOR MANAGEMENT

New techniques were needed to organize and control the growing space agency and its diverse flight projects. To meet these needs, on December 29, 1959, Associate Administrator Richard Horner announced the creation of a Space Exploration Program Council. NASA's leaders expected the new Council to establish the management mechanism for implementing space flight projects, and to reconcile in a timely fashion the differences between Headquarters and the field centers managing the projects. With Horner as Chairman, the Council would consist of only key center directors and Headquarters personnel: William Pickering of JPL (deep space missions), Harry Goett of Goddard Space Flight Center (earth orbiting missions), von Braun of the new Marshall Space Flight Center at Huntsville (launch vehicles) and, in NASA's two offices now sharing operating responsibilities, Flight Programs Director Abe Silverstein and Launch Vehicle Director Don Ostrander.[11] At the first meeting in Washington, D.C., on February 10, 1960, Council members confronted the problem of managing NASA's Agena B flight program.

Headquarters could not go along with the Agena Survey Team's recommendation for a single flight project manager at one field center,[12] but two alternative approaches, Horner announced, had been examined. The first called for controlling NASA's flight project activity by two coordinating committees chaired by Headquarters personnel, each with representation split between Flight Programs and Launch Vehicles along lines that approximated the existing Headquarters organization. In the second instance, a single committee or "Steering Group" would direct the effort. This group would consist of a Headquarters Chairman, but with representation from each of the three affected field installations—JPL, Goddard, and Marshall. Since no other forms of organizing NASA's unmanned space flight projects were candidates for consideration, Horner recommended acceptance of the single steering group, together with appropriate subcommittees and technical panels, as the least complicated and more desirable management mechanism.[13] Of the two choices offered, the assembled field center directors opted for the single committee, though doubtless without enthusiasm, since all preferred field center control over individual flight projects.

Various members at the meeting certainly wondered how the Air Force would be fitted into the committee management scheme. Someone pointed out that with a committee as final authority, von Braun could expect problems in coordinating Agena procurement through the Air Force representatives stationed at the Lockheed plant in Sunnyvale, California. After discussion all around, attendees agreed that a resident project engineer would be assigned to Sunnyvale from the Huntsville staff to take care of matters involving the Air Force there.[14]

Other Air Force personnel concerned with Ranger would be appointed to appropriate technical panels of the new coordinating committee.

In due course, on February 19, 1960, NASA Headquarters established an Agena B Coordination Board. Chaired by William A. Fleming, Silverstein's technical assistant in Flight Programs, the Board was to resolve "all technical problems arising in the execution of the missions within its area."[15] In point of fact, this interagency Coordination Board did not solve any of the important problems of flight project management, but created new ones. In short order, whenever a dispute arose over what participating organization was to do exactly what part of the job, be it providing trajectory calculations for lunar missions or tracking equipment during launch operations, it was labeled a technical problem and referred to the Board for resolution. Month after month in 1960, questions of roles in and jurisdiction over mundane tasks appeared and were carried forward on the Board's agenda. As time went by, molehills assumed mountainous proportions for Ranger's schedules and costs.[16]

Burke's project office at JPL, and similar offices at Huntsville, the Air Force office in Inglewood, California, and in Silverstein's shop at Headquarters, felt the effects directly. Each of these project-related organs held responsibility for some of the technical aspects or for overall guidance of Ranger. Inside or outside of the Agena B Coordination Board, none of them possessed the necessary authority to do its task.

ORGANIZING FOR RANGER: HUNTSVILLE AND THE AIR FORCE

Perhaps the most severe immediate problem resulting from this unusual division of authority and responsibility surfaced in spring and summer 1960 between the former rivals in missilery, the von Braun team at Huntsville and the Air Force headquartered in Inglewood. The Huntsville staff was primarily occupied with developing for NASA the Army-originated Saturn series of super launch vehicles large enough to carry man into interplanetary space. To care for and nurture the Agena and Centaur launch vehicles inherited from the Air Force, von Braun appointed a long-time friend and coworker in the vineyard of rocketry, Hans Hueter, to head a Light and Medium Vehicle Office. To manage Agena Systems, Hueter obtained Friedrich Duerr. Educated in the classics, Duerr was an electrical engineer from Munich who had joined the Peenemünde rocket research center in 1941, where he designed the electrical checkout and firing equipment for the A-4 (V-2) missile. Like von Braun and Hueter, he had come to the United States under "Project Paperclip"[17] in 1945, had remained with the Army's missile program, and eventually transferred to NASA. Duerr was a cultivated man in the old world tradition. "Why," an impressed JPL engineer recalled, "he even *spoke* Latin" (Figure 12).

Hueter and Duerr were to procure, first, all Agena B vehicles used by NASA and, second, all of the Agena's ground support equipment and the launch-to-injection tracking and instrumentation. The second task would be worked out in a

Fig. 12. Marshall Agena Systems Manager Friedrich Duerr

panel of the Agena B Coordination Board, with the Air Force and with Ostrander's newly formed launch Operations Directorate at Cape Canaveral. By agreement between NASA and the Air Force on the important first task, however, the Huntsville office was expected to confine itself to *supervising* the procurement of Agena B vehicles;[18] the Air Force Ballistic Missile Division in Inglewood would handle the *actual* procurement. Though NASA provided the funds, the Air Force would thus administer the contracts and direct Lockheed and General Dynamics-Astronautics. The purpose was to minimize interference between NASA's space program and the Air Force's own high-priority military satellite projects that used the same Atlas-Agena launch vehicles.

This agreement went down hard at Huntsville. Accustomed to the tradition of "in-house" development of missile and space systems,* the German-Americans now found themselves responsible for work they could control only through the

---

*"In-house" refers to the Army practice of undertaking research and development of a weapon system at a military arsenal. After building and testing the equipment, the design would be turned over to an industrial firm for serial production. The Air Force, on the other hand, preferred to contract for the complete package—research, development, and production—directly with an industrial firm.

Air Force as a second party. Worse, it was a party that had bested them in the struggle for jurisdiction over the Army's intermediate-range ballistic missile a few years before, and disdained their arsenal procurement practices as obsolete. Von Braun and Hueter requested a clarification of roles in May. Launch Vehicle Director Ostrander replied that Marshall "will provide complete day-to-day technical, administrative and financial supervision of the industrial contractors participating in the various launch vehicle programs."[19] But when directives began to be dispatched from Huntsville to Inglewood and Sunnyvale under this mandate, countercomplaints from the Air Force arrived at NASA Headquarters.[20]

By summertime, Ostrander was obliged to express the equation anew both to von Braun and to Major General Osmond J. Ritland, the Air Force Commander in Inglewood. Von Braun's group held *responsibility* for the planning and execution of NASA Agena projects, while the actual *implementation* of the projects was the task of the Air Force. General Ritland, happy to agree with Ostrander's clarification,[21] had detailed Major John E. Albert, in April, as the Air Force representative on NASA's Agena Program.[22] Albert assumed responsibility for all Air Force technical matters in Ranger, an assignment which rapidly led him to the peripatetic schedule of the NASA, Air Force Agena, and Ranger team in offices scattered from coast to coast. Quartered in Inglewood, he came in time to sign his correspondence and orders as Air Force "Director of Ranger."[23] But despite the misleading title, Albert's solid expertise, direct approach to technical problem-solving, knowledge of Air Force "rules of the road," and his genuine interest in Ranger soon proved indispensable both to Burke at JPL and to Duerr at Marshall (Figure 13).

The Ranger-Agena group in the Light and Medium Vehicle Office in Huntsville, however, numbered no more than four men, including Robert Pace, Duerr's "resident project engineer" assigned to the Lockheed plant. That number would prove insufficient to "supervise" the far-flung Agena-related activities at Lockheed and Cape Canaveral, much less in Inglewood, where, above the level of Major Albert, Ranger and other NASA projects using Agena tended to be viewed as impediments to the timely prosecution of Air Force work in space. Newly embarked on a supervisory mission for NASA, with ill-defined responsibilities and authority, understaffed in the crunch to get on with Saturn, Hueter and Duerr found themselves unable to command further support for Agena affairs at the Marshall Space Flight Center.

ORGANIZING FOR RANGER: JPL AND HEADQUARTERS

At JPL, Pickering's reorganization had provided separate Lunar and Planetary Program Offices. The Ranger Project reported to the lunar organization. To meet the special demands anticipated in space flight operations, new technical divisions also appeared: Space Sciences, which coupled scientific experiments with spacecraft; Systems, which integrated spacecraft engineering and ran test

Fig. 13. Air Force Ranger Manager John Albert with James Burke

operations; and Telecommunications, which handled spacecraft tracking, command, and control. JPL leaders reasoned that a small staff in each of the flight project and program offices could draw on the personnel in the technical divisions for the support they required. Project Ranger depended upon all of the JPL line technical divisions, new and old. By 1960, direct divisional support of Ranger had grown rapidly from 200 to nearly 600 engineers and technicians. Burke's project office counted two men and one secretary.

Under the JPL management structure, Cummings and Burke allocated the funds, planned, scheduled, and assigned Ranger tasks, and reviewed the progress of these efforts. They did not, however, possess direct supervisory authority over divisional individuals, groups, or sections. The divisions carried out the design and development of the spacecraft, scientific experiments, and tracking net and flight operations. They also directly monitored the various Ranger contracts. Each division chief and his line subordinates not only supervised but selected and placed engineers assigned to Ranger. Division managers could and did substitute key individuals at their discretion. By August 1960, as the Mariner planetary projects got up to speed at the Laboratory, personnel turnover on Ranger reached critical proportions. Cummings complained bitterly to all of JPL's division managers that the "alarming rate of loss to the Lunar Program of the talented personnel who had originally been assigned...and the breaking in of new and less qualified personnel has been detrimental to the program...and incompatible with the priority...established for the Ranger Project."[24] The rate of turnover slowed, but the divisions' closely guarded personnel prerogatives remained unchanged.

Key among the supporting divisions was the new Systems Division. Directed by Harris M. Schurmeier, Burke's Caltech classmate and friend, this division held three major flight project functions: systems analysis, which included flight trajectory, orbit, and the overall analyses to establish midcourse and terminal maneuvers; systems design and integration, which included spacecraft preliminary design, subsystem integration, and design studies; and operations, which covered spacecraft assembly and checkout, qualification and performance testing, quality assurance, and spacecraft launch and flight operations. This single division contributed the core of most engineering cadres for Ranger and JPL's other space flight projects; other technical divisions supplied additional specialized talents.[25] At the outset, in February 1960, Schurmeier appointed Gordon P. Kautz as the Systems Division's Project Engineer for Ranger. NASA's new spacecraft for deep space exploration would be, Schurmeier assured him, a challenging assignment. In fact, during the next five years, Ranger would become his life.

Kautz, a mechanical engineer, had graduated in 1941 from Fenn College (now Cleveland State University) in Cleveland, Ohio. Like the entire student body of this small engineering school, Kautz secured his diploma by means of the Cooperative Plan, alternating work and study for a B.S. degree. Employed with various eastern firms during the next fifteen years, he encountered Caltech's Jet Propulsion Laboratory for the first time in 1955. Impressed by the competence and spirit of those he met, he resigned his position and was hired at JPL. Soon after the Ranger Project Office was formally established in October 1960, Burke named him Assistant Project Manager (Figure 14).[26] As Burke's deputy, in the absence of a concrete agency-wide project structure, Kautz found ways of tracking Ranger developments across the board, including the major contractors involved. He possessed a jugular instinct for potential trouble areas; Burke, on receiving unexpected phone calls from Kautz, often associated his scratchy Ohio voice with bad news.

Together, Burke and Kautz formed the linchpin of Project Ranger at JPL. Their Ranger organizational arrangement, which was characteristic of all of the deep space missions at the Laboratory, was assumed to combine the best of two worlds. It avoided large, ephemeral, "projectized" offices while maintaining permanent operating capabilities in the individual technical division, thereby approximating the separation of disciplines found in university departments. However, the small project office, short-handed, without direct supervisory control

Fig. 14. JPL Ranger Project Assistant Manager Gordon Kautz

over personnel assignments or turnover, had to rely entirely upon the performance of the divisions. If something suddenly needed additional attention on short notice, the office staff, somehow, had to supply it. The work week of the Ranger "Spacecraft Project Manager" and his deputy quickly rose to sixty hours and more.

But in Pasadena the project staff and division personnel at least knew and understood one another. The same could not as yet be said for those groups beyond its confines, in the larger world of Project Ranger. By institutional default, Burke and Kautz found themselves enmeshed in an agency-wide policy question meant for higher levels than that of the project: Where did the field center Spacecraft Project Manager's authority end and that of the NASA Headquarters offices, Marshall, and the Air Force begin? When pressed by Cummings at JPL for an answer, Silverstein at NASA insisted that the matter simply was not crucial;

with everyone anxious to make Ranger a success, he explained, these questions would work themselves out.[27]

In July, Silverstein notified the Laboratory that his Lunar and Planetary Programs Division within the NASA Flight Programs Office would be the group directly responsible for the conduct of the project, and JPL's point of contact in guiding Ranger developments.[28] The Lunar and Planetary Programs Division originally had been formed in the autumn of 1959, as NASA Administrator Glennan began emphasizing a moon flight program. It was headed by Edgar M. Cortright, formerly from NACA's Lewis Aeronautical Laboratory. Silverstein and Newell had moved this exceptionally capable aeronautical research engineer from meteorological satellites and charged him with forming the new office (Figure 15). Cortright promptly assembled a team,[29] and so far as NASA Headquarters was concerned, the Ranger project structure was now complete.

Fig. 15. NASA Lunar and Planetary Programs Chief Edgar Cortright

The project management issue, nevertheless, was hardly solved. Cummings at JPL, intent on comprehending the position and roles of the numerous agencies, offices, committees, and panels involved, prepared what is probably the only organization chart ever made of the early Ranger project (Figure 16). Whatever the merits in capturing the elusive project organization on paper, the exercise offered small comfort to those obliged to live inside that framework. When Burke and Kautz operated beyond the confines of JPL, they lacked authority; diplomacy and appeals to reason and national honor proved the only available means for gaining compliance with requests for action. And these proved to be dull instruments with which to cut the hard schedules established for Ranger.

Launch Vehicles: Management Problems Materialize

The managerial faults were rapidly exemplified in the launch vehicle situation. Upon visiting Inglewood and Sunnyvale as early as February 1960, JPL engineers asked for details about Agena to support spacecraft development. They found that "official requests for and answers to NASA inquiries" were supposed to be sent through circuitous Air Force-to-NASA Headquarters channels. "The Inglewood and Sunnyvale people seemed perfectly willing to give us unofficial answers to our questions, but stressed that these could only be considered tentative answers." Incredulous, the JPL visitors asserted: "This situation could result in some serious bottlenecks in our program unless the future working relationship can bypass this long route."[30] But that long route and working relationship had already been agreed upon between NASA Headquarters and the Air Force in Inglewood. During 1960 it could be shortcut to a limited extent by those working at the engineering level, but it was not to be bypassed at Lockheed on task assignments, change orders, or schedules.[31]

In March 1960 NASA and the Air Force agreed on Lockheed's task. The firm would provide three major launch vehicle components: the Discoverer[32] model of the Agena B, made standard for NASA use; a spacecraft adapter to fit the forward end of the Agena to which the Ranger spacecraft would be fastened; and an over-the-nose metal fairing, or shroud, modified from a standard version used for Air Force space missions, to protect the spacecraft during ascent through the atmosphere. A letter contract for sixteen Agena Bs and this associated equipment, issued through the Air Force on April 12, 1960, permitted work to proceed.[33]

Herschel J. Brown, Corporate Vice President and General Manager of Lockheed's Missile and Space Division, selected Harold T. Luskin to manage the NASA Agena program. In the charged, often rapacious atmosphere that characterized space affairs in the early 1960s, Harold Luskin coupled technical expertise with integrity and decency; he commanded nearly universal respect—even affection—among members of the aerospace engineering fraternity. An American astronautical pioneer in his own right, he had as a young man contributed a chapter on spacecraft temperature control to the 1946 Rand earth satellite study. Now he would have a hand in the NASA unmanned lunar program (Figure 17).

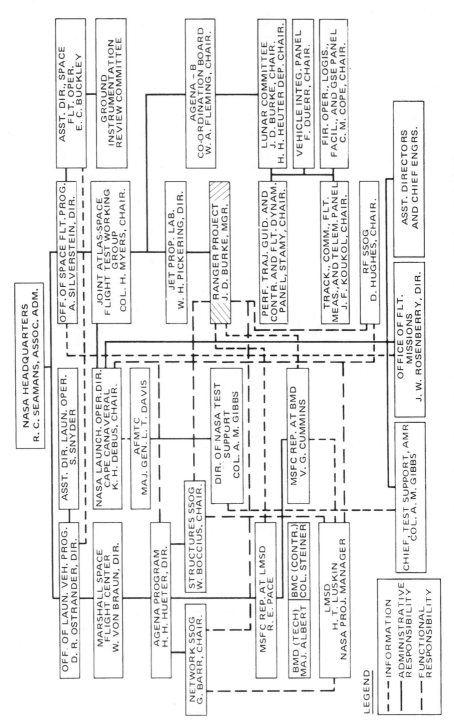

Fig. 16. 1960 Ranger Project Organization (JPL Organization Chart R-5, August 15, 1960)

Fig. 17. Lockheed Agena Manager Harold Luskin

However meritorious the selection of Luskin, the job did not come with a viable organizational base at Lockheed. NASA Agena work remained part of the Air Force satellite activities at the huge plant in Sunnyvale. His people few in number and physically scattered, Luskin found himself in a position all too familiar to Duerr in Huntsville and Major Albert in Inglewood: he had to meet a set of special NASA requirements amid larger, ongoing design and production efforts in which equally strong managers resisted change—from detailing personnel to allocating office space—that might interfere with their own programs. At the same time, established NASA–Air Force command complications were built into the job. The Air Force intermediaries attached to the Lockheed Air Force Plant Representative's Office stood between Luskin and NASA personnel, insulating the NASA effort from the higher-priority, classified Air Force programs. Representatives from JPL and Marshall found themselves unable to obtain Lockheed schedules and other vital Agena information, or even to discuss certain questions, because they lacked what the military liked to call a "need-to-know." NASA should submit its requirements, Air Force personnel insisted, and they would see that the necessary Agena Bs and other equipment fitted together and were delivered at Cape Canaveral in time for launch.[34]

If Luskin needed any more difficulties, he got them in a major contract change. Late in July NASA Headquarters determined that it had firm mission requirements for only nine Agena Bs instead of sixteen.[35] On August 4, 1960, the space agency notified Lockheed of this reduction in numbers. Negotiation of a definitive contract, already well underway, halted. Agreement on a final contract, expected in August, would not be realized until February 1961. Among a growing number of Air Force satellite projects at Lockheed, the NASA work amounted to a very small piece of the action; whatever institutional leverage Luskin possessed with sixteen Agenas was reduced precipitously by an order for only nine. Luskin soon found that he could not command the needed Agena support for NASA in Sunnyvale.

First hardware deliveries testified to the dismal situation. In September JPL received a mockup of the forward portion of the Agena, the spacecraft adapter and nose fairing. Project engineers immediately returned all of the articles to Lockheed because they detected "inconsistencies in design and construction."[36] Burke called a meeting among representatives of the Air Force, Marshall, and Lockheed a few weeks later to evaluate the status of all three components, including the mechanization required to separate the nose fairing and the spacecraft from the Agena. During the meeting it became apparent that numerous deficiencies still remained.[37] Worse, any delay in applying corrective measures would jeopardize the entire Ranger schedule.

Under pressure from Burke at JPL and the Hueter–Duerr team in Huntsville, the Air Force agreed to a Board of Inquiry to investigate conditions in the NASA Agena B program at Lockheed. Inglewood assured Hueter that top management at Lockheed's Missile and Space Division would be directed to take whatever "positive corrective action" might be required to maintain the NASA Agena B program schedule.[38] Even before the Board convened, Burke, Hueter, and Albert J. Kelley, Ostrander's representative in charge of the NASA Agena B program at Headquarters, had no doubts about what was required. Among other things, establishing an immediate NASA need-to-know at Lockheed was imperative.[39] Around Christmas, Hueter, writing for von Braun, reported the findings of the Board of Inquiry to Herschel Brown at Lockheed. Hueter candidly told Brown that the NASA program required more and better qualified engineers, increased authority delegated to Luskin as program manager, improved coordination among Lockheed design groups, and approval of direct engineer-to-engineer contact with NASA organizations together with a NASA need-to-know.[40]

With the consent of the Air Force, Brown vigorously pushed through the needed reforms. Luskin, whose NASA Agena office had been given its own quarters and elevated to project status at Lockheed a few weeks before the findings were released, acquired the personnel and authority he needed. On February 6,

1961, the Agena contract was signed; then, on February 14, the Air Force approved direct contact between NASA and Lockheed personnel in the new *NASA Agena B Launch Vehicle Program Management Organization and Procedures* agreement signed by the Deputy Administrator of NASA and the Air Force Commander of the Air Research and Development Command. No longer an Air Force foster child, NASA's Agena work would become a full-fledged member of the Lockheed "family" with a legitimacy all its own.

With the new year, events in the launch vehicle department appeared definitely on the upturn. Luskin's shop delivered the reworked spacecraft adapter and nose fairing to JPL in January 1961. The equipment performed satisfactorily in "match-mate" tests with the spacecraft[41]—a hopeful sign that engineering and management problems at the Sunnyvale plant were truly a thing of the past.

REORGANIZING FOR RANGER

All the while, officials elsewhere had been working to reorganize the entire NASA Ranger management structure, still frustratingly awkward, along the single project manager lines originally recommended by the Agena Survey Team. At its core, the Ranger management problem remained one between the NASA installations in the field and the Headquarters offices in Washington. Everyone liked to agree with Silverstein that there was universal dedication to move ahead with Project Ranger as rapidly and efficiently as possible. But events did not match either expectations or good intentions. "Each didn't understand the other's methods," Burke explained, "and there wasn't an agreed management under which we could dispute points and have them resolved by higher authority."[42] As conditions on the project deteriorated in 1960, JPL continued to press the question at Headquarters: Would authority for the technical control over flight projects be delegated to the responsible space flight center, or would it continue to be exercised by means of a Headquarters-chaired coordinating committee—attendant difficulties notwithstanding?

The smouldering discontent was brought directly to Administrator Glennan's attention on Bastille Day, July 14, 1960, at the third meeting of the Space Exploration Program Council. At JPL, Pickering had consulted with his lunar and planetary program directors preparatory to the meeting in Washington. Cummings, mincing no words, urged his superior to "denounce the continual fumbling over responsibilities and authority which has resulted in confusion and delays in getting organized..." The Agena B Coordination Board "should be dissolved and program directors clearly delegated full power to establish any management arrangement which has not already been determined."[43] At the Space Exploration Program Council meeting, Pickering, joined by his fellow center directors Harry Goett and von Braun, urged the "liberation" of NASA flight projects from management by committee. But despite the problems encountered by the committee in question, the Agena B Coordination Board, Glennan and his Headquarters cohorts rejected the recommendation.[44]

But the Headquarters contingent was by no means unanimous. However desirable they might regard a Headquarters-chaired coordinating board, the Board itself had obviously failed to resolve many of the disputes over the task assignments and responsibilities of the various project organizations in a timely fashion. Glennan, moreover, opposed increasing the staff at Headquarters to handle project details[45] and, contrary to Silverstein, he held no brief against the decentralization or delegation of authority. In the days that followed the July 14 meeting, Glennan permitted Albert Siepert, Director of the NASA Office of Business Administration and an attendee at the meeting, to consider alternatives for managing NASA flight projects, including direct delegation of the task to NASA space flight centers.

The fortunes of centralized control declined further on September 1 when Richard Horner, a proponent and architect of the Agena B Coordination Board, resigned his position at NASA to return to the business community. As his replacement for the agency's general manager, Glennan selected Robert C. Seamans, Jr., a former professor of aeronautical engineering at MIT and the manager of RCA's Airborne Systems Laboratory. Accustomed to delegating authority as a matter of course, Seamans chose to involve himself at NASA primarily with agency-wide management questions; Siepert now received active encouragement, and the support of Glennan, in drafting a formal agency proposal for flight project management.

During the next four weeks Siepert and an associate, Jack Young, prepared *A NASA Structure for Project Management*. Completed shortly before NASA's Fourth Semi-Annual Staff Conference—where agency policy, advance planning, and management issues were to be reviewed, discussed, and, it was hoped, resolved—Siepert's brief called bluntly for decentralization. A few days before the conference convened in Williamsburg, Virginia, Glennan circulated the document to all those invited to attend. The management proposal, he declared, elaborated three general concepts "with which I agree":

> 1. That the approaching volume and magnitude of NASA projects will make it impossible to follow our present ad hoc approach to project management problems; a basic concept of project management is needed.

> 2. That NASA intends to fix at the Space Flight Centers responsibility for the execution of NASA projects.

> 3. That NASA needs to establish a method which assures a closer link between technical project direction and associated business interests.[46]

The proposal specifically recommended approval of a NASA General Management Instruction that defined project terminology and fixed the authority and responsibility for flight projects at the space flight centers. Marshall Space

Flight Center, providing launch vehicles and launch operations under Ostrander's office, would support a project manager located at JPL or Goddard. Silverstein's office would allocate the funds, establish objectives, and review project developments. In the event of unresolved disputes, Associate Administrator Seamans—not a coordinating committee—would decide the issue. On October 19, 1960, NASA's leaders took up Siepert's proposal and endorsed it.[47]

On January 19, 1961, one day before leaving office with the rest of the Eisenhower administration, Glennan signed NASA General Management Instruction 4-1-1, *Planning and Implementation of NASA Projects.* Judged by many to be the single most important management concept generated and adopted by NASA in its formative years, the instruction abolished the Agena B Coordination Board method of project management. It delegated to space flight centers direct authority and responsibility over their assigned tasks. Its central feature involved what was termed a Project Development Plan. As a controlling, written instrument signed by all participating organizations, the development plan came to be viewed at JPL as the "contract" or statement of commitment and understanding between the Headquarters offices and all NASA field installations having responsibilities for a system or major part of a project.[48]

Coming about the same time as the resolution of Luskin's difficulties at Lockheed, the new mandate for project control cleared the managerial way for Ranger. Burke, now officially named Ranger Project Manager by NASA Headquarters,[49] was eager to get on with the assignment. But an entire year had elapsed since the Agena Survey Team first recommended decentralized project management. And though a sensible organizational structure was indispensable, achieving it had cost project officials valuable time that might otherwise have been spent on the technical details of the spacecraft and launch vehicle. Their development remained on schedule, at least on paper; however, only twenty-four months remained in the lifespan planned for Project Ranger. Official schedules pegged the flight of Ranger 1 in July 1961, a scant six months away. Time was running out—fast!

Chapter Three

SPACE SCIENCE AND THE RANGER
MACHINE

ALL through the managerial fracas, JPL engineers had been pushing ahead with the design of the Ranger spacecraft. The origins of the evolving Ranger machine went back to March 1958, when Major General John B. Medaris, Chief of the Army Ordnance Missile Command, had authorized JPL to evaluate a new launch vehicle based on the Jupiter and eventually known as Juno IV. Paper analyses concluded that Juno IV could inject several tens of kilograms into deep space trajectories. That prospect was enough for JPL Director Pickering. Eager to leapfrog beyond the moon and begin exploring the inner planets before the Soviets, he instructed Daniel Schneiderman, in charge of JPL's payload design group, to define the technology and preliminary concepts for a Juno IV planetary spacecraft, one that would weigh some 134 kilograms (300 pounds) and could be guided to an encounter with Mars.[1]

A PLANETARY MACHINE FOR SPACE SCIENCE

John Small, Chief of Mechanical Engineering, assigned a few engineers to work with Schneiderman on the new spacecraft design: James Burke, Walter Downhower, Marc Comuntzis, and John Casani. For some weeks this group devoted considerable time to the problem of radio communications. To communicate adequately from planetary distances, the spacecraft would require a high-gain antenna—in conventional terms, a narrow-beam "dish"—mounted and hinged so as to point continuously at the earth. There, sensitive receivers, powerful transmitters, and very-high-gain antennas would complete the circuit. All the while, the spacecraft dish antenna would have to be kept pointing in the right direction through an appropriate method of stabilizing the attitude of the spacecraft itself. Early American spacecraft, such as the Explorers and Pioneers, had been stabilized by spinning the vehicle along its roll axis. For flights to the planets, JPL engineers deemed it necessary to have complete control of the spacecraft in all three axes, roll, yaw, and pitch. This would ensure precise pointing of the experiments and the antenna, and maximize solar power collection and thermal control (Figure 18). With full attitude control, the flight trajectory of a planetary spacecraft could also be refined by igniting a rocket engine on board in a "midcourse maneuver." A small rocket would be able to compensate for minor guidance errors introduced by the launch vehicle, thus permitting the spacecraft to approach more closely or even hit a celestial target.[2] In June 1958 a Pickering-appointed JPL review team approved these preliminary design concepts.[3]

Though the Juno IV program was canceled in October 1958,[4] the JPL martian spacecraft continued to evolve, now to be used in Project Vega. In addition to the features of a high-gain antenna and full attitude stabilization, engineers designed the spacecraft so that its longitudinal axis would point continuously toward the sun (except during midcourse or terminal maneuvers), since it was uncertain whether the earth could be "seen" by onboard sensors at planetary distances. This decision simplified the problem of maintaining thermal equilibrium on the spacecraft and permitted the use of solar cells on fixed panels as a primary source of electrical power.[5] With the Vega launch vehicle, JPL's Schneiderman and his colleagues also had more ample weight figures with which to work. Assuming a parking orbit technique, in which the spacecraft is ultimately launched into the solar system from earth orbit, Vega calculations yielded spacecraft weights of 360 kilograms (800 pounds) deliverable to the vicinity of the moon, and approximately 205 kilograms (450 pounds) to Mars or Venus.[6]

But at the start of the space program, the additional spacecraft weight was inseparable from the expensive and unproved Vega launch vehicle. The higher launch costs would ultimately mean fewer flights. The unproved launcher would also likely fail in some early misions; estimates of reliability for the individual Vega stages ran at about 0.5. Each flight could be expected to have less than half a chance of succeeding even if one were to neglect any failures that might occur in the complex spacecraft. To enhance reliability, therefore, members of the JPL design group decided to use a single spacecraft design repetitively. They identified and combined functions common to all flights, with the resultant basic unit termed a spacecraft "bus." Here the design would vary as little as possible for each mission. The bus would provide electrical power, communications, attitude control, command functions, and a midcourse maneuver capability. To the bus would be added the scientific instruments and associated equipment that together comprised a mission package.[7]

Since nearby equipment could adversely affect many kinds of scientific experiments (limits on fields of view, radio or magnetic interference, etc.), tall structures and extendable booms were called for. Like a chrysalis, the craft with its solar wings would be folded into a compact and rugged package inside the conical aerodynamic nose fairing of the launch vehicle, then open out in space while cruising to Mars or Venus. By mid-1959 these preliminary plans had been

THE PLANES OF MOVEMENT THROUGH WHICH THE SPACECRAFT PASSES ARE ILLUSTRATED BELOW. THE ATTITUDE OF THE SPACECRAFT IS ESTABLISHED AND MAINTAINED BY COMMANDS EXECUTED BY THE ATTITUDE CONTROL SYSTEM IN CONJUNCTION WITH OTHER SYSTEMS OF THE SPACECRAFT. THE ATTITUDE CONTROL SYSTEM REGULATES THE MOVEMENTS OF THE SPACECRAFT IN THESE THREE PLANES AND ACCOMPLISHES THESE MOVEMENTS BY IMPARTING SMALL AMOUNTS OF THRUST TO THE SPACECRAFT BUS BY THE DISCHARGE OF NITROGEN GAS FROM COLD–GAS JETS POSITIONED ON AND ABOUT THE HEXAGONAL FRAME OF THE SPACECRAFT. THE DESIRED POSTURE OF THE SPACECRAFT IS DICTATED BY ITS SPATIAL RELATIONSHIP WITH THE SUN AND EARTH. IT MUST KEEP ITS SOLAR PANELS FACING THE SUN FOR SOLAR POWER AND THE HIGH–GAIN ANTENNA FACING EARTH FOR COMMUNICATION. THE SUN AND EARTH SENSORS, PART OF THE ATTITUDE CONTROL SYSTEM, CONTINUOUSLY GENERATE SIGNALS WHICH ARE COMBINED WITH INFORMATION FROM THE RATE GYROS OF THE AUTOPILOTS. THE LATTER SUPPLY THE INFORMATION DEFINING THE RATE OF MOVEMENT. THIS COMBINED INFORMATION IS TRANSLATED BY THE ATTITUDE CONTROL SYSTEM INTO COMMANDS TO THE APPROPRIATE COLD–GAS JETS OR COMBINATION OF JETS. THIS CYCLE IS REPEATED UNTIL THE DESIRED ATTITUDE OF THE SPACECRAFT IS ESTABLISHED AND SUBSEQUENTLY CONTINUES TO FUNCTION MAINTAINING THIS POSTURE.

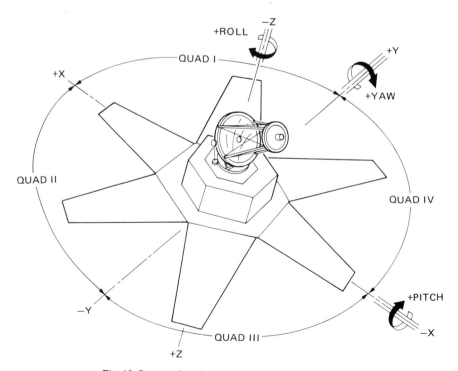

Fig. 18. Spacecraft Attitude Stabilization in Three Axes

communicated to NASA and described to Congress.[8] They would provide the basic model for the Ranger machine (Figure 19).

THE VEGA-RANGER: WHERE PLANET AND SKY SCIENCE MEET

The JPL Space Sciences Division, meantime, drew up a program of experiments for the Vega flights. Created in the reorganization that followed JPL's transfer to NASA, this division drew together JPL experimenters in one group, prepared instruments for flight missions, and processed the information returned to earth for release to the experimenters.[9] As chief of the new division, Pickering named Albert R. Hibbs, an articulate, exceptionally bright young physicist with a Caltech Ph.D. One of the founders of the Lunar and Planetary Exploration Colloquia and a participant in the Explorer and Pioneer IGY projects, Hibbs was already well known within the community of space scientists and engineers.

When Glennan decided to emphasize the lunar objective in July 1959, neither Hibbs nor his JPL engineering colleagues abandoned the martian spacecraft. They preferred to stick with it even though on a 66-hour flight to the moon batteries could suffice in place of solar panels, and a high-gain antenna was unnecessary for communicating over a distance of 400,000 kilometers (a quarter million miles). Adapted to lunar missions, the high-gain antenna, instead of being used for long-range, narrowband communication, would now be used for relatively wideband transmission such as television at lunar distances. The bus and passenger concept, three-axis attitude stabilization, and solar power, its designers reasoned, could be used to develop the technology required for the planetary flights postponed to 1962.[10]

The question of just what scientific experiments the Ranger machines would carry was a matter for decision not only by JPL but by NASA Headquarters, particularly Silverstein's office of Space Flight Programs. Silverstein had awarded the responsibility for determining what scientific instruments would ride into space to his quiet assistant director, Homer E. Newell. A crew-cut, 44-year-old scientist-administrator, Newell was the son of an electrical engineer, who had helped him provision a chemical laboratory in their home in Holyoke, Massachusetts. His mother, an accomplished musician, inspired dinner-time exchanges concerning events of the day. Newell, who earned his Ph.D. in mathematics from the University of Wisconsin, joined the Naval Research Laboratory after World War II. He was soon appointed head of the Rocket Sonde Branch, and then, a few years later, Superintendent of the Atmosphere and Astrophysics Division. Together with James Van Allen, Lloyd Berkner, and other members of the ad hoc Upper Atmosphere Rocket Research Panel, he directed America's postwar assault on the physics of the upper atmosphere, seeking answers to theoretical questions involving earth–sun electromagnetic relationships. Revealing the type of science he preferred, Newell assured the readers of his first book that "the ultimate technical strength of the nation is to be found in the continuing accumulation of fundamental knowledge" (Figure 20).[11]

VEGA PAYLOAD (MARS MISSION) EXPERIMENTS

Measure density and composition of solar corpuscular radiation, cosmic rays, Mars gravity field and reflected light from Mars.

Main experiment might be a spectrophotometer to measure the characteristics of spectral light reflected from Mars in order to determine the possibility of the existence of life on Mars

Fig. 19. Vega Spacecraft Model

Fig. 20. NASA Space Flight Programs Assistant Director for Space Sciences Homer Newell

In October 1958, after serving as Science Program Coordinator for the IGY Vanguard Satellite project, Newell moved to Silverstein's office in the fledgling NASA. He promptly established procedures for shaping the agency's space flight program. In arrangements worked out in succeeding months, the Space Science Board of the National Academy of Sciences contracted to provide advice on long-term goals to NASA. While specific research objectives and operating policy were of course prerogatives reserved to the space agency, Newell recognized that the success of the research program would ultimately depend upon the support it could command within and outside the agency, especially from those scientists engaged in its work. He elected to employ small contingents of qualified scientists at Headquarters and the field centers. In accord with the precedent of the IGY and sounding rocket programs, he chose to delegate the bulk of NASA's basic space research, primarily to university scientists. NASA would tap the best scientific talent available in the nation, and the best scientific talent, Newell hoped, could be expected to provide vocal and articulate support for NASA's science program.[12]

Newell's office made known the opportunities for conducting space research from NASA satellites and space probes to university scientists directly by correspondence, in announcements at annual meetings of the various science societies, and by means of special conferences convened for that purpose, including the first national Conference on Space Physics sponsored jointly by NASA, the National Academy of Sciences, and the American Physical Society in April 1959.[13] Numerous proposals for experiments—both solicited and unsolicited—arrived in short order. In fact, Newell received far more proposals than could be accommodated on authorized NASA space launch vehicles. Ad hoc space science working groups assisted in selecting and developing the most promising of these experiments for NASA flight missions. Composed of well-known university and NASA scientists appointed by Newell, one group dealt with space radiation, another with magnetic fields and plasmas in space, and the third considered experiments for lunar exploration after Robert Jastrow and Harold Urey convinced Newell of its importance.[14]

By the fall of 1959, the Headquarters working groups and Hibbs' scientists at JPL had agreed upon a rough priority of experiments to be carried on the six authorized Vega lunar flights. Since the experiments of sky scientists did not require a precise trajectory, the first two Vega test missions would be instrumented to measure more closely the fields and particles between the earth and the sun. Representations by planetary scientists ensured that the lunar flights to follow would be confined exclusively to investigating the moon rather than the interplanetary medium. This division of available vehicles proved congenial to the interests of both science parties.

In November Hibbs submitted a precis of these deliberations to the JPL Vega Project staff. Planning of the sky science experiments for the first two launches, he informed Cummings and Burke, had reached an advanced stage. Measuring solar corpuscular radiation—the solar atmosphere—and magnetic fields was the objective preferred by these scientists. Second preference had been assigned to investigating the neutral hydrogen cloud that surrounds the earth, third preference to measuring cosmic rays, and fourth, to measuring micrometeorites. Those identified as experimenters were James Van Allen of the State University of Iowa, and others from the Naval Research Laboratory, Goddard Space Flight Center (GSFC), Los Alamos Scientific Laboratory (LASL), JPL, and Caltech. The later lunar rough landing missions would carry the seismometer being developed cooperatively for NASA by Caltech and Columbia University. Additional lunar experiments would be designated in the near future, upon completion of the Vega spacecraft design.[15]

But Glennan's cancellation of Vega occurred just as the spacecraft design neared completion. After Silverstein then ordered the five-flight Atlas-Agena B Ranger Project, JPL Director Pickering and Vega Program Director Cummings faced a choice: considering only the spacecraft, JPL could begin anew on the design of a less complicated lunar machine for use on the Atlas-Agena; or it could proceed with the planetary craft, building on the work already accomplished. Pickering and Cummings opted for the planetary vehicle. If that course was risky, it nevertheless offered a rapid development of planetary flight technology and experience, nearly on Vega schedules. Besides, the time and money already invested, and the experience accumulated, would be saved. Moreover, if pushed in a concerted fashion, the planetary machine might well return valuable data about the moon in advance of the Soviets. And by using the Vega planetary craft for the lunar missions, NASA could also accomplish the scientific program already agreed upon by the ad hoc working groups at Headquarters and Al Hibbs' Space Sciences Division at JPL.

On December 28, 1959, representatives of NASA Headquarters met with Pickering and his senior staff to review plans for Ranger. Pickering recommended pursuing the Vega spacecraft and flight mission plans in Project Ranger. The savings in time and money appealed to everyone, and so did the possibility of surpassing the Soviets.[16] Headquarters personnel authorized JPL to proceed in Project Ranger with what had been the Vega planetary spacecraft and flight sequencing. By implication, they also sanctioned Vega's scientific experiments—sky science ventures on the first two Ranger engineering test flights, planetary science on the three lunar rough-landing missions. Newell assured his associates at NASA that the five Ranger missions appeared to be "well thought out from the scientific point of view."[17] Early in 1960 Headquarters officially approved the list of experiments for all five flights (Table II).[18]

Capable of performing sky science experiments and of exploring the distant planets, the Ranger spacecraft would be far more complex than a vehicle designed for lunar missions had to be. The commitment to develop such a machine demanded a major technological advance—from small spin-stabilized earth-orbiting vehicles weighing tens of kilograms to a completely attitude-stabilized planetary machine weighing hundreds of kilograms. The first earth satellites had been equipped with small radios to transmit data to stations several hundred kilometers below. The Ranger spacecraft would be able to accomplish tasks automatically in response to a preset program, to process and transmit diagnostic and scientific information, and to receive and respond to commands sent to it from the earth—not just at lunar distances but at interplanetary distances of millions of kilometers. Whether the decision to pin the fortunes of lunar exploration on the development of such a machine within a little more than a year was wise or not, officials at NASA and JPL recognized the task as a high-risk venture—one sure to tax the technology and skills both in and out of the Laboratory in Pasadena.

Table II. The Original Ranger Space Science Plan

Ranger test flights	
Experiment	Agency and scientist
Solar corpuscular (plasma) detector	JPL: M. M. Neugebauer, C. W. Snyder
Photoconductive particle (trapped radiation) detectors	State University of Iowa: J. A. Van Allen
Rubidium vapor magnetometer	GSFC: J. P. Heppner
Vehicle charge	Not specified
Triple-coincidence cosmic ray telescope	University of Chicago: J. A. Simpson
Cosmic-ray integrating ionization chamber	Caltech/JPL: H. V. Neher, H. R. Anderson
Lyman alpha scanning telescope (hydrogen geo-corona)	Naval Research Laboratory: T. A. Chubb
Micrometeorite dust particle detectors	GSFC: W. M. Alexander

Ranger lunar flights	
Experiment	Agency and scientist
Single-axis passive seismometer[a]	Caltech: Frank Press
Capsule temperature measurement	Columbia University: M. Ewing
Maximum deceleration at impact measurement	
Gamma-ray spectrometer[b]	U.C. San Diego: J. R. Arnold JPL: A. E. Metzger LASL: M. A. Van Dilla
Television camera[b]	Experimenter Team representing various agencies to be assigned

[a] Capsule to survive hard landing.

[b] Spacecraft bus — destroyed on impact.

CREATING THE RANGER MACHINE

On February 1, 1960, Dan Schneiderman issued the design concepts and criteria for the Ranger spacecraft.[19] Since the Vega third-stage vehicle had employed a hexagonal truss and six longerons, the Ranger spacecraft possessed hexagonal symmetry (Figure 21). JPL experience in electronics packaging on the Corporal and Sergeant missiles suggested the use of modular construction of electronic equipment to withstand high levels of vibration. These modules would be packaged in plain rectangular boxes, one bolted to each of the six sides (Figures 22 and 23). Above the 1.5-meter (5-foot) diameter hexagon would rise a "tower"

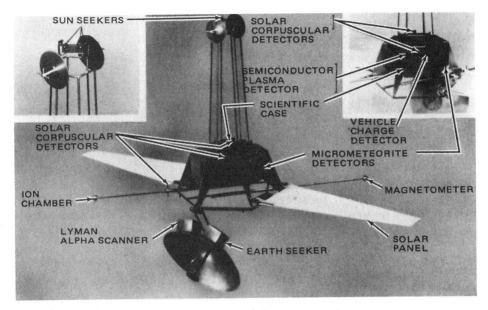

Fig. 21. Ranger Spacecraft Preliminary Design

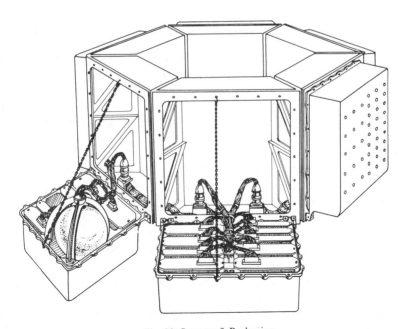

Fig. 22. Spacecraft Packaging

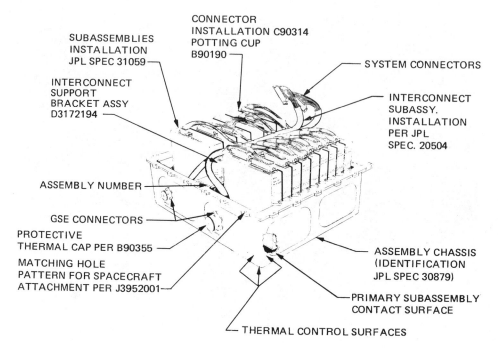

SUBASSEMBLIES
INSTALLATION
JPL SPEC 31059

CONNECTOR
INSTALLATION C90314
POTTING CUP
B90190

SYSTEM CONNECTORS

INTERCONNECT
SUPPORT
BRACKET ASSY
D3172194

INTERCONNECT
SUBASSY.
INSTALLATION
PER JPL
SPEC. 20504

ASSEMBLY NUMBER

GSE CONNECTORS

PROTECTIVE
THERMAL CAP PER B90355

MATCHING HOLE
PATTERN FOR SPACECRAFT
ATTACHMENT PER J3952001

ASSEMBLY CHASSIS
(IDENTIFICATION
JPL SPEC 30879)

PRIMARY SUBASSEMBLY
CONTACT SURFACE

THERMAL CONTROL SURFACES

Fig. 23. Typical Assembly Package

that supported the fixed low-gain antenna and various scientific experiments. Conforming to design concepts that called for tall structures, the tower would become a permanent feature on all early JPL lunar and planetary spacecraft. Below it, two solar panels would be attached to opposite sides of the bus; a 1.2-meter (4-foot) diameter high-gain dish antenna would be stowed directly beneath the spacecraft. Both the solar panels and the dish antenna would be hinged at the frame, designed to swing out from the bus after the spacecraft had been placed on its trajectory in space.

This spacecraft was to be used on the first two test flights, which, with their sky science purpose, were designated Block I. The three further flights, which would photograph the moon and deposit seismometers on its surface, were designated Block II. The three Block II spacecraft would incorporate the midcourse engine and maneuver capability, and a tower to support the seismometer capsule and its retrorocket. Prior Atlas-Agena studies had yielded an estimated weight of 360 kilograms (800 pounds) for the lunar spacecraft, comparable to the weight-lifting performance expected of Atlas-Vega.

Systems Division Chief Schurmeier created a Spacecraft Design Specifications Book to guide the JPL design activity and to describe the mission objectives, design philosophy, design restraints (weight, schedule, power, etc.), and functional requirements (what each subsystem was intended to do, and their interactions).

NASA, through Cummings and Burke at JPL, would provide project objectives and guidelines; the Systems Division would integrate the complete design and test the assembled spacecraft. Other JPL technical divisions would prepare Ranger's detailed design, procure parts, fabricate components, and test each subsystem. To conserve time and quickly demonstrate the technology for planetary missions, changes in the spacecraft design were to be minimized.

As a surrogate of man in the cosmos, the Ranger spacecraft was to serve a four-fold purpose: first, to deliver its scientific cargo to a celestial target within certain tolerances, then position the experiments, perform the proposed scientific program, and transmit the results back to earth. Priorities for Project Ranger, released by Cummings on February 17, 1960, conformed to this progression. In descending order of importance, he committed JPL to (1) develop the spacecraft technology, (2) maintain schedules, (3) establish industrial support for NASA-JPL planetary flight missions, and (4) support science.[20] Although included on the test flights, sky science thus initially held a low position in the scheme of project activities as Cummings and Burke sought to create the technology first, then pursue lunar investigations. In authorizing the eight experiments for the engineering test missions, Silverstein agreed that sky science was not to interfere with the creation of the spacecraft technology.[21]

But NASA had specified scientific investigations of the moon as the objective of the Block II Rangers—to acquire more knowledge about our celestial neighbor.[22] Approved in early 1960, the planetary experiments included a television camera to return closeup photographs of the moon, a seismometer to be deposited upon it, and a gamma-ray spectrometer to determine the chemical composition of the surface material.[23] The midcourse trajectory and terminal attitude maneuvers to be incorporated in these spacecraft would position the television camera to take pictures of the moon, and permit release of the seismometer capsule just prior to impact. However, in a compromise between the scientific objective stipulated by Headquarters and the necessity to create the required technology, technology remained the clear emphasis for the engineers charged with Ranger's prosecution. As the project got underway, the priorities established at JPL revealed the essential purpose of all five Ranger flights to be the development of "basic elements of spacecraft technology required for lunar and planetary missions."[24]

Under direction of the Systems Division, design of the two test spacecraft moved ahead smoothly, and was completed by May 1960 (Figure 24).[25] To reduce complexity to manageable proportions, the engineers divided the spacecraft system into functional subsystems that corresponded to the responsibilities of the line divisions at the Laboratory. The Telecommunications Division, headed by Eberhardt Rechtin, handled the spacecraft command, telemetry, and communications subsystem; Eugene Giberson and the Guidance Control Division looked after attitude control, power, and central control; Geoffrey Robillard's Propulsion Division was responsible for spacecraft pyrotechnic squibs and actuators, and—on

Block II—midcourse propulsion; the Engineering Mechanics Division, led by Charles Cole, developed the thermal control, structures, electronics packaging, and cabling for the spacecraft; and Hibbs' Space Sciences Division dealt with the scientific instruments and their control data automation subsystem. Apportioned so that one individual could comprehend an entire subsystem, each one was assigned to a "cognizant engineer" by the respective division chief; the cognizant engineer, in turn, further subdivided the work and guided subsystem development.

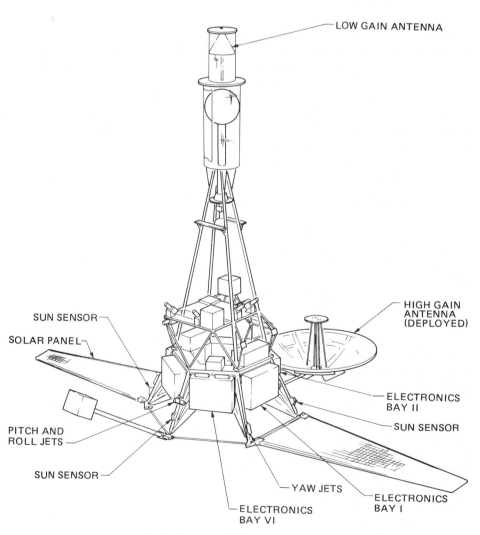

Fig. 24. Ranger 1 and 2 Spacecraft Design

The tasks to be accomplished by the respective Ranger spacecraft may be seen most readily as a set of increasingly complex phases of operation, as shown in Table III.

Ranger had to be designed to operate in free fall and the hard vacuum of outer space. Lacking the protection of the earth's atmosphere, it would be bombarded by radiation from the sun across the entire energy spectrum. Ranger would also demand 100–150 watts of power to operate a host of electronic components. This energy could be supplied continuously by the 1.86 square meters (20 square feet) of solar panels weighing 18 kilograms (40 pounds) or, for up to two days, by a 54-kilogram (120-pound) silver–zinc storage battery. But the electrical power consumed by its onboard instruments would reappear all over the spacecraft in the form of heat that had to be carried away if the instruments were to continue functioning efficiently. After considerable effort, this problem was solved: the final design minimized direct solar energy input and matched it with heat radiated to space without involving any active elements of the spacecraft in this task. Thus, heat levels would be controlled passively—by conduction and radiation—through the use of aluminum and magnesium alloys in the structural elements, by various coatings such as reflective paint and gold plating of electronic boxes, and by carefully distributing heat sources throughout the spacecraft. The electronic boxes themselves were vented so that all equipment operated in the vacuum of space.

Telemetry would determine and record the health of the spacecraft. In all of the subsystems, near-continuous measurements of voltage, current, temperatures,

Table III. Progression of Ranger Spacecraft Technology

Mission phases	Blocks	
Launch (spacecraft as passenger)	I[a]	II[b]
Postlaunch (active period, spacecraft separation, solar panels and dish antenna extended, acquisition of sun and earth)	I	II
Cruise (in free fall, oriented to sun and earth)	I	II
Midcourse maneuver (active period, attitude change, engine burn, reacquisition of sun and earth)		II
Cruise (as above)		II
Lunar operations (active period, terminal attitude maneuver, TV, seismometer capsule launch)		II

[a]Flights 1 and 2.
[b]Flights 3, 4, and 5.

and pressures were to be sampled, channeled through a telemetry coder, and transmitted by radio to receiving stations on earth, where engineers could evaluate the data. But maintaining the health of the spacecraft, so dependent on temperatures, required keeping the vehicle properly oriented in space. That orientation would be determined by information received from photoelectric eyes, and through a working nervous system which included its own electronic brain and gyroscopic sense of balance that directed the motion of the spacecraft by expelling cold nitrogen gas. Especially heavy activity in spacecraft attitude control would occur during initial acquisition of the sun and earth, in the midcourse turns commanded by a central controller with the "muscle" provided by the cold gas jets, and at midcourse motor burn when a special, powerful autopilot would steady the spacecraft attitude by means of vanes in the rocket exhaust.

For its lunar mission, Ranger's brain would be a miniature central computer and sequencer. This element began as a complex alarm clock on the Block I spacecraft; it was to time and trigger events after launch and enable Ranger to convert from its role as a rocket passenger to that of independent spacecraft (i.e., open the solar panels, activate the solar alignment, and point the dish antenna at the earth). With the midcourse maneuver phase added to refine the lunar trajectory for Block II spacecraft, engineers coupled a computer to the sequencer. Unlike the latter, the computer could receive such numerical information from earth stations as the direction and size of rocket thrust required, then initiate the maneuver on command.

The central computer and sequencer would receive its information from the command subsystem using decoding components in the spacecraft and encoding components on the ground. Other spacecraft activities would also be initiated by communications passed through this system. The radio link was to consist of a powerful transmitter on earth and a very sensitive receiver in Ranger. The same radio system would contain a spacecraft transmitter that conveyed the diagnostic telemetry and scientific experiment data to the earth receivers. Ranger's radio would also operate at radar frequency, and the signal loop—which could carry command, telemetry, and scientific data all at once—would provide accurate two-way doppler tracking at the ground receivers. Before sun acquisition or during the midcourse maneuver, commands to or data from the spacecraft would pass through the low-gain antenna atop the tower; in the stabilized cruise mode and at lunar encounter, the more powerful high-gain dish antenna beneath the spacecraft was to be used.[26]

Totaling 270–320 kilograms (600–700 pounds), Block I spacecraft represented a complicated array of interdependent subsystems. The decision to fly scientific experiments on these two engineering test missions precluded adding many redundant or duplicate engineering devices such as a second central controller or attitude control components. The design did include a backup low-power transmitter with independent battery power supply; nevertheless, to a greater extent than might have been preferred by its designers at JPL, the craft depended upon successive functioning of unique components in series; that is to say, the proper operation of each subsystem largely determined the operation of its neighbor. A failure in the attitude control or power subsystems, for example, would have adverse effects throughout the spacecraft. To compensate, engineers planned more redundant features for the lunar spacecraft. Among other changes to the bus, they expected to add a second set of attitude control gas jets and a separate gas supply, in addition to the backup low-power transmitter.

Of all the physical changes, a modified tower above the Block II spacecraft was the most noticeable. The JPL-designed hydrazine monopropellant midcourse rocket engine, another addition, would be tucked away out of sight beneath the hexagonal bus. The tower supported the seismometer capsule and its retrorocket; aimed at depositing the seismometer on the moon, the entire assembly was a major subsystem in its own right. Like the spacecraft in inherent complexity, the capsule subsystem had to be designed and developed at the same time as the bus. Short on manpower and facilities, and responding to the wishes expressed at NASA Headquarters to engage industry in the space program, JPL immediately sought out an industrial firm to design and fabricate this science subsystem. Two weeks after formal NASA approval of Project Ranger in February 1960, three companies received contracts for competitive design studies of the capsule. On April 15 they returned proposals to JPL for evaluation. Ten days later NASA Administrator Glennan announced that the Aeronutronic Division of Ford Motor Company had been selected to build the seismometer subsystem.[27]

The Aeronutronic design was novel, its schedule ambitious. The firm proposed to fabricate, assemble, test, and deliver by September 1961 the required number of 134-kilogram (300-pound) capsule subsystems for a cost of $3.6 million. The design mounted the capsule above a solid-propellant retrorocket; a radar altimeter would signal separation and firing of the capsule's motor at a specified distance above the lunar surface. The full capsule, which was then to separate from the motor, incorporated a crushable outside shell or impact limiter. Inside, a spherical metal survival package floated in fluid to distribute and dampen the structural loads at impact, and to allow erection of the package to local vertical by moon gravity after the capsule came to rest. In addition to tiny batteries, the survival package would contain the single-axis seismometer already being developed by Caltech's Seismological Laboratory and the Lamont-Doherty Geological Observatory at Columbia University. Erection to local vertical on the moon would permit the sensitive axis of the seismometer to be positioned correctly, and allowed deployment of a modest directional transmitting antenna.[28]

The Aeronutronic firm appointed Frank G. Denison manager of its Lunar Systems, the group formed to develop the capsule subsystem. A Caltech graduate and former JPL Section Chief, Denison was well known at the Laboratory. He would report to Burke and Kautz in the Ranger Project Office, and to Schurmeier's Systems Division, which would integrate and test the capsule subsystem in the assembled spacecraft. In June, Denison selected and JPL approved the subsidiary contractors to fabricate the major components of the capsule subsystem: Hercules

Powder Company, solid-propellant retrorocket; Ryan Aeronautical Company, radar altimeter; and Rohr Aircraft, capsule support structure (Figure 25).

The Ranger spacecraft design had reached an advanced stage, with the scientific content agreed upon and the principal contractors selected. Before, in the minds of the participants at NASA and JPL, the name Vega had been associated with a launch vehicle. In the months to follow the name "Ranger," besides acting as a project designation, came to apply increasingly to a specific spacecraft genre. Its design concepts and functions had grown from Juno IV in 1958 via Vega in 1959 to become Ranger in 1960. A machine capable of planetary missions appeared destined to inaugurate NASA's program of lunar exploration.

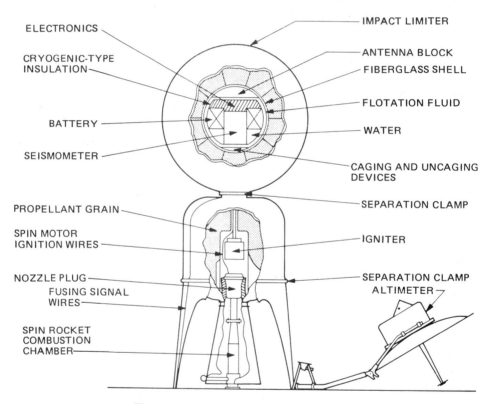

Fig. 25. Ranger 3, 4, and 5 Superstructure Design

Chapter Four

RANGER'S LUNAR OBJECTIVES IN DOUBT

I N 1960 Jim Burke found himself enmeshed in the project-wide organization and management imbroglio, and drawn increasingly away from spacecraft affairs at JPL. More and more time was required to define and hammer out project requirements and responsibilities among the participating organizations. Solving such questions as who would provide needed launch equipment, the type, and when and where it should be delivered often proved as frustrating for the project manager as trapping quicksilver with one's fingers. But the recognized risks in the management side compounded other, less obvious technical risks associated with the new and unproved spacecraft.

A Difference in Weights and Measures

Even as NASA and JPL selected contractors to make components for the spacecraft in mid-1960, questions of weight threatened plans for the Ranger Block II final design. The difficulty suddenly appeared in the form of revised Atlas-Agena performance figures. Original preinjection trajectory calculations—from launch to the second burn of the Agena—had been obtained by NASA from General Dynamics-Astronautics, the Atlas contractor, at the beginning of Project Ranger. They had been used at JPL in determining spacecraft weights and for computing postinjection trajectories—from the second burn of the Agena to the end of the mission. On July 11, Lockheed submitted new ascent trajectory figures through the Air Force Ballistic Missile Division in Inglewood, and these figures contained a significant discrepancy: they specified 34 kilograms (75 pounds) less weight available for the Ranger lunar spacecraft.[1] To determine the reason for the difference in figures between General Dynamics and Lockheed, Burke ordered an immediate investigation and, pending the outcome, directed Schurmeier's Systems

Division to take all reasonable measures to lighten the spacecraft short of removing planned equipment.[2]

The uncertainty over spacecraft weights could not have occurred at a worse time. A firm launch schedule for Ranger had just been issued by NASA, committing the two engineering test vehicles to fly in July and October 1961, and the three lunar missions in January, April, and July of 1962.[3] Far more disturbing than that, planning for the final design of lunar spacecraft to meet this schedule was well underway, and could not continue in an orderly manner if the engineers did not know the weights they had to work with. The test spacecraft needed only some 300 kilograms (675 pounds), and were not adversely affected, but the lunar spacecraft needed the entire 360 kilograms (800 pounds) that the Atlas-Agena B had been projected to carry to injection. The capsule subsystem alone weighed 135 kilograms (300 pounds). Nevertheless, the critical disparity in trajectory figures was not soon resolved.

The trajectory problem, like other launch vehicle management problems in 1960, was nourished by the still unsettled state of project management. At the outset, NASA and the Air Force Ballistic Missile Division had failed to designate a single organization to perform all of the trajectory computations for Ranger missions. With the new performance figures disputed, Lockheed's participants on the Agena B Coordination Board urged that they be given this function. JPL representatives Cummings and Burke, reluctant to delegate this important work and skeptical of the Lockheed calculations, opposed such a move. The question remained in the hands of the Board as summer turned to fall, and fall to winter. Meantime, without a single set of confirmed trajectories, the allowable weight of the Ranger lunar spacecraft remained in limbo.[4]

While the question of Ranger trajectories deepened the paralysis of the Agena B Coordination Board, JPL lightened the bus chassis for the lunar spacecraft as much as possible. Magnesium replaced aluminum in covers, tubes, and fittings. Engineers shaved the thickness of structural members, and directed that minimum size wire and lightweight insulation be used in the power subsystem. As an extreme measure, Cummings ordered holes drilled in the electronic boxes to save a further few ounces, and parts of the craft began to resemble Swiss cheese. Projections of the total weight expected for this machine nevertheless remained above the limit of 325 kilograms (725 pounds) specified by Lockheed, and by fall Denison had added his own reservations to the weight question. The Aeronutronic capsule subsystem, he informed Burke, would probably exceed the 135-kilogram (300-pound) limit.

Believing the Lockheed calculations questionable, Burke delayed ordering any major changes in the final design. To members of the Agena B Coordination Board, he made the point abundantly clear in October: the Lockheed data could have been revised downward "somewhat arbitrarily by the Air Force," he declared, "and may not represent the true capability of the [Atlas-Agena] vehicle." Discord over who would prepare the complete lunar trajectories, he continued, made it impossible for JPL to obtain a set of definitive trajectory calculations and to clarify the divergence in performance figures previously computed for the Atlas-Agena B by General Dynamics and Lockheed.[5]

Despite prodding, two more months elapsed before the Ranger participants agreed even to a statement of work for preparing trajectories. Signed by JPL, the Air Force, Lockheed, and the NASA Agena office in Huntsville, the document was issued on December 14, 1960. It called for the preinjection trajectory to be computed by Lockheed in accord with JPL-furnished specifications. The postinjection trajectory would be prepared by JPL. Finally, a third party, the Space Technology Laboratories (STL) would integrate the two trajectories and generate accurate firing tables under a separate Air Force contract.[6]

This compromise did not actually provide the "definitive trajectory calculations" JPL needed to determine a final weight for Ranger. Moreover, STL could not promise the final set of trajectories for six months—hardly in time for the launch of Ranger 1—and that was contingent upon receiving the JPL and Lockheed pieces of the work. Decisions on the Block II final design and weight could not be postponed much longer. On January 11, 1961, JPL engineers agreed on the placement and way of mechanizing the last outstanding item on the lunar spacecraft. The low-gain antenna would be located above the survival capsule, mounted on a boom to be swung away from the spacecraft, permitting ejection of the capsule subsystem near the moon.[7] For Burke in the Ranger Project Office, a further delay on a decision of final weight of the lunar machine would amount to a decision not to meet the Ranger 3 launch date in January 1962. Time had run out.

Burke and Cummings faced two uncomfortable choices. They could slip the lunar flight dates month-to-month at the possible expense of seeing the Soviets win more lunar laurels—while hoping for an increase in weight from the definitive set of STL trajectory calculations—or they could freeze the spacecraft design at the lightened weight specified in the Air Force-Lockheed figures regardless of any allowable increase that might eventually be found, thus permitting the project to remain on schedule. The Soviets helped Burke and Cummings make up their minds when, on February 12, 1961, they launched a 450-kilogram (1000-pound) spacecraft on a flyby trajectory past Venus. This spacecraft possessed three-axis attitude control, oriented solar panels, a high-gain dish antenna, and made use of an earth parking orbit—all of which Ranger was supposed to have demonstrated (Figure 26). Though the mission was destined to fail, the Soviet attempt at a deep space mission that JPL claimed as its own keyed Ranger participants to a "pinch-hitter's" state of mind.

At the Laboratory, Jim Burke nailed a picture of his Soviet competitor, Venera 1, to the wall directly in front of his desk. Next to it he pasted an old proverb: "The better is the enemy of the good." He would call that aphorism to the attention of those proposing design changes that involved major delays. Design changes would continue to be minimized, delays proscribed. And while there was some sentiment in Silverstein's office to postpone the Ranger flights, it was strongly resisted at JPL. Pickering, Cummings, and Burke were saying with former NASA Administrator Keith Glennen that schedule "postponements should be

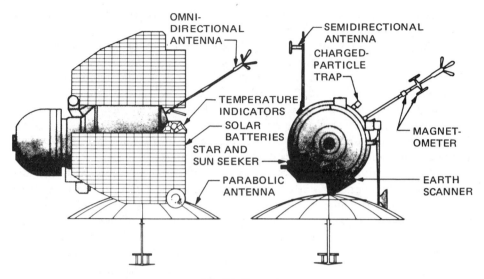

Fig. 26. Venera 1

viewed with concern, not regarded as normal procedure."[8] In the face of the Soviet challenge, flight schedules for Project Ranger remained firm. The lunar spacecraft would necessarily have to shed more weight.

Four days after the launch of Venera 1, though still distrusting Lockheed's performance figures, Burke notified the Systems Division that "...we must begin removing items from the spacecraft." Original plans to include a redundant set of attitude control gas jets had already been abandoned. Now Burke instructed Schurmeier to remove the backup low-power transmitter and its battery, as well as a set of engineering instrumentation gyroscopes (to be used to determine initial sun–earth acquisition rates) and their associated equipment on all the lunar spacecraft. In addition, he asked that the Systems Division "review the entire instrumentation schedule and remove a portion of the [engineering telemetry] equipment so as to save weight at the expense of creating a higher-risk situation," and "continue to insist that the capsule, which is currently overweight, must meet its weight goal of 135 kilograms (300 pounds) or not fly" on Ranger.[9]

Burke announced the decision to other project participants a few weeks later at the ninth meeting of the Agena B Coordination Board. At an anticipated 330 kilograms (730 pounds), Ranger 3 would still be somewhat heavier than available performance figures allowed. Nevertheless, Burke insisted, "JPL believes that the allowable weight may turn out to be more like 355 kilograms (790 pounds)."[10] In April 1961, still holding to this opinion, the Ranger Project Office "froze" the final design of the lunar spacecraft at the reduced weight.[11] Additional equipment would be stripped from these machines only in the event that STL confirmed the available trajectory calculations. No more equipment would be removed on the strength of suspect figures.

In fact, as Burke had surmised, no weight reduction campaign had ever been necessary. On May 28, shortly before assembly of Ranger 3 began, STL issued the definitive trajectory calculations for Ranger missions. To his superiors at JPL, Burke glumly reported that an additional 74 kilograms (164 pounds) were available for Block I and 52 kilograms (116 pounds) for Block II spacecraft, resulting in confirmed total weights of 371 kilograms (824 pounds) and 378 kilograms (841 pounds), respectively. This increase in allowable weight was, he explained, simply "unexploitable at this late date."[12] When assembled in the fall, Ranger 3 weighed in at 327 kilograms (727 pounds). Rangers 1 and 2, already at Cape Canaveral, each weighed 304 kilograms (675 pounds). For these particular planetary spacecraft, all options for improving reliability through redundant features had been irretrievably lost. The road to Ranger, recognized as a high-risk avenue from the management standpoint in 1960, had been paved with still graver technical risks in 1961 because the spacecraft was lightened to meet an erroneous weight limitation.[13]

PREPARING FOR THE TEST FLIGHTS

Between June 1960 and August 1961, while the Ranger officials in Pasadena wrestled with the various problems of project management and spacecraft weights, Rangers 1 and 2 were fabricated and tested. In performing these tasks, JPL followed the pattern of missilery that relied on preliminary engineering flight missions to test the entire space and ground components of the system.[14] For these first Ranger flights, however, JPL employed three versions of the spacecraft to validate the final design: a spacecraft mockup, a thermal control model, and a proof test model.[15] Engineers used the mockup, constructed in July 1960, to confirm the mechanical aspects of the spacecraft, including cabling and the layout of equipment. The thermal control model was then assembled and tested in a 2.7-meter (6-foot) diameter vacuum chamber, where conditions in outer space could be approximated. Though lacking active electronic components, this model held equipment containing resistors and other equivalent heat-producing sources, and perfected the design for passive thermal control.

The proof test model would be as nearly identical to the actual flight articles as possible but subjected to tests above the performance and stress levels expected during the actual flight of Rangers 1 and 2. As the name implied, the proof test model was used to shake down the design. All of the unforeseen and undesired characteristics that appeared here in testing could be isolated, identified, and corrected before assembly of the flight machines began. This particular test program, though more severe, essentially duplicated that planned for the flight Rangers. But most important were the qualifying tests of each component and subsystem undertaken by the JPL technical divisions, and systems tests conducted by the Systems Division (Figure 27).

Fig. 27. Ranger Block I Proof Test Model and Nose Shroud Mockup

Although each spacecraft test was important, the systems tests were crucial. Missiles had previously been tested at the subsystem level at JPL, then assembled and launched at White Sands in a test of the "system." The idea of testing the total spacecraft system prior to flight was new. This preflight check corresponded as nearly as possible to all of the mission phases planned for Ranger, and was divided roughly in two: mission sequencing and environmental tests. First, by means of elaborate test consoles, cables, and radio-frequency links, engineers exercised the proof test model, flight spacecraft, and scientific instruments through their complete sequence of operations from launch to end-of-mission. Commands would be sent to the space machine, events monitored against expected performance, and discrepancies rectified by repair or replacement of parts (Figure

28). Second, environmental testing meant evaluating the spacecraft in the specific conditions encountered at launch and in space operations. Here, the spacecraft was first mounted on a large "shake table." Such a table, weighing hundreds of kilograms, vibrated the spacecraft at various amplitudes both in horizontal and vertical planes, approximating conditions to be expected during ascent atop the Atlas-Agena B. To test its thermal design and performance in a vacuum, the spacecraft, less solar panels, was suspended and operated inside the 2.7-meter (6-foot) diameter vacuum chamber. As in the other qualifying tests, its conductors monitored or commanded appropriate functions by means of wire cables and test consoles.

Thorough testing of the spacecraft, however, required facilities beyond those available in 1960. A missile assembly building, used in JPL's programs for Army Ordnance, doubled as the spacecraft assembly and mission sequencing test area. Next door, another small building housed the shake table and small vacuum chamber to be used for the environmental tests. With modification, these facilities could serve the immediate needs of Project Ranger in makeshift fashion; they would, nevertheless, be inadequate to accommodate other NASA lunar and planetary spacecraft scheduled to be assembled at the Laboratory in 1961 and 1962.

Fig. 28. Ranger 1 in Systems Test Complex

With the test procedures established in 1960, NASA and JPL officials completed plans for the necessary facilities. By July, contracts had been issued for the construction of (1) a spacecraft assembly facility to be used for assembly and functional tests; (2) an environmental laboratory, for shock tests and testing on larger shake tables and in the small vacuum chamber; and (3) adjacent to the environmental laboratory, a cylindrical building to house an eight-story, 7.5-meter (25-foot) diameter, vertical vacuum chamber in which the complete spacecraft could be tested on simulated missions to the moon and planets. Contractors completed the first two of the new facilities in mid-1961, in time to accommodate the Block II spacecraft. But the large vacuum chamber, because of problems encountered with the artificial solar heat and light sources, would not be placed in operation until November 1962. None of the five Ranger spacecraft could be tested there. Thus, the importance of Ranger's first two test flights was emphasized. They would serve as engineering missions in the vacuum of outer space to provide equivalent data on flight performance that could not otherwise be obtained in available ground test complexes.

In Silverstein's Office of Space Flight Programs, JPL's progress on the entire project, including Ranger testing, was monitored and evaluated against NASA guidelines by Oran W. Nicks. Born and raised on a ranch in the southwest in the dust and depression of the 1930s, Nicks took an intensive two-year course at Spartan College of Aeronautical Engineering in Tulsa in 1942 and 1943; then, after service with the Army Air Forces, worked his way to a degree in mechanical engineering at the University of Oklahoma in 1948. First with North American Aviation on the West Coast, later with Chance-Vought Aircraft in Texas, he held a succession of increasingly responsible positions. In March 1960 he joined NASA as Chief of Lunar Flight Systems (Figure 29). Circumspect, disciplined, and practical, Nicks possessed an instinct for fathoming complex situations rapidly and accurately. The technical advances required to achieve Ranger's lunar objectives, he believed, warranted more cautious reflection and a less open display of confidence and self-assurance by his JPL colleagues. Technical prospects aside, from past experience he could be sure of one thing. Regardless of their Caltech credentials, the informality of JPL project personnel and the casual attire they sometimes sported in meetings at Lockheed and Cape Canaveral provoked unfavorable reactions among the colonels and generals of the Air Force who were expected to make room for NASA's Ranger.

Beyond different backgrounds, James Burke and Oran Nicks differed in their attitude towards schedules and testing. For Burke and his associates, meeting schedules and costs—even if it meant lightening the spacecraft and removing tardy or overweight scientific experiments—assumed first importance. Reliability would be achieved through the sound design of the hardware. Testing necessarily had to be compressed to fit the NASA schedules within available facilities.[16] Nicks, less inclined to accept the swift schedules established in response to the Soviet challenge, was more disposed to slip flight dates, even at the risk of increasing costs, if he believed it was required to ensure that scientific experiments got on

Fig. 29. NASA Lunar Flight Systems Chief Oran Nicks

board, or more adequately to test and qualify spacecraft components.[17] And if Burke reasoned in 1960 that the JPL project office ought to direct Ranger activities in the absence of a viable project management structure, Nicks became convinced that that office had to be more responsive to directives from NASA Headquarters, which, after all, paid the bills and was ultimately accountable to the Congress.[18]

Nicks might remain skeptical and differ with Burke on the test philosophy and importance of schedules, but actual experience in qualifying Rangers 1 and 2 for flight further heightened confidence in the soundness of the systems test procedures and the spacecraft design among the project engineers—heightened it at least until the question of sterilization arose.

WHEN AND HOW TO STERILIZE SPACECRAFT

With the first earth satellites in orbit late in 1957, Detlev W. Bronk, the president of the National Academy of Sciences, openly expressed interest in a subject heretofore discussed quietly among biologists. Would the Chairman of the Earth Satellite Panel of the U.S. National Committee for the IGY, he inquired, act as chairman of a planning committee to organize a symposium on the prospects for biological research in space, and of detecting possible low-order life forms on other celestial bodies in the solar system? He would indeed, and the Academy was joined by the American Institute of Biological Sciences and the National Science Foundation in sponsoring a symposium held in 1958.[19]

Before planning had progressed very far, however, one member of the life sciences fraternity expressed a formidable reservation. In a private memorandum circulated in January 1958, Professor Joshua Lederberg, Chairman of the Department of Medical Genetics at the University of Wisconsin, cautioned that earth organisms aboard spacecraft that landed on celestial bodies might reproduce, making it impossible forever to discover and examine indigenous extraterrestrial life forms.[20] The issue was debated in the National Academy of Sciences, then before the International Council of Scientific Unions (ICSU), which established a special committee to study the question. That group adopted sterilization as an important policy, and urged both the United States and the Soviet Union to implement measures to avoid introducing earth organisms on other bodies in the solar system.[21]

Thus, in September 1959, even before Project Ranger officially began, NASA Administrator Keith Glennan received a letter from the National Academy of Sciences advising the space agency to adhere to the ICSU policy and sterilize United States space probes.[22] If he did not foresee all of the ramifications spacecraft sterilization might entail, Glennan did appreciate its importance for science and the prestige of the nation. With his approval, on October 15 Silverstein issued an initial guideline to all NASA field centers. "As a result of deliberations," the directive read, "it has been established as NASA policy that payloads which might impact a celestial body must be sterilized before launching." Center Directors were informed that "of the several means of sterilization proposed, NASA considers the use of ethylene oxide in its gaseous phase as the most feasible agent at this time."[23]

As a target of biological interest, the moon was considerably more doubtful than any of the inner planets. Scientific opinion, nevertheless, was divided.[24] If there were ice or water trapped somewhere beneath the surface, microorganisms, insulated by the material above, might exist there. On that chance NASA Headquarters included lunar spacecraft under the terms of the directive. First in line for detailed application of the NASA sterilization techniques, as a prototype unmanned planetary machine, was Ranger.[25]

Problems of weight and mechanical "bugs" were one thing for Ranger's engineers. Unfamiliar worries about real live bugs was something else again. To sterilize surgical instruments, one simply boils them in water and keeps them in an autoclave until ready for use. But how would they sterilize a 326-kilogram (725-pound) spacecraft and keep it sterile from Pasadena to the moon? To be sure, bathing a spacecraft in toxic ethylene oxide gas would drastically reduce the number of organisms on exposed surfaces; however, the gas would not necessarily reach bacteria between joints or trapped in electrical potting substances. While this treatment might decontaminate a machine, it could not guarantee sterility. JPL investigators suggested "one chance in ten (perhaps a hundred) of a viable organism remaining on the probe [as] an acceptable infection tolerance."[26] To attain that goal it would be necessary to heat the components, or perhaps the entire machine. Even then full sterility might not be achieved.

Without firm specifications on the degree of spacecraft sterility desired, JPL established procedures believed to be consistent with expectations of reliable equipment performance. George L. Hobby, a research biologist in the Space Sciences Division, was placed in charge of this work. Together with Burke, in April 1960, he formulated plans to sterilize the three lunar spacecraft.[27] The plans called for a three-step operation embracing fabrication, assembly, test, and transportation. All components, including the lunar capsule subsystem components, would be assembled, then subjected to heating at 125°C (257°F) for 24 hours. This was a compromise of sorts, because at temperatures and times above these levels, electrical equipment often failed, while below these levels some organisms were found to survive. The sterilized components would then be assembled and the machine tested in a segregated area in the new spacecraft assembly facility. All surfaces joined during assembly were to be cleaned thoroughly with alcohol. Finally, shipped to Cape Canaveral in a controlled environment and after completing final prelaunch tests, the entire spacecraft would be "soaked" in ethylene oxide gas inside the Agena nose fairing. Theory held that once the machine was rendered essentially sterile, any subsequent contamination would be on its surface and exposed to the lethal gas.

On May 25, 1961, a date coinciding with the distressing resolution of final allowable weights for the Block II spacecraft, Cummings submitted JPL's suggested program for sterilizing these vehicles to NASA. It conformed to the plans made in 1960 but noted that exceptions would need to be made for certain heat-sensitive parts.[28] NASA Associate Administrator Seamans approved this program in June 1961 in time for the assembly of Ranger 3.[29] Without question, sterilizing spacecraft would be an expensive, new, and unusual demand upon the project—and another burden for Burke, who, in listing 16 competing requirements for Ranger, put reliability first in priority, sterilization fourteenth.[30]

SPACE SCIENCE AND THE ORIGINAL RANGER MISSIONS

Meantime, further difficulties for Burke cropped up in reconciling the experimental ambitions of space scientists with the engineering requirements of the Ranger machine. In accord with NASA General Management Instruction on the "Selection of Scientific Experiments," April 15, 1960, Headquarters exercised final authority, through a Space Sciences Steering Committee, over what experiments would fly on NASA vehicles. Subcommittees, arranged by discipline, with membership drawn from scientists at Headquarters, field centers, and the universities, evaluated and recommended experiments to the Steering Committee for authorized flights. The Committee—with Newell as Chairman—reviewed and approved these recommendations, then submitted the proposed experiments to Silverstein for final approval. Once the experiments were selected, field centers assisted in fabricating and qualifying instruments, integrating them in a spacecraft, and processing the data returned from them. The experimenters were thus expected to work closely with the field center space science division and project office in realizing flight objectives, and in analyzing and publishing the findings.[31] But the management instruction did not specify the precise relationship between the experimenters and the field centers, neither for NASA flight projects generally nor for JPL and Project Ranger in particular.

Burke and the engineers assigned to Project Ranger at JPL could agree with Newell, Hibbs, and Schilling on the ultimate role of Ranger as a scientific mission—specifically, one to investigate the moon for planetary science. Burke dealt directly with JPL's space sciences chief Hibbs on matters of science and the spacecraft; Hibbs, in turn, worked out details with the individual experimenters and supported them in designing, fabricating, and qualifying their instruments for flight. Burke pledged Hibbs his full support and commitment to the lunar goals; Ranger would fly for planetary science.[32] But Burke elected to insulate himself as much as possible from the interested experimenters and focus his attention upon the project organization and technology. To him, the demands of organization, of developing the Atlas-Agena vehicle, the spacecraft, the tracking net, and flight operations procedure all had to be met first so that the scientific goal could be realized. The technology, including even the technology to accommodate the experiments—had to take priority over the science.

Burke and his fellow engineers, all the same, could by no means ignore the science on Project Ranger. The complement of scientific instruments affected nearly every aspect of the Ranger machine, from the center of gravity to calibrated magnetic characteristics. The space science instruments occupied volume, added weight, consumed power, produced heat, and demanded data handling and instrumentation that might otherwise be given over to redundant engineering subsystems and telemetry in the spacecraft. The project team did plan for and deal with these complications day to day as the spacecraft design evolved. Their efforts, however, required a delicate balancing between project science and engineering. They proceeded without disruption only so long as NASA's basic list of approved experiments remained unaltered. But the scientists themselves were not easily dissuaded from proposing alterations to a list on a mere scrap of paper, especially since, as the first, large, completely stabilized spacecraft scheduled by NASA for launch into deep space, Ranger offered rare research opportunities for experimenters.

On April 13, 1960, shortly before Burke froze the final design of the two test vehicles, a delegation of scientists from the Los Alamos Scientific Laboratory and the Sandia Corporation paid a visit to Pasadena. They came to explore prospects for placing a Vela Hotel experiment on board the Ranger Block I spacecraft. Vela Hotel was the code name for a project managed by the Atomic Energy Commission (AEC); it sought to develop a satellite-borne X-ray and gamma-ray monitoring system that could detect above-ground nuclear explosions. In theory, its successful operation hinged upon the absence of a natural background source of certain X-rays. Experiments first had to be conducted to determine whether the sun was a source of microsecond bursts of such X-rays, and the AEC scientists were eager to fly their experiments at the earliest opportunity.[33]

Informed of their mission, Burke received the scientific delegation at JPL with grave misgivings. He showed his visitors the Ranger spacecraft, described the flight system, and gave them material concerning the scientific experiments and telemetry system. Patiently, he explained that science for Rangers 1 and 2 had already been approved by NASA, and that "no additional experiments could be accommodated in the program as now planned." The Los Alamos and Sandia scientists were, of course, "welcome to examine any results [obtained] with our instruments."[34] Far from dampening interest among his AEC guests, Burke's cautious briefing substantiated their expectations of the spacecraft's capabilities. They understood that Ranger, pointed at the sun as a major reference in space, was just right for Vela Hotel. NASA Headquarters determined science policy; Headquarters might consider more science for Ranger.

Two days later Army Brigadier General A. W. Betts, the Director of ARPA, dispatched a letter on the subject to John F. Clark, Schilling's deputy for the Planetary Science Program. Betts moved quickly to his point: "Since both of these [Ranger] vehicles are scheduled to penetrate the regions of space of prime interest to Project Vela Hotel (outside the trapped radiation fields), it is requested that as much space, weight, and telemetry as possible be made available so that detailed planning and design may be initiated immediately."[35] Clark discussed the issue with Schilling, Newell, and Silverstein, and General Betts received a qualified answer. NASA asked that the AEC submit a proposal explaining how Vela Hotel would be incorporated in the Block I spacecraft. Should the experiment turn out to be impractical, it might then be refused. On May 3, 1960, the Los Alamos Scientific Laboratory and the Sandia Corporation complied. Recognizing the design of the NASA spacecraft to be well advanced, the proposal aimed at minimizing interference. The AEC experiment would consist of a primary battery power source, two radiation detectors, a data handling and logic module, and the necessary packaging and shielding.[36] All in all, with some reshuffling of equipment on Ranger, adding the experiment appeared to be entirely feasible.

In early June Cummings summarized JPL's evaluation of the proposal for Silverstein's Lunar and Planetary Program Director, Edgar Cortright. The 5.4-kilogram (12-pound) Vela Hotel experiment, Cortright learned, was technically compatible with the test spacecraft.[37] Within the Laboratory, Burke notified affected division chiefs that "it is probable we will be requested to incorporate the experiment." He directed them to begin altering the design for just such an eventuality. Taking these steps now, he declared, "will ease the pain of incorporating it considerably."[38] On June 29 Silverstein approved the new experiment on Rangers 1 and 2.[39]

Besides Vela Hotel, there was one other change in the complement of scientific instruments planned for Rangers 1 and 2.[40] NASA dropped plans for the vehicle charge experiment in May, and substituted a small engineering friction experiment in its place. Conceived by members of JPL's Materials Research Section in the Engineering Mechanics Division, it would measure the coefficient of friction in the vacuum of space of various metal discs rotated at a few revolutions per minute by a small electric motor.[41] Engineers expected the information derived from this experiment to prove valuable in the design of bearings and gear surfaces for future space machines. But it was not, strictly speaking, a "scientific" experiment. At JPL, Burke and Hibbs gave preference in design, operations, and data handling only to scientific instruments. Accordingly, they established the following priority among experiments to guide project engineers: (1) solar plasma detector, (2) magnetometer, (3) trapped radiation detector package, (4) ionization chamber, (5) cosmic-ray telescope, (6) Lyman alpha scanner, and (7) micrometeorite detector.[42] Last, and not listed, was Vela Hotel. The entire array of experiments for the Ranger engineering test flights was now complete (Figure 30).[43]

If the experience with Vela Hotel proved "painful" to Burke, it also left him skeptical of Headquarters' true commitment to Ranger schedules and of its appreciation for the engineering demands of the project. To Burke, Newell's science staff appeared entirely too willing to hazard launch schedules and defer demonstration of the spacecraft technology in favor of still more science. "This difference in viewpoint," he explained to JPL Deputy Director Brian Sparks, "adds vigor to the technology-science controversy and encourages undisciplined efforts by our own [JPL] science people to get Headquarters to order us to wait for them if necessary."[44] Presiding over the birth of a new spacecraft, Burke began to view his scientist-clients as overly enthusiastic members of the family—waiting anxiously outside the delivery room with bat and glove, eager to play ball with the infant machine before it was able to crawl, let alone run.

A few weeks later, in August 1960, Burke's suspicions were aroused again. Effects of the discrepancy in ascent trajectory calculations on the weight of the lunar spacecraft had just turned that design effort into an engineer's nightmare. What was more, contract negotiations had broken down at Lockheed over the reduction in NASA's order for Agenas. Project schedules appeared threatened across the board. At that moment scientists at the Los Alamos Scientific Laboratory and Sandia Corporation produced a new recommendation: they wished to fly a Vela Hotel experiment on Rangers 3, 4, and 5.[45]

The Ranger Project Manager received the document with incredulity. Vela Hotel had nothing whatsoever to do with the moon. Not a single planetary scientist associated with the lunar Rangers would endorse the new proposal if it meant that his own lunar experiment might be compromised, much less if it were eliminated. Having seen Rangers 1 and 2 reworked to accommodate the AEC experiment, Burke had no desire to repeat that exercise on the trouble-plagued Block II spacecraft. Several other experiments, in fact, had been recently suggested for these lunar machines by scientists at the Goddard Space Flight Center. Should they now be combined with Vela Hotel and approved by NASA, the ensemble would consume 9.5 precious kilograms (21 pounds), not to mention power and telemetry support.

After conferring with JPL's Lunar Program Director Cummings, Burke wrote to Cortright at Headquarters and told him that the problems of weight,

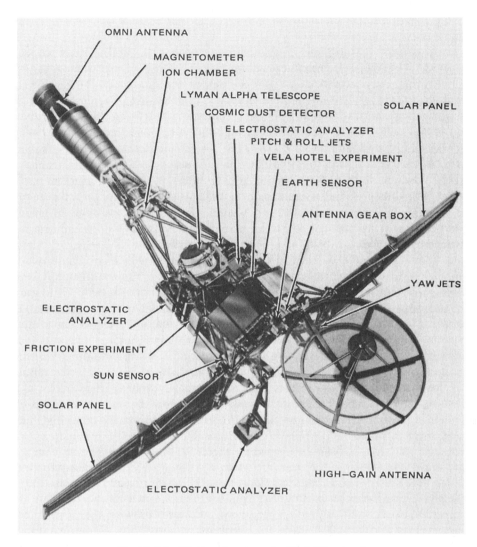

Fig. 30. Scientific Experiments on the Block I Spacecraft

packaging, power, and space prohibited accepting this AEC proposal.[46] Cummings also urged that the Goddard science proposals not be considered for flight on the lunar Rangers.[47] Los Alamos Scientific Laboratory received a copy of Burke's communication to Cortright. Made aware of the seriousness of the situation, and failing to obtain support among other mission scientists, the scientists in New Mexico withdrew their proposal.[48] For the rest, Cortright, the aeronautical engineer, concurred with Cummings. Further experiments would not be considered on the lunar flights that involved the important midcourse and terminal maneuvers.

Engineers qualified the original lunar experiments approved by NASA's Space Sciences Steering Committee without serious delays. The 3.6-kilogram (8-pound), single-axis seismometer developed under the guidance of Frank Press, Director of Caltech's Seismological Laboratory, and Maurice Ewing, Director of Columbia University's Lamont-Doherty Geological Observatory, was basically a magnet suspended in a coil by a spring, and restrained radially so that it responded only to motion parallel to its axis. Tested to withstand a 3000-*g* impact force, floated in a viscous fluid inside the survival sphere, it would assume a vertical position in response to lunar gravity after the capsule came to rest.[49] Although its designers did not anticipate significant tectonic activity on the moon, the detection of microseismic activity would significantly contribute to man's understanding of the thermal character of the lunar body, as well as indicate the existence of crust, core, or both, and the density distribution with depth.[50]

James R. Arnold, a chemistry professor at the University of California at San Diego, first suggested the gamma-ray spectrometer for Project Ranger.[51] After approval by NASA, Arnold developed the instrument with Ernest C. Anderson and Marvin A. Van Dilla of the Los Alamos Scientific Laboratory, and Albert E. Metzger of JPL. The heart of this device, a scintillation detector capable of measuring natural radioactivity emanating from lunar surface material, was mounted on the end of a telescoping 1.7-meter (6-foot) boom. Uranium, thorium, and potassium emit gamma-rays (very energetic X-rays) in radioactive decay. They are also among the elements that are enriched towards the surface of a planetary body by differentiation if the temperature below the surface has exceeded the melting point of the rock any time in the body's history. By detecting these gamma-rays, it would be possible to determine the degree of differentiation that had occurred on the moon's surface—and, therefore, whether the surface material had formed by meteoritic accretion or volcanic eruptions. The instrument, turned on four hours after launch and extended on its boom shortly after the midcourse maneuver, would establish background counting rates (caused by proximity to the spacecraft) to a high degree of accuracy, and also provide the first direct measurement of an interplanetary or cosmic gamma-ray flux, if one existed. It would continue operating until the lunar seismometer capsule ejected at 24 kilometers (15 miles) above the lunar surface, at which time the spacecraft was expected to tumble, throwing its high-gain antenna out of lock with receiving stations on earth.

The 5.9-kilogram (13-pound) television camera completed the list of experiments first approved for the lunar mission. Activated 4000 kilometers (2500 miles) above the moon, the camera would transmit high-resolution pictures of the approaching surface until the lunar capsule was ejected. JPL awarded the Radio Corporation of America the contract to develop the vidicon sensor, based largely on that firm's experience in the field, particularly with the cameras for the Tiros meteorological satellite. The Space Sciences Division at JPL fabricated the f/6 aperture optical telescope with a 102-centimeter (40-inch) focal length, while RCA produced the slow-scan vidicon with its deflection and focus coils, and the supporting electronics package.[52] Under the best of viewing conditions on earth, the largest optical telescope could obtain a surface resolution of 300 meters (1000 feet) on the moon. The last of Ranger's anticipated 100 pictures, taken at 47 kilometers (29.5 miles) altitude, would achieve a resolution of approximately 3 meters (10 feet) or better with a 200-line system, and provide planetary scientists the first closeup pictures of the lunar surface. In October 1961, NASA's Space Science Steering Committee appointed the experimenters who would analyze and interpret the data to be returned by the television camera: Gerard P. Kuiper, an astronomer and Director of the Lunar and Planetary Laboratory at the University of Arizona; a geologist, Eugene M. Shoemaker of the U.S. Geological Survey; and a chemist, Nobelist, and prime instigator of NASA's unmanned lunar program, Harold C. Urey of the University of California at San Diego.[53]

One more scientific experiment added to the lunar missions in early 1961 caused hardly a ripple of concern in Burke's project office. Not so much added as devised, it involved using in a new way the existing radar altimeter, which was designed to function as a range meter to trigger separation of the lunar capsule assembly from the spacecraft bus at a predetermined height above the moon. No plans existed to telemeter this radar data from the spacecraft to the earth. Walter E. Brown, head of the Data Automation Systems Group in JPL's space Sciences Division, examined the radar altimeter in March 1961 and prepared a memo urging that the Ranger encoder be modified to permit telemetering the radar data to ground stations. He averred that "the correlation between the optical [TV] data and the Ranger 3, 4, 5 radar reflectivity data will provide information about the dust or surface layer." Specifically, such information on the properties of the lunar surface could include its density, conductivity, and the thickness of the layer, useful information for scientists and the Surveyor soft-lander scheduled to follow Ranger.[54]

An associate of Brown's, Harry Wagner, soon formulated a relatively simple method for feeding the radar data into the telemetry system in such a way that it would not interfere with information being transmitted from the gamma-ray spectrometer. That satisfied James Arnold, who of course had no desire to turn off his experiment in favor of data from the radar altimeter. With all hands pleased at this outcome and with no weight or space penalties threatening the spacecraft, Burke approved the necessary modifications to the telemetry system.[55] Members of Newell's Space Sciences Steering Committee agreed as well, and NASA accepted the radar reflectivity measurement as a new experiment a few months later, naming Brown principal investigator (Figure 31).[56]

By the end of 1961, at NASA Headquarters and at JPL, Rangers 3, 4, and 5 were set as lunar missions expressly for scientific purposes. But the production of Atlas-Agenas, the trajectory squabbles, the miscalculated weight of the lunar spacecraft, and the unusual testing and sterilization procedures fully occupied James Burke. He thus viewed the redetermination of the scientific experiments for

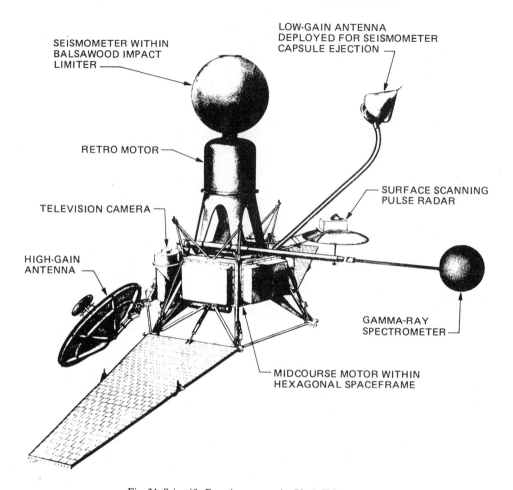

Fig. 31. Scientific Experiments on the Block II Spacecraft

Rangers 1 and 2 as a gratuitous and vexing exercise. And he was more than ever determined to insulate Ranger from what he considered to be "undisciplined efforts" by the scientific community to interfere with established schedules and with the efforts needed to perfect the spacecraft bus, launch vehicle system, and flight operations.

Chapter Five

SUPPORTING THE FLIGHT
OPERATIONS

A LL the while, other JPL engineers were conceiving and building an essential
part of Ranger's operational system—the part for tracking the spacecraft's
position, retrieving its engineering and scientific information, and recording that
information on the ground. Termed here collectively as support of flight
operations, this element, like the spacecraft, was intended for NASA's use in
subsequent deep space flight projects. It came in time to consist of three distinct
components.

First there was the Deep Space Network composed of the radio tracking,
telemetry, and command stations at different points around the earth, a control
center from which to direct the activities of these stations, and the earth-based
communications network that tied the stations and control center together.

Second, there was the mission operations function; this involved, among
other things, computing the space trajectory, receiving and analyzing telemetry
data for science and data on the condition and performance of the spacecraft, and
generating the commands necessary for Ranger to complete its mission
successfully. This activity and the Deep Space Network shared a common control
center at JPL.

Third, there were launch operations that involved the preflight testing of
the launch vehicle and spacecraft at Cape Canaveral, the actual launch itself, and
the immediate tracking support downrange from Cape Canaveral that was needed
before Ranger entered its trajectory in space.

THE DEEP SPACE NETWORK

Ranger's designers recognized and accepted the limited opportunities for
flight testing and the potential for improving the reliability of a spacecraft by

flying a single configuration repetitively. They also recognized that the requirements for planetary space flight would be more severe than those for lunar flight; orienting the design toward the planetary goal was a logical, forward-looking choice. This same reasoning had led to the JPL "bus and passenger" concept, where the functions common to the spacecraft in general were separated and maintained from flight to flight, with changes confined to those "passenger" items peculiar to a given mission. The ground-based systems to support flight operations were thought of in the same way. The design and components for each tracking station were to be made to a single, interchangeable standard. New technology would be introduced only after it had been successfully demonstrated at a test site, and after special recording equipment for a given space mission had been installed at the stations for use during the lifetime of that project. As conceived and implemented, the Deep Space Network* would support all of NASA's unmanned deep space flight projects.

Withal, the spaceborne and earth-based segments of the Ranger system would operate to a single purpose. Scientific information concerning the moon or the interplanetary environment was to be the end product. The scientific experiments would act as the sensing elements, the launch vehicle and spacecraft as the sensor positioners. The Deep Space Network, in turn, would perform the function of position indicator and sensor output recorder. The Network would also permit more accurate measurement of the position and motions of objects in space and allow the efficient transmission to a spacecraft of a command signal to alter its position. At lunar and planetary distances, very sensitive receivers and powerful transmitters that could be pointed directly at the spacecraft would be required on earth, since the weight and power of the spacecraft continued to be restricted by the modest size of the available launch vehicles. In 1958, when ARPA approved the first Pioneer lunar probes, Eberhardt Rechtin and his associates, Walter Victor, Henry Richter, William Sampson, and Robertson Stevens among them, turned their attention to these new demands.

Eberhardt Rechtin was the man ultimately responsible for providing the radio tracking, telemetry, and command of the Ranger spacecraft. Pickering's former student in electrical engineering, he had graduated from Caltech a class behind Burke and Schurmeier in 1946. When he came up from campus to accept a job at the Laboratory four years later, he held a PhD cum laude in electrical engineering. Positive, forceful, energetic, and enthusiastic, Rechtin quickly established a reputation for his ingenious solutions to technical problems. "He could grasp the ramifications of a complex system and foresee benefits and problems far down the road," a colleague remarked, "when many had difficulty seeing the road itself." In 1958 Pickering named Rechtin Chief of the Telecommunications Division. Two years later, as the programmatic aspects of

---

*Originally known as the Deep Space Instrumentation Facility, the name was later changed to the Deep Space Network when more functions were added. The latter term is used throughout the text to avoid unnecessary confusion.

deep space tracking became evident, he was appointed Program Director of the Deep Space Instrumentation Facility* as well (Figure 32).[1]

Under circumstances where a spacecraft far from earth rose and set each day like other celestial bodies, two choices for a deep space radio tracking network presented themselves: First, a single station could be constructed in the United States, and the spacecraft could be interrogated during a single period each day when it came into view. Second, a network of three stations, approximately 120 degrees apart in longitude, could be constructed around the globe. While the second choice would involve possible international political complications, it meant that the spacecraft's transmissions could be received and the craft itself kept in view continuously. Before NASA's creation, in the early months of 1958, Rechtin, his colleagues in the Telecommunications Division, and the engineers charged with

---

*The Deep Space Instrumentation Facility was composed of the radio tracking, telemetry, and command stations around the world. It did not include the ground-based communications network which linked them to the control center at JPL. All three elements, the stations, communications network, and the control center, eventually made up the Deep Space Network.

Fig. 32. JPL Deep Space Instrumentation Facility Director Eberhardt Rechtin

design of the Juno IV spacecraft, recommended the latter course, a course which ARPA, and subsequently NASA, approved.[2]

This decision eased the difficulty in the design of the thermal control system for the spacecraft, and it helped make more reliable the performance of electronic parts aboard the space machine. Continuous operation of the electronics afforded engineers a constant heat distribution pattern with which to work in designing the thermal control system. Turned on and left on, the electronic components were also found to possess a longer lifetime than equipment operated intermittently. Three radio tracking stations, moreover, when fitted out with both receivers and transmitters, added flexibility to the system's ability to detect malfunctions and to trouble-shoot any time a spacecraft encountered difficulty. Designers of the Soviet deep space radio tracking system selected the first approach, the single tracking station which would view their craft for a single period each day. Operated intermittently, their interplanetary space machine was designed with an active thermal control system (using convection), with most of the electronics hermetically sealed and pressurized inside a tank (see Figure 26). Early failures of the spacecraft on deep space missions caused the Soviets to move their design and operation more closely to the American model, and to build a second tracking station—though both remained confined within the broad borders of the Soviet Union.[3]

Members of a JPL team began examining sites for the global deep space stations in 1958–59. They found the first one in March 1958, near Goldstone Dry Lake in Southern California's Mojave Desert, 160 kilometers (100 miles) east of JPL. The choice considerably facilitated acquisition of the land, since this area lay inside the Army's Camp Irwin military reservation.[4] Team members selected the remaining sites in 1959: Island Lagoon, another dry lake bed, near the Woomera Test Range in East-Central Australia, and, for the last station, a shallow valley near Johannesburg, in the Republic of South Africa. The U.S. Department of State worked out arrangements with the respective governments for NASA to use these overseas sites and for the construction and staffing of the installations.[5] Reporting to the national cooperating agencies, but technically and operationally to the American network management, the indigenous staff at the Australian and South African stations would furnish competent and enthusiastic support to this international venture (Figure 33).

Commercial radio stations use as much as 50,000 watts of broadcast power, but Ranger's small transponder would radiate only 3 watts. To acquire and sort out this signal from among the random galactic background noise and man-made radio signals bouncing around in the earth's atmosphere, the antennas at the tracking sites had to be very large. They also had to be steerable—to point at the spacecraft. Finally, they had to be equipped with extremely sensitive electronics to receive the signal from space, and have sufficiently powerful transmitters to send commands to the spacecraft. Two of Rechtin's colleagues, Robertson Stevens and William Merrick, quite literally "lifted" an antenna design from the radio astronomers to meet these demands. The standard dish was 26 meters (85 feet) in

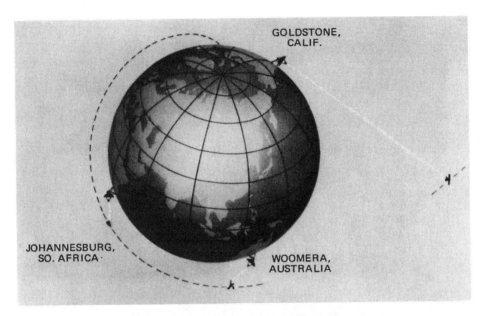

Fig. 33. Station Locations in the Deep Space Network

diameter, with an equatorial or polar mount based on astronomical requirements. Both the dish and mounting designs suited the needs of deep space tracking, where rates of movement were significantly less than a degree per second, and where a distant spacecraft looked more like a star in the sky than did such rapidly moving objects as missiles or low-altitude earth satellites (Figure 34).[6]

The astronomical antenna, nevertheless, had to be modified for two-way deep space communications. The engineers added a device that would automatically point the antenna at the spacecraft. Once contact was made using a precalculated ephemeris, the radio signal from the spacecraft itself would be used to drive the antenna servo pointing mechanism. The parabolic surface of the antenna would collect and focus the spacecraft signal upon a "feed" (signal collector) above the center of the dish, which in turn would send it to a series of sensitive amplifiers that boosted its intensity and processed the signal. To command or interrogate the spacecraft with the transmitter, the process was reversed.[7]

In a further modification, the transmitter and receiver on earth were coupled to a diplexer. The diplexer permitted simultaneous transmission and reception via a single antenna, as shown in Figure 35. Earth-to-space transmission, known as "uplink" to its practitioners, was provided by a 10-kilowatt transmitter operating at 890 megahertz. The Ranger transponder received this radio signal, phase-locked with it, stripped off the uplink command data, multiplied and

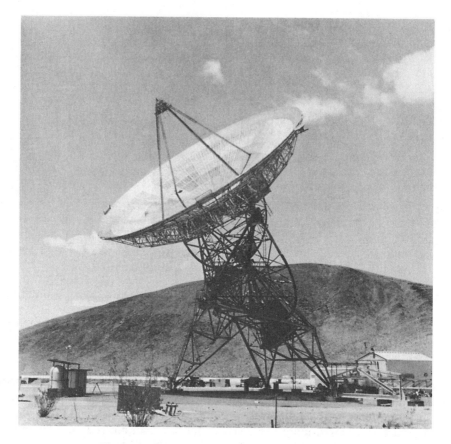

Fig. 34. Twenty-Six-Meter Radio Antenna at Goldstone

modulated the basic downlink signal with engineering and scientific telemetry,*
and retransmitted it from space to earth at 960 megahertz. This mechanization
made it possible to determine the velocity and direction of the spacecraft by
comparing the frequency transmitted to the spacecraft to the frequency received on
the earth to obtain the doppler shift. The trajectory in space could be accurately
determined from the observed "two-way" doppler shift and the position of the
antenna with respect to the earth and the spacecraft.[8] The large, steerable,

---

*Telemetry, the technique of transmitting instrument recordings over a radio link, was pioneered
during World War II. As rockets became more powerful and flew to greater altitudes and ranges,
engineers and scientists found it increasingly difficult to gather data about the vehicle and any
experiments it carried. Most rockets could not be recovered intact after flight testing with the result
that much valuable information was lost. With the advent of radio telemetering systems, however,
various functions of the rocket's performance, such as fin position, combustion pressure, skin
temperature, missile attitude, and critical voltages in electronic gear, could be monitored in flight and
the information relayed by radio to ground receiving stations, where it was recorded for analysis.

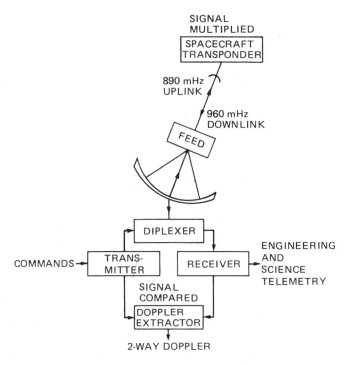

Fig. 35. The Deep Space Network–Spacecraft Link

parabolic antennas and continuous two-way communications would become the salient features in all NASA space missions and in military space communications for command and control.

NASA approved, funded, and monitored Deep Space Network developments after that agency began to function in late 1958. Edmond C. Buckley, Silverstein's Assistant Director for Space Flight Operations, supervised the effort from the Office of Space Flight Programs. Later, when James Webb established the tracking and operations function as a separate program in 1961, Buckley would become Director of the Office of Tracking and Data Acquisition. A graduate of Rensselaer Polytechnic Institute, personable and articulate, Buckley brought to NASA many years of NACA experience in the development of the Wallops Island Launch Range, where he had been responsible for the tracking and instrumentation associated with free-flight research. Buckley and Rechtin respected and liked one another, and got on well together. No major disagreements marred the planning and creation of the Deep Space Network.

The ground communications network connected all of the deep space tracking stations, the tracking radars downrange from Cape Canaveral, and the control center at JPL. Each station was tied to the control center by two teletype circuits and one voice circuit. Over these circuits came the administrative messages by phone and the time-labelled tracking, telemetry, and scientific data in digital form suited to the computers housed at JPL.[9] The Goldstone station, JPL, and Cape Canaveral would be linked on trunk tie lines in the United States. The small tracking antennas downrange from Cape Canaveral, in Puerto Rico, and on Antigua and Ascension Islands, were connected by means of a submarine cable, as was the deep space station in Woomera, Australia. But while one could talk with colleagues in Australia directly via a submarine cable, such a direct link to South Africa did not exist in the early 1960s. Phone calls and teletype messages to the Johannesburg station had to pass by submarine cable to England, then by trunk line to Spain. From North to South Africa, across an entire continent, the calls were patched together with high-frequency radio links and land lines. Both phone calls and teletype information from Johannesburg to JPL returned in the same manner. A good connection over the radio segments in Africa depended on stable ionospheric conditions, and disruptions during certain times of the day were common.

On July 4, 1961, NASA declared the Deep Space Network, including the deep space stations, the ground communications system, and the flight control center at JPL, to be operational.[10] As preflight shakedown tests for Ranger 1 began, however, flight controllers encountered an unforeseen problem with the voice circuits. They found the volume on these lines decreasing precipitously as the date of launch drew near, making it extremely difficult for the controllers at JPL to hear their counterparts in other parts of the net. Checks of the circuits failed to reveal clues to account for this mysterious occurrence. Then one astute observer interrupted his conversation and called out "operator." Several voices answered simultaneously. The volume loss had resulted from an increasing number of operators "listening in" as the moment of launch approached. Thereafter, routings were employed that largely eliminated this difficulty, but it did little to improve the poor quality of the high-frequency radio links across Africa.[11] With commercial communication satellites as yet unavailable to handle this international traffic, the radio links would have to be tolerated until 1968, when a submarine cable connecting South America with South Africa was finally completed.

Space Flight Operations

"Space flight operations" would become the art of conducting, on the ground by remote control, a mission being performed by a vehicle in space. It would include determining and correcting the flight path, analyzing telemetry from space, specifying and issuing commands to the spacecraft, and recording and processing the scientific information from the mission. This specialized activity would also be performed in a single flight control center that served both as the "command post" and as focus of the deep space stations, the tracking stations downrange from Cape Canaveral, and the ground communications net that tied these elements together around the world.

In May 1960, detailed studies of a system that could receive, evaluate, and process the data returned to earth, and would permit control of the maneuverable Ranger spacecraft in space, began at JPL. The plans were completed in November, with the aggregate effort designated the Space Flight Operations System. The system consisted of the equipment, computer programs, and groups of technical and operational personnel required to support Project Ranger. Since the activity involved all of the technical divisions in a combined effort, Systems Division Chief Schurmeier was charged with staffing and directing the new function. He would appoint a Flight Test Director to implement the system. The Test Director would report to him on system matters in general, and to Burke as they affected Ranger in particular. Together, they would work out the details of space flight operations.[12]

Just as the Ranger spacecraft was broken down into subsystems, space flight operations were likewise divided into functional components, each reporting to the Flight Test Director. There was to be a Deep Space Station Operations Manager, a Central Computing Facility, a Scientific Data Group, a Data Reduction Group, and a Spacecraft Data Analysis Team. The groups would consist of the cognizant engineers and representatives from the technical divisions who had participated in the development and testing of Ranger's subsystems. They would receive the engineering telemetry and determine and analyze the state of Ranger's health. At the end of 1960 NASA and JPL officials decided to house Space Flight Operations in a control center at JPL, along with the computers and the main technical staff. The building containing the computer facility would be modified to incorporate the operations rooms needed for the Ranger flights. Systems Division Chief Schurmeier named Marshall S. Johnson as Flight Test Director. Johnson had been employed in 1957 as one of JPL's growing collection of computer wizards. The job of Flight Test Director called for securing the cooperation of all of JPL's technical divisions involved in the conduct of mission operations, a task of no mean proportions. Johnson's knowledge, persuasive talents, and demonstrated willingness and ability to tackle any assignment—or anybody obstructing an assignment— marked him as a natural choice. One year later, in October 1961, he was named Chief of a new Space Flight Operations Section as well.[13]

The flight control center took shape in 1961 prior to the flight of Ranger 1. A large room was cleared next door to the computers to serve as the operations center. Blackboards and pinboards, used to post current flight information, lined the walls. Desks and phones were installed for the flight operations teams. Perpendicular to the operations center, another room housed the Science, Command, Spacecraft Data Analysis, and Trajectory Groups. With time, Johnson and his crew integrated more rooms, added consoles containing closed-circuit television (allowing each controller to view incoming data in the adjacent rooms), and replaced the blackboards with rear projection screens, where information of a more general nature could be displayed for all to see. Johnson somehow managed to overcome the normal bureaucratic inertia and allegiances, forced people to face problems early, worked around the inevitable man-made complications, and improvised the combination of hardware, "software,"* and people that made space flight operations and the flight control center a functioning reality (Figure 36).

The new control facility, makeshift at best, was recognized as inadequate to deal with several space flight missions operating simultaneously, a condition projected for post-Ranger lunar and planetary flights. In February 1961, JPL Director Pickering formed a committee to evaluate these future requirements. The committee recommended construction of a new building, devoted exclusively to deep space flight operations. In July, NASA Headquarters approved the plans.[14]

This building, the Space Flight Operations Facility, would be completed on schedule and placed in operation in early 1964. As Chief of the Space Flight Operations Section, Johnson oversaw the design and construction of the new control center. Seemingly never wearing a coat, dressed in open shirt and red suspenders, he could be found at work any time. His idiosyncrasies made legends at the Laboratory. NASA Space Sciences Director Homer Newell was among those

---

*Software means computer programs and operating routines.

Fig. 36. JPL Space Flight Control Center, 1961

to acknowledge Johnson's efforts at the dedication ceremony: "Planner, cajoler, threatener, union arbitrator, designer, budget parer, trouble shooter," Newell declared, "you name it and he had to do it—and he did."[15]

Both the original flight control center and the one that replaced it swelled the number of new installations created expressly for NASA's deep space missions. And all of them, including the Environmental Test Laboratory, the Spacecraft Assembly Building, the Deep Space Tracking Stations, and the Space Flight Operations Facility, were to be employed first in Project Ranger. So, too, were the procedures and facilities devised for launching NASA's Atlas-Agena B vehicles at Cape Canaveral in Florida.

Launch Operations

NASA's global deep space network and space flight operations could control Ranger once the spacecraft had been injected into its space orbit. But the United States Air Force had first to launch Ranger from the Cape Canaveral Missile Test Annex in Florida, a facility controlled from the Missile Test Center at Patrick Air Force Base, situated a few miles away at Banana River. In 1960 Major General Leighton I. Davis commanded both installations. Davis was a spit-and-polish West Point graduate and pilot who made it plain that the new space agency, like the Army before it, was welcome at Cape Canaveral as a tenant, subject to the Air Force regulations that governed all operations there.[16]

The Air Force group responsible for actually launching the Rangers from Cape Canaveral was the 6555th Aerospace Test Wing. The 6555th oversaw the checkout of the assembled vehicles on the pad, as well as the launch operations proper. This arrangement was made explicit in July 30, 1960, when NASA issued the *Agena B Launch Program Management Organization and Procedures*. The same document set forth the organizational framework that NASA had selected for its launch operations. Thus, as with launch vehicle procurement, NASA depended on the Air Force both for the facilities and launch support for Project Ranger. And if the organization to procure the vehicles was complex, the one created to launch them was equally confusing. JPL's Lunar Program Director Cummings and Ranger Project Manager Burke would have to "manage" Ranger launchings through the Air Force launch organization, two NASA Headquarters offices, several subsidiary offices, and the Agena B Coordination Board. This problem would become one of the most difficult with which they had to deal in 1960 and 1961.[17]

General Ostrander's Office of Launch Vehicle Programs comprised the first headquarters group. In March 1960, Ostrander had created a Launch Operations Directorate with offices at the Marshall Space Flight Center in Huntsville, Alabama, and at Cape Canaveral in Florida. The German-American Kurt H. Debus, a longtime associate of Wernher von Braun, directed the NASA launch organization. Composed of elements of the Army Missile Firing Laboratory, the Debus Launch Operations Directorate acted "as the directing field agency for

NASA in all matters concerned with NASA launchings at the Atlantic Missile Range..." Quartered at the Cape, Debus would have direct access to General Davis and, through intermediaries, to von Braun in Huntsville. To assist in coordinating activities and to plan and order the range support equipment needed for Atlas-Agena launches, Debus established a subsidiary office at Cape Canaveral called the NASA Test Support Office.[18]

Launch operations also involved Silverstein's Office of Space Flight Programs. Silverstein did not choose to let details of launch operations escape his attention, and in March 1960 he established an organization of his own at the range, known as the Office of Flight Missions. This group served as Silverstein's representative at the Cape, reported directly to him, and acted as the central point of contact for the flight project offices, including Project Ranger. In Silverstein's view, it would further serve to "coordinate range support requirements" and provide "administrative and logistic support to all of the NASA elements."[19]

By June 1960, JPL hands at Cape Canaveral complained vigorously that this bifurcated arrangement was unworkable. "Effective action [here]," one declared, "will have to await resolution of the administrative maze."[20] The launch program management organization and procedures document issued in July settled the question of authority. Although it did not reduce the number of groups in the multifaceted organization, it at least defined what each one was and how it related to the others. Meantime, Burke directed Luskin's Agena organization at Lockheed to prepare an engineering plan specifying the tracking, telemetry, data processing, data transmission, and computational facilities and equipment required for the support of Ranger. Eventually known as the *Program Requirements Document,* it contained the needs for the launch hardware and software, hammered out among the participants.

Resolving these and other launch details dragged on throughout 1960, as proposals bounced back and forth between the various operations offices and the technical panels of the Agena B Coordination Board. Agreement, and the necessary subsequent action, were painfully slow. Under the existing procedures, engineers from Debus' Launch Operations Directorate could act as "monitors" of NASA's launch operations, but the 6555th Aerospace Test Wing "supervised" the actual work. In the presence of hesitancy and confusion, Air Force representatives bluntly informed NASA: "Just tell the Air Force what you want, and we will put it in orbit for you."[21]

Like Hans Hueter and Friedrich Duerr in the Light and Medium Vehicle Office at Huntsville, the Launch Operations Director Kurt Debus chafed in a role where he held responsibility for activities he could not control. As the new year began, NASA launch operations remained in the hands of the Air Force. NASA Headquarters might not wish to press for more direct control over launch operations, since such a move "could only be looked upon by the Air Force as a lack of confidence on the part of NASA," as one official later commented,[22] but the organization formed in response to that need seemed hardly adequate.

From a low point at the end of 1960, matters improved in early 1961. The Agena B Coordination Board was abolished in January, and in February Burke learned that the Air Force had authorized procurement of the mobile antennas needed to track the Atlas-Agena downrange from Cape Canaveral. At the same time, the Air Force made available Launch Complex 12 and Hangar AE at Cape Canaveral for NASA's Project Ranger. Headquarters had also moved to alter the cumbersome organization for launch operations. The new NASA Administrator James Webb, Deputy Administrator Dryden, and Associate Administrator Seamans argued forcefully that the Air Force might properly furnish the launch vehicles as the procuring agency, but that NASA, as the using agency, should be wholly responsible from launch through the completion of a mission. A NASA-Air Force agreement of July 17, 1961, recognized NASA as a joint tenant with the Air Force at Cape Canaveral and called for the eventual replacement of the 6555th Aerospace Test Wing by NASA's own launch groups over a period of time.[23] And shortly thereafter, in early 1962, the entire NASA launch organization was overhauled. Meantime, however, the first Atlas, Agena, and Ranger had been delivered to Cape Canaveral.

Chapter Six

TEST FLIGHTS AND
DISAPPOINTMENTS

BY mid-1961, project personnel were preparing for the launch of Rangers 1 and 2. These missions were to demonstrate the engineering performance of the spacecraft and its flight support operations, provide operating experience with the Atlas-Agena B launch vehicle, and measure, by means of sky science instruments, fields and particles along the selected trajectories.[1] Each spacecraft would be launched into a low earth orbit from Cape Canaveral. Then, upon second burn of the Agena stage, each would be injected into a highly elliptical orbit with an apogee (or high point) of approximately one million kilometers (620,000 miles) and a perigee (low point) of a few hundred kilometers. Of course, with no rocket engine aboard, no midcourse maneuver could be made. But, swinging far out into space beyond the orbit of the moon, with an operating life expectancy of five months, the flights of Rangers 1 and 2 promised a great deal: for engineers, a first opportunity to command and exercise this new breed of spacecraft in deep space; for sky scientists, new knowledge of the space environment from measurements conducted over a prolonged period (Figure 37). Indeed, by 1961, just enough scientific data on the magnetosphere and outer space phenomena had been returned by NASA's early Pioneer space probes and Explorer satellites to hint at the fascinating electromagnetic interactions occurring between the earth and the sun.

PLANNING THE ASCENT

The Atlas booster, which would deliver the spacecraft into near space, was America's first intercontinental ballistic missile. Developed in San Diego for the Air Force in the mid-1950s by General Dynamics-Astronautics, it measured 18 meters (60 feet) in length, and towered five stories high. Fueled with liquid oxygen

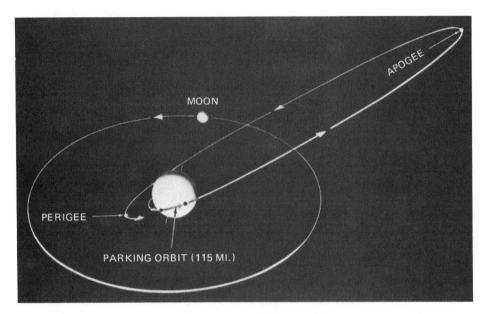

Fig. 37. Space Trajectory Selected for Rangers 1 and 2

and kerosene, it weighed 117,000 kilograms (130 tons). The rocket engines, supplied by Rocketdyne Division of North American Aviation, produced 163,080 kilograms (360,000 pounds) of thrust at sea level—equivalent to the combined power of six Boeing 707 jet liners. To increase the weight carried into space, the Atlas designers had radically lightened the stainless steel skin and structural members. The structure was so light, in fact, that it could not support its own weight. To provide the needed rigidity, the vehicle was pressurized internally, like a balloon. The engineers also determined to ignite all three main engines at launch since they could not be certain whether liquid propellant rocket engines could be ignited at very high altitudes and at very low pressures (see Figure 5).

The Agena booster satellite, stacked atop the Atlas, had been designed by the Lockheed Aircraft Corporation's Missiles and Space Division. The model B to be used in Project Ranger employed propellant tanks of increased capacity over an earlier model, and a Bell Aerosystems rocket engine that could be ignited in space, shut down, then reignited to burn a second time (Figure 38). This 7248-kilogram (16,000-pound) thrust engine consumed two noxious liquid chemicals known as unsymmetrical dimethyl hydrazine and inhibited red fuming nitric acid. The Agena B would hurl Ranger into its ultimate deep space trajectory.[2]

There were two methods by which Ranger could be sent into its deep space path: the direct ascent and the parking orbit. In the first of these, the spacecraft would be launched upward through the earth's atmosphere, directly into orbit toward its celestial target. In the parking orbit trajectory, the upper stage of the launch vehicle and the spacecraft would first be placed into an earth satellite orbit;

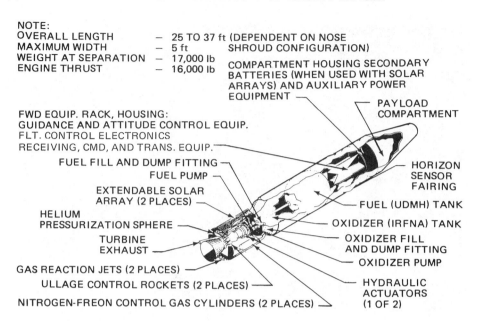

NOTE:
OVERALL LENGTH   — 25 TO 37 ft (DEPENDENT ON NOSE
MAXIMUM WIDTH   — 5 ft   SHROUD CONFIGURATION)
WEIGHT AT SEPARATION — 17,000 lb
ENGINE THRUST   — 16,000 lb   COMPARTMENT HOUSING SECONDARY
                                      BATTERIES (WHEN USED WITH SOLAR
                                      ARRAYS) AND AUXILIARY POWER
                                      EQUIPMENT

PAYLOAD
COMPARTMENT

FWD EQUIP. RACK, HOUSING:
GUIDANCE AND ATTITUDE CONTROL EQUIP.
FLT. CONTROL ELECTRONICS
RECEIVING, CMD, AND TRANS. EQUIP.

FUEL FILL AND DUMP FITTING
FUEL PUMP

EXTENDABLE SOLAR
ARRAY (2 PLACES)

HELIUM
PRESSURIZATION SPHERE

TURBINE
EXHAUST

GAS REACTION JETS (2 PLACES)
ULLAGE CONTROL ROCKETS (2 PLACES)
NITROGEN-FREON CONTROL GAS CYLINDERS (2 PLACES)

HORIZON
SENSOR
FAIRING

FUEL (UDMH) TANK
OXIDIZER (IRFNA) TANK
OXIDIZER FILL
AND DUMP FITTING
OXIDIZER PUMP

HYDRAULIC
ACTUATORS
(1 OF 2)

Fig. 38. Agena B Satellite Configuration (For Project Ranger, the Agena Solar Panels Were Removed and a Spacecraft Adapter Plate Was Substituted for a Portion of the Forward Equipment Rack.)

then an additional rocket thrust would inject it onto its deep space orbit. In both approaches the planner could alter the launch azimuth (direction) of the ascent from Cape Canaveral to accommodate some of the changes in position of the earth and a celestial target. But the parking orbit trajectory offered still further advantages for deep space missions, and the "dual-burn" feature of the Agena made a parking orbit feasible.

Dual burn meant that trajectory planners could alter the point of launch above the moving earth and could, therefore, aim at a wider variety of points on the lunar target. Thus, by varying the launch azimuth at liftoff *and* the coasting time of the Agena in an earth parking orbit, it would be possible to compensate continuously for celestial motions, and expand the time allowable for the launch interval (called the "launch window"). This increased the chance of getting a launch away despite any countdown problems that might occur in the first operation of these large and complex vehicles.* To simplify the operations of the first two Ranger test missions, and to ensure that the Agena second burn would take place within view of the downrange tracking station on Ascension Island in the South Atlantic, predetermined values were established for the launch azimuth and the dual burn and parking orbit. For the lunar flights, however, both the

---

* Were a launch window missed because of delays, the flight had to be postponed for an entire lunar month, until the target point and lighting conditions (necessary for photography and solar power) presented themselves again in the same geometry.

variable launch azimuth and parking orbit coast times would be employed, continuing Ranger's progression from the simple to the complex in design and operation.

At liftoff, under autopilot and programmer control on the desired launch azimuth, the Atlas would begin a gentle arc, tilting toward the horizontal above the earth. With data of the vehicle's performance fed into a large digital computer at Cape Canaveral after the two outboard engines had shut down and separated, steering commands for the sustainer phase of operation could be generated and transmitted over the earth-to-Atlas radio link. Then, upon shutdown of the Atlas sustainer engine, the Cape computer would predict the performance required of the Agena booster. Two commands (parking orbit coast time and second-burn velocity-to-be-gained) would be radioed to and stored in the Agena. Another command would shut down the two small vernier engines on the Atlas when the proper velocity and heading were attained.

Upon separating from the Atlas, the Agena's aerodynamic nose fairing that covered and protected Ranger from the flow of the wind would eject automatically. The Agena B engine would then burn once to establish the parking orbit around the earth. After the proper coasting period, the engine was to be reignited, injecting the vehicle onto the final transfer trajectory. All of these latter events would take place under the command of the Agena's guidance and control unit (composed of gyroscopes, accelerometers, and infrared horizon scanners), without further radio control from the ground. With the final trajectory established, the Ranger spacecraft would be mechanically separated from the Agena by a set of springs; a few moments later, several thrusters aboard the Agena were to be fired, carrying the booster safely away from Ranger and the moon, onto an altogether different trajectory (Figure 39).

Apart from the proper execution of the planned ascent, the success of the Ranger missions would also depend upon the spacecraft performing its appropriate routine in space, upon the proper operation of the deep space stations and the JPL flight control center, and upon how well Ranger's engineers and scientists had performed their preflight tasks.

PREPARING FOR LAUNCH

As expected, systems tests conducted with the Ranger Block I proof test model at JPL revealed a number of subsystem discrepancies and electrical interferences that had to be corrected on the flight hardware. However, by mid-1961 the first spacecraft had passed through assembly and the qualification tests on time and without serious incident. In May, Oran Nicks examined Ranger 1 and its complete test history. Although short on redundant features, the design appeared sound. Nicks authorized shipment to Cape Canaveral.

The allowable launch period for Ranger 1 extended from July 26 through August 2, 1961; its daily launch window extended from 4:53 to 5:37 am EST. For

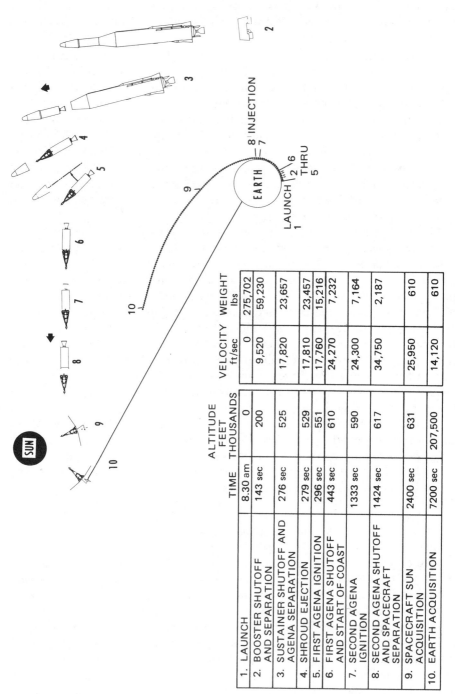

	TIME	ALTITUDE FEET THOUSANDS	VELOCITY ft/sec	WEIGHT lbs
1. LAUNCH	8.30 am	0	0	275,702
2. BOOSTER SHUTOFF AND SEPARATION	143 sec	200	9,520	59,230
3. SUSTAINER SHUTOFF AND AGENA SEPARATION	276 sec	525	17,820	23,657
4. SHROUD EJECTION	279 sec	529	17,810	23,457
5. FIRST AGENA IGNITION	296 sec	551	17,760	15,216
6. FIRST AGENA SHUTOFF AND START OF COAST	443 sec	610	24,270	7,232
7. SECOND AGENA IGNITION	1333 sec	590	24,300	7,164
8. SECOND AGENA SHUTOFF AND SPACECRAFT SEPARATION	1424 sec	617	34,750	2,187
9. SPACECRAFT SUN ACQUISITION	2400 sec	631	25,950	610
10. EARTH ACQUISITION	7200 sec	207,500	14,120	610

Fig. 39. Ranger Ascent Sequence

the first time all of the hardware for a Ranger mission was brought together, and all entered the last round of inspections and flight acceptance tests. Once the countdown began, the project members would also for the first time participate in an actual launch operation. In the weeks immediately preceding launch, Burke and Schurmeier, together with Air Force Major Jack Albert, Marshall's Agena chief Hans Hueter, and other participants, spent an increasing amount of time at Cape Canaveral ironing out bureaucratic differences and seeing to final details.

The participants reconciled differences of opinion, and the Ranger vehicles and flight operations complexes finished their test cycle in June and July. Moved to Launch Complex 12 at the Cape in late June, Agena B 6001 was stacked atop the Atlas 111D. Ranger 1, attached to an adapter plate and sealed inside the nose fairing, was then connected to the forward end of the Agena, and engineers began combined systems tests of the assembled vehicle. The ensemble satisfactorily passed what was termed the Joint Flight Acceptance Composite Test on July 13, and was judged ready for launch. The Deep Space Network, meantime, underwent net integration tests involving the deep space tracking stations, the JPL control center, the JPL command center in Hangar AE, and the tracking stations downrange from Cape Canaveral. Although encountering some problems, especially with communications to South Africa,[3] these tests were also deemed successful. The control center received the tracking, engineering, and scientific telemetry data from all of the stations in the correct format and proper sequence.[4]

To permit tracking Ranger in its initial parking orbit above the earth, a small, mobile tracking antenna had been moved to the Johannesburg station. Though it could not track radio signals at great distances, this 3-meter (10-foot) dish antenna possessed a beamwidth of 10 degrees—ten times as wide as the large, permanent antennas—and it could swing at a rate of 10 degrees per second—ten times as fast as the big dishes could be moved. A small group led by Earl Martin of JPL readied the mobile antenna and prepared to acquire and track the rapidly moving Ranger above South Africa, as the Agena reignited its engine to inject the spacecraft onto its deep space trajectory. Information provided by the mobile tracking antenna and the tracking stations downrange from the Cape would enable the big dishes in the Deep Space Network to locate and follow Ranger 1 as it moved out and away from the earth on its space trajectory (Figure 40).[5]

On July 26, 1961, everything was ready; at least everything that had been thought of and tested by myriad engineers and scientists in a dozen organizations around the world.

A LEARNING EXPERIENCE

The first countdown for the launch of Ranger 1 began on the evening of July 28, three days late. Trajectory information required by the Range Safety Officer was delayed, at a cost of one day; a guidance system malfunction in the Atlas booster consumed another; and the third was lost when engineers found that the guidance program to be fed into the Cape computer contained an error. The

Fig. 40. Three-Meter Antenna at Johannesburg

countdown proceeded normally into the early morning hours of July 29. Power interruptions occurred eighty-three minutes before launch, requiring momentary holds to permit all stations to check and recover. Then, at 5:02 am EST, twenty-eight minutes before launch, Cape Canaveral, aglow with lights and the hum of launch activity, was plunged into darkness. Commercial electrical power had failed, much to the increased consternation of launch officials when they learned why: inadequate allowance had been made for changes in cable sag caused by variations in temperature on the new power poles recently installed at the Cape. The launch was delayed yet another day.

The second countdown was canceled on July 30 when engineers discovered a leak in Ranger's attitude control gas system. After the faulty seal had been repaired and the spacecraft returned to Launch Complex 12, the third countdown commenced on July 31, only to be terminated once again when a valve malfunctioned in the liquid-oxygen tank on the Atlas booster.

Early in the fourth countdown on August 1, ground controllers turned on a spacecraft command applying high voltage to the scientific experiments for calibration purposes. Immediately all stations reported a major spacecraft failure.

An electrical malfunction had triggered multiple commands from the central clock timer, and Ranger 1 "turned on" as it had been programmed to do in orbit. The explosive squibs fired, solar panels extended inside the shroud, and all the experiments commenced to operate.

Project engineers disengaged Ranger 1 from the Agena and hastily returned it to Hangar AE. Meantime the launch was rescheduled for August 22, the next available opportunity. Subsequent tests and investigations determined the activating mechanism to have been a voltage discharge to the spacecraft frame; although engineers suspected one or two of the scientific instruments, they could not determine the precise source of the discharge with certainty. In the days that followed they replaced and requalified the damaged parts, and modified the circuitry to prevent a recurrence of this kind of failure.[6]

If the frustrating delay in the launch of Ranger 1 embarrassed NASA and JPL project officials, it also vexed Burke for quite another reason. As he viewed the matter, the ultimate loss of one month was the direct result of an unorthodox power discharge among the sky science experiments—cargo that was incidental to and conflicted with the primary engineering test objectives. Several of the experimenters, moreover, had already complained to him and to JPL Space Sciences Division Chief Al Hibbs about another kind of conflict between the instruments themselves. J. P. Heppner, the experimenter for the rubidium vapor magnetometer, announced that he found the friction experiment generated a magnetic field in excess of the background field for which his magnetometer had been designed.[7] Since the friction device was actually an engineering experiment, he argued that space science be served by removing it from the spacecraft. For the sake of future bearing and gear designs in space machines, J. B. Rittenhouse and his coexperimenters likewise insisted that the friction experiment be retained. More certain than ever that the cargo should not be allowed to further jeopardize or delay the test flight, Burke told Hibbs and the experimenters concerned that, at this crucial point, no changes could be made in the science or engineering content of the Ranger Block I vehicles. Under existing schedules, each experimenter had to obtain the best data he could from his own instrument.[8]

In the evening of August 22, 1961, the fifth countdown for the launch of Ranger 1 commenced on schedule at Cape Canaveral. The weather, although clear, was still warm and humid. At 6:04 am EDT, amid incandescent flame and the roar of the Atlas engines, Ranger 1 lifted from Launch Complex 12 and rose into the dawn sky, leaving in its wake a luminous, spectacular trail visible over the length of the Florida peninsula (Figure 41).

Like other launch officials anxiously awaiting the first return of data from the downrange tracking stations, Burke, now the formal mission director, moved to the JPL command post in Hangar AE. As the minutes passed, early returns were spotty and disconnected. Both a Department of Defense tracking ship in San Juan Harbor and the station on Ascension Island, he learned, had fought for control of the launch vehicle's radio beacon, with the result that data from the tracking ship were not immediately available. Nevertheless, what little information did arrive

Fig. 41. Launch of Ranger 1

seemed encouraging. Both the Atlas and Agena appeared to have performed normally, and an earth parking orbit had been achieved. First returns from the Mobile Tracking Station in South Africa, however, were unclear. The spacecraft had appeared over the horizon five minutes later than anticipated. As Ranger 1 streaked eastward across the Indian Ocean, data from the mobile station suggested that the Agena B had not ignited a second time.

Ninety minutes after launch, Burke had solid—and bad— tracking news from one of the big dish antennas at Walter Larkin's Goldstone tracking facility in California. The Agena second burn definitely had gone improperly. With a perigee of 168 kilometers (105 miles) and an apogee of 501 kilometers (313 miles) altitude, Ranger 1 was stranded in a near-earth parking orbit—a flight profile for which neither it nor the deep space tracking net was designed. Disappointed at this outcome, Burke remained temporarily at the Cape, and conferred with Marshall Johnson by phone at JPL to see what could be done to achieve at least some of the mission's engineering objectives.

With Ranger 1 in a near-earth orbit, the deep space stations around the globe attempted a program of low-altitude satellite tracking. But the high angular velocity and the poor "look" angles near the horizon on many of the orbital passes severely taxed the capabilities of the large antennas (Figure 42). The polar-mounted antennas at the Woomera station in Australia and the Johannesburg station in South Africa were closed down for a portion of the flight. In the days that followed, the small dish of the mobile tracking antenna in Johannesburg, and one of the large dish antennas at Goldstone which used an azimuth–elevation mounting, performed the lion's share of the tracking.[9]

In space, meantime, Ranger 1 struggled to meet the unusual requirements imposed upon it by force of circumstance. Telemetry data arriving at the JPL control center disclosed that the spacecraft had separated properly from the Agena, had deployed its solar panels and high-gain antenna, and that the attitude control system had stabilized the machine in three axes. With its "top hat" low-gain antenna pointed at the sun, the solar panels were generating power, and Ranger 1 could receive and act on commands sent to it by the Goldstone transmitter.

But, designed for deep space flight, Ranger 1 could not meet the sustained demands of a near-earth satellite orbit. As it passed into the shade of the earth every ninety minutes, the spacecraft lost solar power and orientation, and had to realign and stabilize itself anew when it reemerged again into the sunlight. This process used the nitrogen gas of the attitude control system at a prodigious rate. On August 24, one day after the launch from Cape Canaveral, the gas supply was exhausted and Ranger 1 began to tumble in orbit. With its solar panels no longer aligned with the sun, the machine was left with nothing but its battery as a source of power. For the next three days telemetry continued to be received from a number of the scientific experiments and on the performance of the spacecraft subsystems. Then, on August 27, the main battery went dead. The telemetry output and all spacecraft functions ceased. Blind and inarticulate, Ranger 1 reentered the denser layers of the atmosphere on August 30 and was consumed in a ball of fire over the Gulf of Mexico.[10]

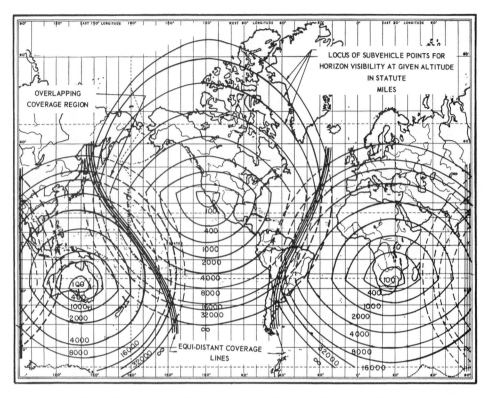

Fig. 42. Deep Space Tracking Coverage as a Function of Spacecraft Altitude

As disappointing as this first flight was, project engineers in Pasadena and their associates at NASA Headquarters could find solace in the performance of the Ranger machine in an unorthodox situation. Even though precious little information had been returned by the sky science instruments[11]—to the greater disappointment of the scientists—engineers considered the soundness of the basic spacecraft design to be demonstrated.[12]

Postflight analyses of telemetry tapes indicated the cause of the engine failure: the Agena B restart sequence had begun at the proper time, but terminated immediately when a malfunctioning switch choked the flow of red fuming nitric acid to the rocket engine. Nevertheless, sufficient oxidizer gas had been expelled to add a velocity increment of approximately 70 meters per second (158 miles per hour), assuring Ranger 1 a slightly higher orbit.[13] Steps were taken to fix the miscreant switch and resolve other tracking problems that had occurred downrange from Cape Canaveral. Perhaps more important, in the wake of the difficulties that had cropped up during the launch, the Air Force, the contractors, NASA, and JPL all made extra efforts to ensure that Ranger 2 succeeded on schedule without any more embarrassing complications.

The Second Lesson

Ranger 2, Agena B 6002, and Atlas 117D had been delivered at Cape Canaveral as the Ranger 1 mission reached its premature conclusion. Preflight preparations began immediately to meet the launch period planned for October 20 to 28, 1961. In the weeks that followed, each of the vehicles completed its exhaustive preflight tests, culminating on October 11 in a satisfactory Joint Flight Acceptance Composite Test at Launch Complex 12. The Deep Space Network, in turn, completed net integration tests on October 18. Project officials made special efforts to preclude the downrange tracking and communications difficulties that had occurred in Ranger's first flight. Again, every facet of Ranger launch and tracking operations was brought to a state of readiness.

The daily launch window established for Ranger 2 was identical to that of its predecessor. The countdown began promptly at 9 pm on the evening of October 19, and proceeded without incident into the early morning hours of October 20. Two hours before launch mission controllers called a brief hold when telemetry indicated irregular performance in the electronics associated with Chubb's Lyman alpha scanning telescope experiment aboard Ranger 2. Officials conferred inside the blockhouse. Since engineers quickly determined that the misbehaving scientific passenger posed no threat to the mission's engineering objectives, Project Manager Burke authorized the countdown to proceed. But little more than an hour later, forty minutes before launch, electrical power to the Atlas guidance package failed. As time in the launch window ebbed swiftly away, the cause of this more serious difficulty could not be readily detected, and Burke, Debus, and Nicks decided to postpone the launch. The difficulty, traced to a faulty splice in the power lead to the Atlas guidance system, was rectified readily.[14] Meantime, another space flight was to take place from Cape Canaveral the following day, and the officials rescheduled Ranger 2 for a second countdown beginning on October 22, a forty-eight-hour delay.

But J. P. Heppner, the magnetometer experimenter from the Goddard Space Flight Center, did not want Ranger 2 to leave just yet. He had approached Burke shortly before the first countdown and had again requested that the friction experiment be removed from the spacecraft. Burke, who viewed himself in a situation analogous to that of the captain of a ship threading its way through a mine field in the middle of a storm, had no patience for a discomfited passenger complaining about the champagne. Burke again refused.[15]

That agitated Heppner more than ever. He had counted on two flights with his magnetometer aboard Ranger. If the friction experiment, as he suspected, did interfere with his measurements on the first mission, there would still have been time to resolve the problem in some fashion before the flight of Ranger 2. Now he could count on one flight only, and the offensive friction device with its interfering magnetic field, moreover, remained on the spacecraft. He stood to lose everything for which he had labored over eighteen long months. T. A. Chubb, though relieved to have an opportunity to fix his own Lyman alpha telescope before another launch attempt, doubtless sympathized with Heppner. To them, project engineers, in their eagerness to flight test the Ranger machine, seemed insensitive to the needs of space science.

Bypassing project management altogether on October 21, in the day between launch attempts, Heppner called and explained the disquieting situation to colleagues in the science group in the Office of Space Flight Programs at Headquarters. The next morning at the range, as preparations for the second countdown neared completion, Burke received a phone call from Flight Programs Director Silverstein. The launch of Ranger 2, Silverstein told him, would be postponed until the conflict between the friction and magnetometer experiments could be resolved.[16]

It was Burke's turn to be exasperated. Besides being annoyed with Heppner's unorthodox maneuver, the Ranger Project Manager cared little whether either instrument stayed or went. The purposeful interference with established schedules and objectives simply threatened the flight. Burke notified Debus and Nicks of the development, then hastily called a conference among the parties involved. JPL Space Sciences Division Chief Hibbs presided at the meeting, in which the scientists argued their respective positions once more. And everyone in the chain of command, from Burke, Cummings, and Pickering at JPL to Nicks and Silverstein at NASA Headquarters, became involved in the decision. As events turned out, neither experiment was removed. Rittenhouse and his associates agreed to fabricate a new shield to cover the friction experiment, reducing its magnetic field to a level which Heppner found acceptable. Working around the clock, technicians completed and installed the shield in time for a second countdown beginning late on October 23, at a cost of another day in Ranger 2's October launch period.[17] But the delay this exercise occasioned was insignificant compared to continuing problems with the Atlas-Agena launch vehicle.

After the second countdown began, in the early hours of October 24, the hydraulic system used to gimbal (swivel) the Atlas vernier engines developed a leak, cancelling the launch. Then, as the third countdown commenced late in the same day, more bad news arrived. Luskin at Lockheed informed Debus that investigation of a recent Thor-Agena B failure at Vandenberg Air Force Base on the West Coast indicated that an Agena engine hydraulic system had malfunctioned. Pending an investigation and solution of this affair, officials agreed that the launch of Ranger 2 should be postponed until the next available opportunity beginning in mid-November.[18] As Burke prepared to return to his office in Pasadena he mused over events of the preceding week. In view of the Air Force Agena failure at Vandenberg, Heppner's obstinacy a few days before might actually have saved Ranger 2 after all.

The Agena hydraulic problem was found and repaired in the intervening weeks, in time for the fourth launch attempt. On November 17 Major Albert, Hueter, Nicks, and Burke gathered again for the countdown of Ranger 2, eager for the satisfaction of seeing their efforts succeed after months of painstaking labor.

Overhead the sky was clear and the temperature moderate—a welcome relief from the heat and humidity of the summer months. Good fortune seemed to be in the air as well, and the count, which began at 8:57 pm EST, proceeded normally with only minor delays (Figure 43). Then, the familiar puff of white smoke and the bellow that quickly swelled to a roar marked ignition of the Atlas engines. The

Fig. 43. Ranger 2 Countdown Progresses Beneath a Full Moon

vehicle strained against its retaining bolts as the thrust reached 360,000 pounds. Amid the flames and clouds of steam from the cooling water circulated in pits below, the launch vehicle was released from Launch Complex 12 at 3:12 am EST on November 18, 1961, and rose into the dark morning sky. Ascending along with it went the audible shouts and silent prayers of the assembled engineers and scientists.

With other project officials, Burke left the blockhouse and drove across the Cape to the JPL command post in Hangar AE to await word from the downrange tracking stations. Once again the first returns were auspicious. The tracking ship in San Juan harbor and the station on Ascension Island reported that the Atlas had performed as programmed, and the first burn of the Agena had placed Ranger 2 in the desired parking orbit. The downrange stations and the communications links also performed flawlessly. With the good news, hopes rose and the tension increased proportionately. Then the mobile tracking antenna in South Africa, which acquired Ranger 2 at 3:44 am EST, showed that the Agena second burn again had not occurred: Ranger 2, like Ranger 1, was stranded in a near-earth parking orbit. Worse, as it turned out, no additional impulse had been obtained in the second-burn sequence, and the orbit of Ranger 2 was even lower and more tenuous than that of its predecessor. Its working lifetime in space would be much shorter.

Dismal though the situation was, a few embers of hope still flickered. Unaware of the true orbit, those present in the command post learned that Ranger 2 had separated from the Agena, and that its central clock timer had initiated the proper series of maneuvers and switching functions. The solar panels, high-gain antenna, and particle analyzer boom had been extended, the commands to acquire the sun and earth issued, and the high voltage to the scientific instruments turned on. But just how much less could be expected from Ranger 2 became apparent during its second and third orbital passes. Acting out its predesigned mission, Ranger 2 began to look for the sun at the moment it entered into the earth's shadow. The earth acquisition command, issued a few minutes later while the spacecraft was still in the darkness, found Ranger unable to "see" the prescribed celestial points of reference. Ranger's attitude control system did not recover, and the spacecraft tumbled in orbit. Tumbling, Ranger 2 could not generate solar power. It used internal battery power to transmit its diagnostic and scientific telemetry to anyone who could track and listen.

Few could. The large and powerful dish antennas at the NASA–JPL deep space tracking stations were unable to keep pace with the high angular velocities and cope with the poor look angles. No backup commands, therefore, could be radioed to the hapless spacecraft. Only the small mobile tracking antenna in South Africa could successfully track and receive telemetry from the spacecraft, and even it would not view Ranger 2 for long. On its fifth pass over that station, some seven hours after launch, Ranger's orbit descended below the South African horizon. No further tracking could be accomplished. Twenty hours after rising above the beach sands of Florida on a stiletto of fire, in its nineteenth orbit, Ranger 2 died. No one heard the final sing-song tones of the analog telemetry on November 19, as the cart-wheeling spacecraft dipped into the atmosphere and quickly incinerated.[19]

THE AFTERMATH

The JPL command post at Cape Canaveral had taken on the cast of a somber requiem when the bad news first arrived. Voices were subdued and emotions held in check as controllers mechanically went about the tasks of receiving and recording available telemetry. The next day, having paid their last respects, project officials began to drift away to their respective offices around the country. Tired and dispirited, Burke was among the last to leave, returning to Los Angeles on November 19 to prepare for the postflight investigation—one that would focus on the performance of the Atlas-Agena launch vehicle.

When development of the vehicle began at Lockheed in 1959, the Air Force had publicly announced its intention to employ the upper stage Agena B in numerous military satellite missions. NASA officials had selected the Atlas-Agena B launch vehicle for Project Ranger in part for the reliability that they anticipated would follow these Air Force flights. But the Agena B had not appeared in the Air Force flight schedules as rapidly as NASA hoped; in fact, because Ranger had held to schedule while other projects slipped, only one other Atlas-Agena B actually had ever been flown before. Long-time Ranger hands recalled the NASA-Agena B development difficulties at Lockheed in 1960. And the more recent miscalculation of the Agena B's performance capability in mid-1961—resulting in the lightweight Block II Rangers—still raised hackles among engineers on every side.[20] Now, two consecutive flight failures on top of the prior troubles generated disturbing speculation. If the Agena B design were sound, either Lockheed was not doing its homework or the civilian space agency had been assigned the costly and ridiculous task of debugging the vehicle for the Air Force.

The implications were quite clear to Major General Osmond J. Ritland, Commander of the Air Force Space Systems Division in Inglewood, California. Two Agena failures in an important NASA space program quite obviously affected the credibility and technical reputation of the Air Force and its contractors, not to mention prospects for individual careers. The Air Force preferred to straighten affairs in its own house before NASA pressed the issue. Two days after Ranger 2 met its fiery end in space, General Ritland fired off a stern memo to Colonel Henry H. Eichel in charge of his launch vehicle office. He ordered Eichel immediately to form an Agena Failure Investigating Board composed of NASA, Lockheed, and Air Force personnel. He expected the board to conduct an investigation and return with its findings and recommendations within nine days: no later than November 30. He wanted all necessary corrective action to be implemented before the flight of Ranger 3.[21] If the General had anything to say about it, every Lockheed Agena B supplied to NASA would meet or exceed performance specifications.

Colonel Eichel, matching his reputation for toughness, moved with dispatch. The investigating board was named and notified; on November 27 it met at the Lockheed offices in Sunnyvale. Lockheed participants reported on analyses of the Agena telemetry tapes during the first day of the two-day meeting. They found the roll gyroscope in the Agena B guidance package to have been inoperative at takeoff, most likely because of a faulty relay in the power circuitry. The Agena's attitude control system had compensated for the roll control failure through gas jetting, but the supply of attitude control gas was exhausted by the time the Agena completed first burn. In a parking orbit without attitude control, the Agena's roll motion had been converted entirely to pitch–yaw or "propeller" motion. Second burn of the Bell Hustler rocket engine had failed since the liquid propellants inside the Agena tanks had sloshed away from the engine intake lines.[22]

On the morning of December 4, 1961, the Air Force formally conveyed these results to NASA. Luskin of Lockheed, Air Force Colonel Henry Eichel, and Major Jack Albert presented the board's case and plans for remedial action in the newly completed NASA Headquarters building in Washington, a block-long marble veneer and glass edifice that contrasted sharply with the turreted and weathered red brick pile of the Smithsonian Institution down the street on Independence Avenue. NASA and JPL officials accepted the findings and corrective measures. Luskin pledged to submit a report on the accomplished changes in early January 1962, in advance of the flight of Ranger 3, as well as review the entire reliability program for the NASA Agena B.[23] When the conference broke up at lunchtime, Luskin, Eichel, and Albert were satisfied that they had solved the Air Force–Ranger launch vehicle problem.

But the representatives of JPL, NASA Headquarters, and the Marshall Space Flight Center reconvened after lunch. In light of past performance, they deemed the approved corrective measures, though necessary and desirable, insufficient. Cortright instructed Hueter to initiate a separate and detailed review of Lockheed field operating equipment and procedures, with emphasis on eliminating any differences in checkout procedure remaining between Air Force and NASA space projects.[24] NASA Headquarters, changing its attitude from customer to consumer, had decided to undertake greater involvement in matters of Agena procurement and preflight testing.

JPL, for its part, also altered preflight checkout procedures for the spacecraft. Based on the experience with Ranger 1,[25] the Systems Division decided that high voltages hereafter would not be applied to scientific experiments during a launch countdown. To accomplish a midcourse maneuver, succeeding Ranger spacecraft would incorporate a fueled rocket engine on the pad. Any high-voltage discharge that ignited Ranger's squibs and other pyrotechnics here could have severe repercussions, possibly destroying the spacecraft as well as the fueled Atlas-Agena B launch vehicle. Burke agreed with this decision and with the rationale behind it. The experiments, after all, would have completed "at least 15 systems tests before [being] committed to flight; all of this after design type approval testing and flight acceptance testing... If the equipment is so suspect that personnel would worry over its condition between the last systems test and launch, then it is not reliable enough to be considered as flight equipment.''[26]

At least the engineers had the lunar flights in which to see Ranger perfected, but the flight of Ranger 2 effectively slammed the door on any substantial dividend for sky scientists. Placed in near-earth orbits, Heppner's magnetometers had been saturated by the earth's magnetic field on both test missions. Van Allen's cadmium–sulphide particle detectors likewise had been overwhelmed by sunlight for the greater part of each mission, as were the light-flash detectors of Alexander's cosmic dust experiment. Though the Lyman alpha telescopes so carefully prepared by Chubb had returned some background measurements, the pictures were mostly smeared, and little was known about the direction in which the telescopes were pointing when the spacecraft tumbled. And so it went. Ironically, among the few instruments to return any data of value were the friction device and the AEC's Vela Hotel experiment in the flight of Ranger 1.[27] But if sky science had lost out on Rangers 1 and 2, at least lunar science, its practitioners assumed, still stood to benefit from Rangers 3, 4, and 5.

Chapter Seven

A NEW NATIONAL GOAL

ON April 12, 1961, a few months before the ill-fated flight of Ranger 2, the Soviet Union launched Major Yuri Gagarin on a one-orbit flight around the world. The new Soviet space achievement brought to a head an increasingly salient debate about the American space program, a debate whose outcome profoundly affected the Ranger Project.

MAN ON THE MOON

The debate centered on whether the U.S. space program should stress manned or unmanned exploration of space and the planets. For some time NASA had been pushing ahead with Project Mercury, which aimed at the earth-orbiting of manned spacecraft, and NASA leaders had long been discussing manned projects to follow Mercury. As early as May 1959, NASA's Goett Committee had recommended that the space agency undertake a manned circumlunar flight mission, tentatively identified as Project Apollo, though the Eisenhower administration had refused to budget anything for it beyond planning studies. But in a report of mid-January 1961, John F. Kennedy's Ad Hoc Committee on Space, composed largely of scientists and chaired by Jerome Wiesner, the incoming President's Special Assistant for Science and Technology, criticized the emerging programmatic emphasis on man in space. By placing the highest national priority on the Mercury manned earth-orbital project, the report declared, "we have strengthened the popular belief that man in space is the most important aim of our nonmilitary space effort." Accompanying national publicity had further exaggerated "that aspect of [the nation's] space activity..." The report argued that NASA needed to increase emphasis on space science, allot more missions for earth applications, develop larger launch vehicles, and organize the entire space program more effectively.[1]

Elsewhere, in NASA's Lunar and Planetary Programs Division, Edgar Cortright and Oran Nicks objected to engineering for man in space at the expense of quality space science. As Project Mercury suggested, just to fly a man into orbit, keep him alive, and return him safely to earth called for remarkable feats of engineering—from complex life support systems to atmospheric reentry and recovery systems. Compared with the effort that went into satisfying these engineering demands, pure science was virtually invisible in Project Mercury.[2] On March 31, the Space Science Board of the National Academy of Sciences added views of its own. In separate reports to the new administration the Board advised that man in space could be an important component of the nation's exploratory program, though it was "not now possible to decide whether man will be able to accompany expeditions to the Moon and planets." The board went on to recommend that planning proceed "on the premise that man will be included." But to the board, man in space meant trained scientists seeking knowledge, not engineer-pilots bent on showing the flag.[3] And the board continued with an endorsement of unmanned scientific space exploration strong enough to make its approval of manned flights seem a mere genuflection to public sentiment.[4]

Nevertheless, the public at large seemed generally unconcerned with the advance of sky or planetary science as such. In 1960, while scrutinizing the proposal for Project Ranger, Congressman Albert Thomas of Houston, the powerful chairman of the House Appropriations Subcommittee on Independent Offices, interrupted the testimony of NASA Associate Administrator Richard Horner to inquire just how far away the moon was from the earth. Horner replied that it was a quarter-million miles distant. Nodding and holding up background data furnished by NASA, Thomas read: "Your justifications continue: 'as our nearest major body in the solar system, basically unchanged for billions of years, the Moon offers unique potentialities for better understanding of historical and contemporary phenomena of the solar system...' We know all about that," Thomas snapped. "If it has not been changed for billions of years, why not leave it alone?"[5]

After Gagarin's feat, Thomas could agree with the sentiments of Congressman James Fulton of Pennsylvania: that NASA should "stop some of this WPA scientific business." Along with other Congressmen, Fulton declared emphatically: "We are in a competitive race with Russia," a race that prohibited a space program aimed merely at serving the needs of science.[6] In the White House, eight days after the Gagarin flight, President Kennedy, sensitive to public sentiment, directed Vice President Lyndon Johnson to take charge of surveying the space program and determining where "we have a chance of beating the Soviets..."[7] In short order, Johnson drew up a report with the aid of, among others, the Secretary of Defense and the new head of NASA, James E. Webb. President Kennedy announced the resulting decision to a joint session of Congress, where, on May 25, 1961, he called for a national program to land a man on the moon and return him safely to earth before the end of the decade.

Amid the eagerness to best the Russians in what was now commonly called the space race, the authorization for Project Apollo whisked through the Congress. The outcome was a clearcut victory for an emphasis on manned over unmanned space exploration, for technological spectaculars, some critics snapped, over sound space and planetary research. Whatever the general accuracy of the critics, the commitment to Project Apollo certainly raised havoc with the scientific planning for Project Ranger.

RANGERS FOR APOLLO

Even before funds for the manned lunar landing were in hand, NASA Administrator Webb, his deputies Dryden and Seamans, and Space Flight Director Silverstein had decided that the unmanned scientific lunar program would contribute to making straight the way for Project Apollo.[8] On the day that Kennedy delivered his special message to Congress, Abraham Hyatt, NASA Director of Program Planning and Evaluation, issued a Master Plan of launches for the conquest of the moon. Instead of missions for scientific purposes, both of JPL's lunar flight projects, Ranger and Surveyor, appeared listed as missions in "direct support" of the manned lunar landing program.[9]

When on June 8, 1961, NASA Deputy Administrator Hugh L. Dryden appeared before the Senate Committee on Aeronautical and Space Sciences, he testified that provision already had been made to strengthen the nation's manned lunar program by augmenting NASA's unmanned lunar flights. "We want to know something about the character of the surface on which the landing is to be made," he explained, "and obtain just as much information as we can before man actually gets there."[10] In subsequent testimony, Silverstein supplied further details. Project Ranger would be increased by four flights to secure closeup pictures and possibly other data concerning the moon's topography. Planetary scientists might obtain information and debate lunar hypotheses at their convenience, but, speaking as NASA's chief engineer and a prospective manager of the manned lunar landing program, Silverstein declared: "It is [now] vital for us to know what the moon's surface is like."[11] The legislators agreed. Several weeks later, Congress appropriated funds for the additional Ranger missions.

Silverstein formally notified JPL of NASA's plans on June 9. Reiterating the justification for extending Project Ranger, in a letter to Pickering he observed that the manned lunar landing created the "need for *early* data on the surface topography and hardness of the moon..." Four more flights would contribute to meeting this goal, "afford a better opportunity for Ranger project success, and make a corresponding contribution to national prestige during the early phases of our lunar program." He instructed the Laboratory to examine improved high-resolution television cameras and other instruments that could return information on the hardness of the lunar surface, such as a penetrometer capsule. JPL's recommendation, together with estimates of the cost and level of effort required, was to be submitted to NASA Headquarters before the end of the month.[12]

High-resolution photographs of selected areas of the moon's surface, JPL Lunar Program Director Cummings informed Edgar Cortright and Oran Nicks in a meeting at Headquarters on June 21, would be the greatest single contribution Ranger could make to Project Apollo. The landing gear of an Apollo vehicle designed for the worst lunar surface conditions, he said, could easily claim 50 percent of the allowable weight. On a more favorable surface, however, the landing gear could likely be reduced to less than 5 percent of that weight. Closeup pictures taken by Ranger spacecraft prior to impact would permit engineers to design Apollo's landing gear with increased confidence. For this purpose, the standard Ranger bus could be modified to support a tower carrying a television subsystem in place of the Block II seismometer capsule. Necessary information on the hardness of the lunar surface could be obtained from the Block II radar reflectivity experiment suggested by Walter Brown in March, from special accelerometers contained in the Block II lunar capsules, from ground-based radar investigation of the lunar surface, and from the Surveyor soft-landing missions now scheduled to follow Ranger in 1964.

To implement this proposal, Cummings continued, JPL recommended contracting for the complete television subsystem, much as the Block II lunar capsule subsystem was designed and developed by the Aeronutronic Division of the Ford Motor Company. For the new missions, however, if the flight schedules projected by NASA to begin in January 1963 were to be met, no time could be spared for the normal process of competitive bidding. A single firm should be selected and a contract awarded immediately. JPL preferred the Radio Corporation of America. Of the five potential sources, RCA alone possessed experience in developing the space television cameras for the original Ranger and Tiros projects. After further discussion, Cortright approved the television as payload for the extended Project Ranger. Nine days later, Cummings learned that Headquarters would approve the proposed sole-source contract with RCA for the television subsystem.[13]

No time was spared. On July 5, 1961, representatives of the Astro-Electronics Division of RCA in Hightstown, New Jersey, met at JPL to define plans for the payload. Photographs of the lunar surface, it was decided, would be taken by a special television system with a shuttered vidicon. The participants assigned top priority to high-resolution pictures. And complete responsibility for the television subsystem, including details of its design, fabrication, and testing, was delegated to RCA. JPL's Systems Division would monitor RCA's performance on the contract with consultant support from the Space Sciences Division. On August 25, RCA accepted the JPL Letter Contract. The company agreed to furnish one proof test model and four flight units of a television subsystem, each consisting of a battery of "three or more" high-resolution TV cameras.[14] Four days later, on August 29, NASA Associate Administrator Robert Seamans publicly announced the expansion of Project Ranger from five to nine flights. The new Ranger missions, he informed the nation, would return high-resolution pictures of the lunar surface as "part of the general acceleration of the program to land an American on the moon by 1970."[15]

This expansion brought with it changes in project staffing and controls. At NASA Headquarters, Lunar and Planetary Program Director Cortright instituted quarterly project review meetings. Supplementing the normal flow of communications, NASA and JPL Ranger personnel would meet—beginning in September and following once every three months thereafter—to evaluate the progress and problems of the project and make plans for any remedial action necessary. NASA's knowledge of and control over project developments, Cortright and Nicks correctly perceived, would thereby be strengthened. At JPL, new personnel assignments were made. Cummings established a Ranger Project Policy and Requirements Document, controlling parts selection, test, and quality assurance requirements primarily for JPL's in-house spacecraft effort—from the design phase through space flight operations. In addition, to assist Ranger Project Manager Burke, he appointed a Ranger Spacecraft Systems Manager, Allen E. Wolfe.[16]

Wolfe had come to JPL in 1952 after military service, a stint with the Army's Ballistic Research Laboratories at the Aberdeen Proving Ground, and a wartime degree in electrical engineering from Caltech. At the Laboratory he had participated in developing the instrumentation for various missiles tested at the White Sands Proving Ground in New Mexico. Taciturn, Wolfe had acquired a well deserved reputation for technical competence and coolness under pressure—a characteristic perhaps most appreciated by those who had occasion to meet him in a sociable game of poker. In the fall of 1960, Systems Division Chief Schurmeier had appointed Wolfe Ranger Project Engineer, replacing Gordon Kautz, who had moved over to the Ranger Project Office as Burke's deputy. Now, in August 1961, as Ranger Spacecraft Systems Manager, his sole responsibility was to guide development of the spacecraft through all phases of assembly and testing, not only for the follow-on missions but for the remaining Block II flights as well (Figure 44).

With the missions approved, the principal contract awarded, and his staff augmented, Burke instructed the JPL technical divisions to place orders for the hardware necessary for Ranger flights 6 through 9.[17] To be known collectively as Ranger Block III, these spacecraft would be subject to design constraints and a test regime similar to the ones imposed on their predecessors. Components for both the spacecraft bus and RCA's television subsystem, for example, would be heat sterilized before assembly, and treated with ethylene-oxide gas afterwards. But one prior difficulty was removed. The engineers knew the allowable weight of the spacecraft; at least 360 kilograms (800 pounds) could be allocated among its various subsystems.

RCA completed design of the television payload in September 1961.[18] Weighing 135 kilograms (350 pounds), the subsystem was composed of three major assemblies: a tower superstructure and thermal shield, a central box containing the principal electronic components, and, located above the box near the top of the tower, six television cameras and their associated bracketry and

Fig. 44. JPL Ranger Spacecraft Systems Manager Allen Wolfe

electronics. Two of the cameras would be equipped with 25-millimeter wide-angle lenses, and four with 75-millimeter telephoto lenses. The four narrow-angle telephoto cameras, positioned for overlapping coverage, would furnish a resolution on the order of 0.1 to 0.5 meter (4 to 20 inches) in the final pictures. All of the cameras were to be operated in sequence under a continuous recycling program from turn-on to impact. While one camera was taking a picture, another would be in a state of "readout"—converting the video data for transmission to earth—while on another the images would be erased preparatory to taking further pictures. Operated by a separate battery power supply, the television system was to be independent of the Ranger power subsystem (Figure 45). Two transmitters, one acting as a backup in the event of any technical malfunction in the other, would ensure transmission of the information to earth. At the Goldstone tracking station, the video data would be recorded on 35-millimeter film and magnetic tape.[19]

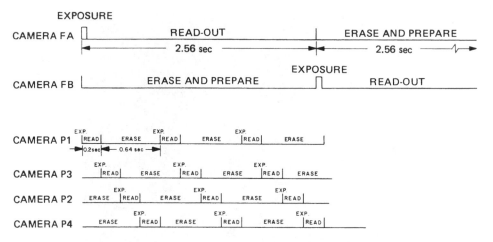

Fig. 45. Ranger Block III Television Camera Sequencing

In appearance, outside of the television package, Rangers 6 through 9 looked virtually identical to their predecessors. Made up of the same subsystems, they featured trapezoidal solar panels, a midcourse propulsion unit beneath the bus, a hinged high-gain dish antenna, and an omnidirectional antenna atop the tower superstructure (Figure 46). To meet the short schedules, and adhering to Burke's dictum, design changes beyond those necessary to accommodate the television package would be minimized. The total weight of 360 kilograms (800 pounds), nevertheless, meant a net gain of some 23 kilograms (50 pounds) for the spacecraft bus—as contrasted with the erroneous figure of weight assigned the Block II lunar spacecraft. For the first time, weight could be used for redundant engineering components. Most notable among them: two backup clocks to enable Ranger to complete a mission even if its electronic brain, the central computer and sequencer, should fail in flight. All of the television cameras were to be pointed so that the normal sun and earth acquisition procedures of the Ranger spacecraft would automatically position them to sweep the moon in most trajectories even without accomplishing the planned midcourse or terminal maneuvers.[20]

On September 19, 1961, NASA Headquarters issued the flight schedule for Rangers 6 through 9; they would be launched in January, April, May, and August of 1963, respectively.[21] The urgency expressed by this schedule, Nicks explained to Cummings, derived from the national commitment to Project Apollo and the delay in the Centaur launch vehicle program, which postponed the flights of the Surveyor lunar soft-lander project. Every effort must be exerted, he advised, "to get the most from existing Ranger developments."[22] Privately, Nicks questioned what "the most" might be. The Atlas-Agena B vehicles had experienced failures with Rangers 1 and 2. The spacecraft and the ground-based Deep Space Network were not yet proved in actual lunar flight—through command and execution of the

Fig. 46. Model of Ranger Block III Preliminary Design

spacecraft midcourse and terminal maneuvers. The most that might reasonably be expected, he confided to Cortright, was one success in Ranger Block II, and perhaps two successes in the Block III flights.[23]

At JPL, Cummings and Burke could concur with Nicks' reserved prognosis. Burke, in fact, in discussion with his counterparts at NASA Headquarters, used the figure of one complete success in each remaining block of lunar flights as a reasonable expectation. Planning had begun at NASA and JPL to add still more Rangers to increase the probability of success, building upon the technology to be demonstrated in these early flights.[24] But more important to both JPL men at that moment were the changed justification and objectives of Project Ranger. The original Ranger Block II, to be sure, would continue in the name of science. The new Ranger Block III, however, was to fly in support of Apollo; engineers had to have a closeup look at the surface upon which Apollo must land. Planetary scientists could draw on these lunar photographs for what they could find in them.

As enunciated by the President, members of the Congress, and the NASA leadership, the new national goal could hardly be misunderstood.[25] Cummings explained its portent and ramifications for NASA-JPL unmanned flight projects to Vice President Johnson on October 4: "Originally our lunar program had been oriented toward scientific and technological objectives," he stated. "Now...the emphasis has been changed so that support of the manned operations is the primary objective, and space technology and lunar science are secondary. We believe, however, that we can accomplish the space science and technology objectives as planned, while at the same time providing essential support to the manned effort."[26]

THE NEW ORDER

The downgrading of space science forced by the commitment to Apollo bothered high NASA officials, including the head of the agency, James E. Webb. The son of a county school superintendent in Oxford, North Carolina, Webb, aged 54, had already distinguished himself in half a dozen careers, including those of aviator, lawyer, administrator, and statesman. He impressed Congressmen with his ability to field questions, remember figures, and overwhelm with information. "Listening to Jim Webb," one of them remarked, "is like taking a drink from a fire hydrant!" Webb had long been familiar with the policy and politics of scientific research and development, not least because before World War II he had been an executive of the Sperry Gyroscope Company and after it, in the Truman administration, the Director of the Bureau of the Budget (Figure 47).

Webb appreciated science qua science, and he understood the scientists' contradictory aspirations to academic independence, peer-group determination of scientific merit, and substantial governmental funding. In later months, prodded to cut back NASA's space science efforts until Apollo was accomplished, Webb would tell the President that he could not run a truncated program. But Webb, who had endorsed the Apollo program to Kennedy, now had his priorities fixed on the manned lunar landing, and he reorganized the space agency accordingly.

The realignment, effective November 1, 1961, was marked at Headquarters by the following salient features:

1. All of NASA's field centers, including JPL and the new Manned Spacecraft Center to be constructed in Houston, reported directly to Associate Administrator Robert Seamans on all institutional matters.

Fig. 47. NASA Administrator James Webb

2. The Headquarters program offices of Advanced Research Programs (Ira Abbott), Space Flight Programs (Silverstein), Launch Vehicle Programs (Ostrander), and Life Science Programs (Charles Roadman), were abolished.

3. Four new Headquarters offices were created: Advanced Research and Technology (Ira Abbott), Space Sciences (Homer Newell, formerly Silverstein's deputy), Manned Space Flight (D. Brainerd Holmes, a new appointee from RCA), and Applications (vacant). A supporting office of Tracking and Data Acquisition (Edmond Buckley) was also formed. These offices would also deal with the field centers on project matters.

Beyond acknowledging the significance now attached to the manned space flight program (which existed previously as a subdivision within Space Flight Programs), this reorganization made Seamans the general manager of the whole agency. JPL's lunar and interplanetary flight projects remained under the functional direction of Newell's office, where Edgar Cortright moved up to become his deputy. Oran Nicks became Director of the Lunar and Planetary Programs Division in place of Cortright.[27]

In the new Office of Space Sciences, the deemphasis of space science as such troubled Homer Newell.[28] As an engineering tour de force, landing a man on the moon might bolster a sagging national economy, reestablish a shaken faith in the supremacy of American technology, or guarantee some short-term advantages in international diplomacy. It definitely promised to dislocate the planning for the scientific exploration of the moon; it even threatened to undermine the raison d'être of NASA's unmanned lunar program already in being or projected. First, Apollo was physically separated from the rest of the lunar program, located in an office for which Newell was not responsible. Second, Newell was expected to represent the interests of space science in NASA, and to push for adding space science in Project Apollo. Third, if he were successful in this endeavor, his own unmanned lunar missions and the remaining experimenters hard at work on their instruments would be in for more trouble. Why rough-land a seismometer on the moon with Ranger, or soft-land a variety of instruments with Surveyor, or return lunar surface samples with Prospector, when Apollo might perform these same functions in a few years' time? Brainerd Holmes and other high officials in the newly formed Office of Manned Space Flight, who would shortly petition President Kennedy for supplemental Apollo appropriations, might have added, "Why, indeed?"

Nevertheless, Newell intended to secure a significant return for space science from all of the automated flights at the earliest opportunity, and to enlarge that return from Apollo as the manned program progressed. He personally had selected the name and emphasis for the organization he now commanded: the *Office of Space Sciences.* Though it might be dwarfed in size, on the charts of the reorganized NASA it would be at least recognizable on a par with the Office of Manned Space Flight. Internally, Space Sciences was reorganized to combine the science and engineering functions so carefully separated under Silverstein. In the new Lunar and Planetary Program Office, Oran Nicks established individual offices for Ranger, Surveyor, and Mariner R, a new JPL project recently approved, to send a modified Mariner and Ranger spacecraft—hence the name Mariner R—to Venus in 1962.

The individual offices included program engineers and scientists.* N. William Cunningham, previously a staff scientist for Silverstein, was named Ranger Program Chief (Figure 48). A wartime radio operator aboard naval aircraft, who had later earned degrees in meteorology and physics at Texas A & M College, Cunningham would guide Ranger developments with Burke in the Project Office at JPL. His deputy and Program Engineer was Walter Jakobowski, a man of

---

*The word "program" was substituted in place of project to avoid confusion with and imply a different function from the Project Offices already existing at the field center level.

Fig. 48. NASA Ranger Program Chief William Cunningham

inquiring temperament well suited to his role as troubleshooter. A Ranger Program Scientist was not immediately named. Charles P. Sonett, a UCLA physicist with a background in sky science and the Chief of Sciences in Nicks' office, served in this capacity in the interim. NASA now had a management structure with which to prosecute Project Apollo, and the Office of Space Sciences had combined science and engineering in a program office for Project Ranger.

James Burke in Pasadena was pleased with the creation of a counterpart Ranger Office at NASA Headquarters. At last, after two years, two individuals at Headquarters were to be concerned with his problems alone. Moreover, from past dealings on the scientific side of Ranger, Burke knew that Cunningham understood the engineering constraints that science entailed and would be quick to foresee other potential problems as well. Burke was even more pleased by the national commitment to Project Apollo. The potential interference of space science activities, he reasoned, would have to be contained during the monumental engineering effort to land a man on the moon and return him safely to earth.[29] That enterprise might well relieve some of the pressure for "more science" in Project Ranger—especially since Ranger now was considered to be supportive of the Apollo goal.

Chapter Eight

THE QUESTION OF SCIENCE AND RANGER

D ESPITE the redefinition of Project Ranger, Burke's problems with the development of the spacecraft were hardly over. Indeed, in 1961, although NASA and JPL had the policy and plans for a concerted attempt at sterilizing spacecraft for deep space missions,[1] the first results on Project Ranger proved truly disquieting. The Aeronutronic Division of the Ford Motor Company, experimenting with the new procedures, encountered increased component failures in the Ranger seismometer capsule subsystem after heating to the JPL specifications. This subsystem, the heart of the planetary science package on the Block II Rangers, demanded far-reaching advances to protect the delicate instruments from the high-impact loading and stresses involved in a rough landing on the moon. Even after deceleration by retrorocket, the capsule would strike the lunar surface with a residual velocity of 61 meters per second (136 miles per hour), with a force of 3000 g. The sterilized seismometer, battery pack, and communications equipment had to survive that shock and operate at least for the two-week duration of a lunar day.

LUNAR SPACECRAFT DEVELOPMENT: STERILIZATION IN PRACTICE

Frank Denison at Aeronutronic had expected a rash of technical difficulties even without sterilization, and the expected development problems did materialize, especially for the radar altimeter that was to trigger separation of the capsule subsystem, the solid propellant retrorocket that would slow the seismometer capsule, and in the capsule itself. In May 1961 Aeronutronic began drop-testing sterilized lunar seismometer capsules encased in their balsawood impact limiters in the Mojave Desert. Five capsules released from an aircraft over both flat and rocky surfaces produced disappointing results. Those that fell on rocks failed to operate.

After modifications, engineers conducted more impact tests in October. The seismometer operated after impact, but malfunctions occurred in the sterilized electronic equipment inside the survival sphere.[2] Prospects appeared slim that the capsule weight and performance requirements could be met in time for the January 1962 launch date of Ranger 3, now only three months away. At NASA Headquarters, dismayed with the progress to date, Oran Nicks ordered Cummings and Burke to strengthen technical direction of the Aeronutronic work, and produce a detailed recovery plan for the time remaining.[3]

On November 6, 1961, Aeronutronic reorganized its Ranger capsule project team. Donal B. Duncan replaced Frank Denison in charge of capsule development. Ranger spacecraft systems manager Wolfe led a contingent of JPL personnel to the Aeronutronic facility in Newport Beach to assist in the recovery effort. For its part, NASA Headquarters issued waivers of heat sterilization for some sensitive electrical components in the survival sphere. In his recovery plan submitted to Nicks on November 9, Burke affirmed that the outstanding technical problems associated with the radar altimeter, retrorocket, and capsule were under control. Tests in succeeding weeks, including impact tests of operating seismometer capsules, supported that observation,[4] but the ability of the seismometer and batteries to withstand the cold temperature of a two-week "lunar night" appeared questionable. And if heat sterilization and glove-box assembly of the lunar capsules inside clear plastic containers had proved a demanding and time-consuming task for Aeronutronic, they were magnified when applied to the Ranger 3 spacecraft at JPL (Figure 49).[5]

Except for the lunar capsule handled separately at Aeronutronic, JPL's technical divisions heat sterilized all of the subsystems for Ranger 3 in a special oven in the new environmental test laboratory. Division Chiefs immediately reported major problems—in particular with electronic components: the heat rendered cabling brittle, reduced the tensile strength of soldered joints, and caused capacitors to leak. After assembly in July 1961, Ranger 3 entered system tests. The results were positively disturbing. "Component failure rates," JPL advised NASA, "are much higher...than those recorded on the proof test model. This is almost the exact reversal of the experience noted on the Ranger 1 and 2 program. The only difference between the Ranger 3 proof test model and the flight spacecraft is heat sterilization." The report continued: "Although no failures are directly traceable to heat damage, it is felt that heat sterilization does shorten the expected life of electronic components and circumstantial evidence seems to bear this out."[6] At the same time, engineers in the Systems Division noted that heating the Ranger 4 bus had resulted in "warping the structure... This, however, can be compensated for by suitable shimming." They had also found a wire in the ring harness "to be parted near a splice. This damage occurred sometime during the sterilization process."[7]

Practical experience hardly encouraged confidence in the new procedure or in its implications for reliability. Granted, the procedures had been devised after a short test program with selected components in 1960; but, surprisingly, no one at

Fig. 49. Sterile Assembly of a Seismometer Capsule at Aeronutronic

JPL or NASA had foreseen the deleterious results that heat sterilization could have for the spacecraft when applied on a full-scale basis. Although lacking firm evidence that this requirement caused the equipment failures, JPL now requested and received more waivers from NASA Headquarters on heat sterilizing certain crucial components. Among them were the capsule batteries, various transistors, capacitors, explosive-actuated switches, and the retrorocket and spin motor propellant, igniter, and squib. More and more it appeared that the moon would have to tolerate some bugs from the earth.

Ranger 3 completed systems tests after all failed parts had been carefully requalified and replaced. NASA accepted the machine for shipment to Cape Canaveral in late November 1961. Cummings informed NASA Associate Administrator Robert Seamans of those parts that had been excluded from heat treatment, and of achievement in the NASA spacecraft sterilization program to date. "The degree of decontamination attained," he explained, "is the result of a major effort which has added substantially to the cost and to the risk of the mission. It is likely that the sterilization procedures have compromised spacecraft reliability; however, there is insufficient data to positively confirm this suspicion.

Though complete sterilization has not been achieved, the total possible contamination level [remaining] is known to be small."[8]

At a minimum, heat sterilization appeared to have affected the life expectancy of many electronic components. After hours of subsystem and system testing, no one could be certain exactly how many more hours of life remained in Ranger 3. The lunar capsule and its retrorocket had just qualified for flight and should properly undergo more tests. Furthermore, the failure of the Ranger 2 mission in November raised questions about the need for improving the Atlas-Agena B launch vehicle to ensure its reliable performance. Existing flight schedules simply did not afford time for this work. Should the launch of Ranger 3 be delayed to provide it?

Burke discussed the matter with Cummings, Kautz, and Wolfe, as well as with others at the Laboratory. The Atlas-Agena, scrutinized and modified, seemed likely to perform properly. Burke and his fellow JPL engineers were also satisfied that the spacecraft would work. The mission was, after all, a lunar flight of only three days, and they could be sure of that much component lifetime. Prevailing sentiment remained heavily disposed to launching Ranger 3 on schedule. Even if the capsule subsystem itself failed to perform precisely as planned at the moon, valuable experience would be gained for the first time with the spacecraft, by the deep space tracking net, and in space flight operations—through exercising the spacecraft in its midcourse correction and terminal maneuvers.[9] Two more flights remained in which to secure the lunar science objective. Feeling an additional urge to make haste, the JPL officials knew that the Soviets had yet fully to demonstrate these important techniques on a deep space mission.

JPL Director Pickering summarized the situation for Seamans on November 15: "The elaborate real-time operations involved in determining the orbit, computing and executing the in-flight maneuvers and commands, and returning the lunar TV data cannot properly be simulated without a flight," he wrote. "We must weigh carefully the value of early experience in these areas against a postulated improvement in the capsule subsystem" that might obtain by delaying the mission.[10] On December 28, 1961, one month after Ranger 2 had burned to a cinder in the earth's atmosphere, in a meeting of NASA and JPL officials at Headquarters, Burke presented the case for launching Ranger 3 on schedule, regardless of the costs to science. He argued forcefully that the advantages in proceeding to launch outweighed the potential disadvantages. The launch vehicle, spacecraft, and deep space tracking net were ready. And should the retrorocket or the capsule electronics fail at the moon, he asserted, NASA still would have the engineering and operational experience gained in a flight there. Oran Nicks, Director of the Lunar and Planetary Program Office, disagreed. Because of the questionable performance of the capsule subsystem, he urged delaying the launch to qualify this key scientific experiment. But Burke's position carried as Cortright, Newell's Deputy Director in the Office of Space Sciences, concluded that Ranger 3 would launch on schedule.[11]

Though science could still be acknowledged as the primary lunar objective, the decision to launch Ranger 3 as planned was a victory for Ranger's designers and engineers in Pasadena. They would have the opportunity to surpass the Russians on the one hand, and to test the Ranger spacecraft and flight support operations through a complete lunar mission on the other. Space science, at least in the form of the lunar seismometer experiment, would have to take its chances. Nicks accepted the edict, but thought it a distortion of the Block II objectives— downgrading space science in favor of engineering.

AFTER THE APOLLO DECISION: WHAT SCIENCE AND WHERE?

On December 1, 1961, the Space Sciences Steering Committee in the Office of Space Sciences reshuffled the experimenters for the visual imaging television package aboard the remaining Ranger flights, 3 through 9. To Urey, Kuiper, and Shoemaker, chosen a few months before in October,[12] the Steering Committee now added Ray Heacock and Edwin Dobies of JPL's Space Sciences Division. JPL Space Sciences Division Chief Hibbs was appointed "convener" of the entire group.[13] Oran Nicks informed JPL Lunar Program Director Cummings that by this change NASA hoped to encourage these scientists to operate together as a "team" from development of the television system at RCA through analysis of the lunar pictures after the mission. He asked Cummings to integrate this scientific cadre directly into the project management structure, and to show that assignment in the Block III Project Development Plan. The blending of scientists and engineers begun in the Office of Space Sciences in Washington was to be extended to the field center project level.[14] At the same time, the Office of Space Sciences also considered the possibility of more science in Project Ranger.

The scientific loss in the first Ranger missions, along with the commitment of the Block III Rangers 6 through 9 to Apollo, had brought Homer Newell numerous complaints from NASA space science experimenters around the country.[15] Even more distressing, perhaps, was the increasing tendency of the aerospace trade journals, the press, and many space agency personnel to represent *all* of NASA's unmanned lunar flights as engineering missions in support of the manned lunar landing.[16] With these conditions straining more than ever NASA's relations with the nation's scientific community, Newell determined to remain true to the principle of inquiring science. JPL, among several other field installations, reported to the Office of Space Sciences, *not* to the Office of Manned Space Flight. As long as Newell had anything to say about it, wherever weight and space permitted, science would be accommodated in NASA's unmanned lunar and planetary projects.[17]

But where to find the spacecraft for deep space missions with sufficient weight and space available to meet the demands of experimenters in 1962–1963? The postponed Surveyor lunar soft lander, experiencing development problems in both the spacecraft and its Centaur launch vehicle, was out of the question. And the head of the Mariner R project at JPL had successfully fought off attempts to add more experiments to this first planetary mission to Venus. But the four Ranger Block III television spacecraft, on which Burke's project engineers were busily programming extra weight and space for redundant engineering features, did remain. Though their objective might be acknowledged as Apollo support, some of the sky science experiments lost on Rangers 1 and 2 conceivably could be added here. Granted, a time limit of sixty-five hours on measurements during a flight to the moon would be a severe drawback. Among those anxious for another change, however, such data were much to be preferred to no data at all.

Out at JPL some months earlier, Burke and Hibbs had collaborated in producing a proposal of their own for three sky science experiments to go along with the primary television package on Rangers 6 through 9. Submitted to Headquarters in August 1961, the three experiments made up a package designed to evaluate the radiation hazards to man between the earth and the moon's surface. Supplementing the return expected from the television system, they would thus support Project Apollo.[18] The Space Sciences Steering Committee evaluated the JPL plan on August 21, and asked Newell's Lunar and Planetary Programs Division to conduct a review and return with recommendations.

On September 26, while the Steering Committee's request was still in Newell's Division, the Ranger Block III Project Development Plan was issued. It specified that the three secondary sky science experiments would be "considered" for inclusion on Rangers 6 through 9. But at JPL Burke and Hibbs assumed that the experiments slated for possible inclusion would be theirs, and they permitted parts procurement and fabrication of these machines to proceed apace so as to meet the established specifications and schedules. Then, early in 1962, because of a shortage of skilled manpower at JPL, Burke and Hibbs decided that the three experiments could not be completed in time for all of them to fly on Rangers 6 through 9. Ranger 6 would incorporate only one experiment along with the television package, a Neher ion chamber. Ranger 7 would incorporate two, the ion chamber and geiger counters. Rangers 8 and 9 would carry three, the first two plus tissue-equivalent dosimeters if they could be prepared and delivered on time. On February 5, Burke reported the change to Nicks at Headquarters.[19]

Burke's news arrived in Washington just when the Office of Space Sciences was deciding upon its objectives in the years ahead, especially the role of space science in the unmanned lunar program. To Newell and his staff, Burke's reduction in the number of sky science experiments for Rangers 6 through 9 was tantamount to another downgrading of space science in favor of spacecraft engineering. The reduction, after all, seemed to be justified only by a requirement of meeting a flight-test schedule for the Ranger machine. Newell and his staff thus saw the issue of whether to approve Burke's sky science experiment prospectus in terms of a single fundamental question. Given existing and anticipated funds, was it better to perfect the spacecraft and launch vehicles, then pursue science, or should the technology and space sciences be pressed simultaneously? Taken up during February, the question was, Nicks later recollected, most difficult to answer.

Charles Sonett, Chief Scientist in the Lunar and Planetary Program Office, pointed out that the Ranger spacecraft bus had been specifically designed to accept a variety of add-on scientific instruments. The space, power, and weight available on Rangers 6 through 9, he maintained, could in fact support more than the few experiments recommended by JPL.[20] The Space Sciences Steering Committee, moreover, had not given final approval to the space experiment recommendations of Burke or scientists elsewhere. If only to relieve some of the pressure from space scientists for more flight opportunities, the committee would likely endorse the addition of some experiments on Rangers 6 through 9. Besides, not one Atlas-Agena launch vehicle or Ranger spacecraft had as yet been proved through a complete mission. No one could be certain which flight might succeed first.[21] Nicks recalled that the scientist, "if he saw a spacecraft launch and work successfully, and it didn't carry anything but engineering telemetry [would have] said: 'I was right,...just look at the free trip I could have had.' On the other hand, if [the scientist] worked very hard and got a good set of instruments on board and it went splash in the ocean, he was frustrated because he wasted all that time."[22]

Newell, Cortright, and Nicks, sure that scientists should at least be given the chance to acquire fundamental knowledge at the earliest time possible, concluded that Rangers 6 through 9 as well as other Office of Space Sciences lunar missions would develop the technology and support the interests of space science and Project Apollo, all at the same time. A few months later Newell announced the objectives of his NASA bailiwick: The Office of Space Sciences would be devoted to the "scientific exploration and investigation of outer space." Its flight missions would "contribute to basic knowledge," then "support manned space flight," and finally, encourage "social, political, and economic growth," in *that* order.[23]

On February 28, 1962, William Cunningham, newly appointed Ranger Program Chief in Nicks' Lunar and Planetary Program Office, picked up the phone and called JPL. He told Burke that the matter of secondary experiments for Rangers 6 through 9 was still open. It would be discussed further by the Space Sciences Steering Committee within the next few weeks. Meantime, JPL would resubmit to the Steering Committee a formal proposal for the secondary experiments it desired.[24]

SCIENCE REASSERTED IN PROJECT RANGER

Burke was first dismayed, then incensed by what he perceived to be a sudden 180-degree turnabout in the objectives for Ranger Block III. Although he had steadfastly emphasized the immediate importance of creating and demonstrating the spacecraft technology, Burke had also acknowledged in Project Ranger the eventual primacy of *planetary* science. The moon, after all, was the ultimate subject of interest.[25] The new Ranger missions had been planned, represented, approved, and thus far prosecuted as a single-purpose endeavor to return pictures of the lunar surface for use in the design of the Apollo lunar lander. The three additional sky science experiments had happily complemented that objective, but three extra experiments were all the project office had programmed to design and pay for. Nevertheless, Cunningham had explained, more such experiments were now to be considered. With all restraint apparently gone by the boards at Headquarters, it seemed to Burke as if all science in Project Ranger was to be redetermined in the laissez faire manner of the Ranger 1 and 2 era.

"The question of what auxiliary experiments Rangers 6 through 9 will carry," Burke tersely notified JPL Deputy Director Brian Sparks, "has been reopened by Dr. Sonett...The timing is bad because JPL schedule and documents show the issue already decided, work proceeding accordingly."[26] Any support that Sparks or Pickering might provide in forestalling these plans would be appreciated. To Cunningham, Burke reiterated the critical problem of timing. Whatever experiments might be considered, whether they had anything to do with the moon or not, the status of these spacecraft had, back in September 1961, been fixed to meet the NASA flight schedule:[27]

Spacecraft	Design Freeze*	Hardware Delivery
Ranger 6	Past	5-21-62
Ranger 7	Past	8-1-62
Ranger 8	4-1-62	10-8-62
Ranger 9	6-1-62	12-3-62

But the planning for more science continued at Headquarters.

On March 13, 1962, NASA's Lunar and Planetary Programs Office evaluated all the proposals for more secondary experiments. A total of nine sky science instruments had been submitted for consideration, including the three recommended by JPL. Of these, only the JPL tissue-equivalent dosimeters were found inappropriate for referral to NASA's Space Sciences Steering Committee. The next day, the Steering Committee tentatively agreed to the entire ensemble of eight experiments.[28] Burke talked again with Cunningham in an attempt to have Headquarters reconsider the issue. Such a large number of experiments, he argued, could be expected to create new weight and power problems on the spacecraft. Adding them now would jeopardize existing schedules, project costs, and the stated NASA flight objectives. Schedules and costs aside, there just wasn't the manpower. Operating under a ceiling of 3500 total personnel at the Laboratory was straining the Space Sciences Division as well as the other technical divisions to the breaking point.[29] The large number of JPL engineers required to plan and integrate the design changes for so many experiments simply did not exist. But Cunningham could not be swayed. Afterward, Burke complained to Cummings: "Cunningham says that all other [NASA] satellites are handled the Sonett way [through the

---

* While Rangers 7, 8, and 9 were planned as identical spacecraft, their designs could be altered slightly to incorporate improvements, based on the dates of hardware delivery.

SSSC]," adding bitterly, "if they wanted to do an exercise like this, they had half of 1961...and now it is pretty late."[30]

On March 20, 1962, Nicks formally advised Burke of the findings of the Space Sciences Steering Committee, and requested him "to take steps to integrate these experiments into the spacecraft bus." If the unsolicited Vela Hotel experiment on Rangers 1 and 2 had aroused the project manager's worst fears, the addition by NASA of eight "trinkets" on Rangers 6 through 9 confirmed them. To Burke, Headquarters had determined upon a course of action for science without regard for its engineering implications, and expected the JPL project office obediently to carry it out. Two more aeronomy experiments, he learned moreover, were also being considered. A final NASA decision on the precise number would be made before the end of the month.[31]

By now news and rumor of these events were abroad at the Laboratory. Burke found himself on the receiving end of distress phone calls and memos from irate JPL engineers. Believing the project office had agreed to intemperate demands, Charles Cole, Chief of the Engineering Mechanics Division, was blunt and to the point: "We wish to remind you," he wrote crisply, "that the engineering problems attendant to the location and installation of science systems on the spacecraft are many (temperature control, structural and mechanical design, and space availability) and the constraints are unusually severe. This Division has not planned for the manpower or funds to carry on either an effort to ascertain the feasibility of incorporating proposed additional experiments or the 'hard design' required for approved experiments."[32] Beset with technical problems in the Atlas-Agena and light-weight Block II spacecraft, and trapped in the crossfire between scientists and engineers on the Block III missions, Burke was emphatically angry.

On March 28, 1962, Burke attended a meeting convened at NASA Headquarters to settle problems of integrating the secondary experiments on Rangers 6 through 9. Lunar and Planetary Programs Director Oran Nicks, his Chief Scientist Charles Sonett, and Ranger Program Chief Cunningham were there along with the Deputy Chief of the JPL Space Sciences Division, Manfred Eimer.[33] Burke, purposefully composed, summoning all the logic and persuasion at his command, suggested that Headquarters simply did not understand the severe problems involved at the project level. With the available manpower and time limits, he argued, so many scientific experiments could be added to the spacecraft only at the expense of engineering reliability, which posed a direct threat to mission success. Eimer followed, explaining that the JPL Space Sciences Division would be forced to subcontract for manpower simply to supervise and qualify these experiments for flight. Such use of inexperienced personnel would also increase the risk in the science area.[34]

For Nicks, incredulous, this was the last straw. Burke's intransigence and continued resistance in the face of Headquarters decisions were inexcusable. The leaders of the Office of Space Sciences had set as a policy the service of pure science. That included flying scientific experiments while at the same time developing the space technology. Good scientific results, precedent already suggested, could be secured even in the event of a partial failure of the space flight hardware, and in any case Ranger had been purposely designed to permit add-on experiments. But perhaps most disturbing, the Lunar and Planetary Programs Director now faced what he considered a direct and crucial challenge: Who, in effect, was in charge of the program, the space agency or the contractor? More to the point as viewed by Nicks, if NASA wanted leopard skin upholstery in the Ranger spacecraft and would pay for it, then that was precisely what JPL-Caltech would provide. Mincing no words, Nicks told Eimer and Burke that the secondary experiments would fly on Rangers 6 through 9. He did not request, but instructed Burke to reexamine the problems and present a plan for integrating these instruments on April 9, at the next Ranger Quarterly Review meeting.[35] Convinced at last that his reasoning could not possibly prevail, Burke submitted.*

Led by Nicks, Sonett, and Cunningham, Headquarters was well represented at the JPL Ranger Quarterly Review meeting. Burke and Eimer presented their plan and reviewed the prospective problems of including the eight sky science experiments on these Ranger missions. Manpower to support this work could be contracted for with a local vendor. The existing flight schedule then could be maintained, they averred, if NASA phased in the new experiments after the launch of Ranger 6, as shown in Table IV.[36]

After discussion all around, NASA personnel agreed to the plan.[37] Burke told Nicks that despite past objections he would "proceed aggressively in the attempt to get them aboard. However," he added, "by placing this demand on the personnel and resources of the Laboratory, we may jeopardize not only other JPL efforts but also the higher-priority [Apollo] objectives within the Ranger Project itself." Unless directed otherwise, the main goal of the JPL project office would "continue to be the production of high-resolution lunar pictures at the earliest possible date..."[38] To help meet that goal, he asked that Headquarters set a priority among the sky science experiments, thus allowing project engineers some basis for treating unforeseen "interference problems."[39] But perhaps because priorities are frowned upon in science, where each area of inquiry is claimed to be equally worthy, Nicks and Sonett refused, saying that they would "review any conflicts and settle each on its own merits."[40]

On April 23, 1962, the Space Sciences Steering Committee gave final approval to the sky science experiments.[41] Aware that Burke and his associates still doubted the wisdom of the decision, Newell personally notified Pickering of the Space Sciences Steering Committee's findings. The Steering Committee, he assured Pickering, had "discussed the relative importance of these experiments to the

---

*Burke, it seems clear, essentially viewed the issue in terms of what it meant to the project rather than what it meant as an institutional matter. Years later he reflected: "The Ranger [Block II] effort already was in serious technical trouble. That was the key thing. And what skilled manpower we could command was busy fixing the technical troubles that existed... But...I just couldn't understand, I couldn't comprehend the thinking which would suddenly load on eight additional 'trinkets' with all their demands and interactions right at the moment when we were perhaps approaching our first [flight] payoff." (Interview of James Burke by Cargill Hall, January 27, 1969, p. 14, 2-1391.)

Table IV. Ranger Block III Sky Science

Spacecraft	Experiment	Agency and scientist
Ranger 6	Neher ionization chamber	Caltech/JPL: H. V. Neher, H. R. Anderson, W. S. McDonald
Rangers 7, 8, and 9	Neher ionization chamber	Caltech/JPL: H. V. Neher, H. R. Anderson, W. S. McDonald
	Geiger counter assembly	Caltech/JPL: H. V. Neher, H. R. Anderson, W. S. McDonald
	Electron flux measurement	Applied Physics Laboratory: G. F. Pieper
	Low-energy solar proton detector	Ames Research Center: M. Bader
	Dust particle detector	Goddard Space Flight Center: W. M. Alexander
	Search coil magnetometer	JPL: E. J. Smith
	Electron-proton spectrometer	University of California at L.A.: T. A. Farley
		Space Technology Laboratory: N. Sanders
	Low-energy proton measurement	Goddard Space Flight Center: G. P. Serbu, R. E. Bourdeau

stated objectives of this follow-on program. The primary objective of these flights is to obtain data of significance to the manned lunar effort, and it is believed that good measurements in cislunar space as well as in the lunar environment will be of great value in this regard. It is certain," he added in closing, "that the Space Sciences program will benefit from these added experiments."[42]

MODIFYING RANGER FOR MORE SPACE SCIENCE

In the weeks that followed, JPL engineers designed a universal bracket for the spacecraft bus capable of supporting most of the experiment sensors and their associated electronics. Three of these brackets would encircle the hexagonal frame of each machine. They subcontracted for the data automation system to receive and process the scientific telemetry for transmission to earth stations.[43] Assisted by the JPL Space Sciences Division, the experimenters set about the design, fabrication, and bench testing of their separate instruments as rapidly as possible, for the first flight units were scheduled to be delivered to JPL at the end of July 1962.[44]

The time element, as it affected preparation of these experiments, remained the source of justifiable concern. Several of the instruments, the Space Sciences

Division cautioned in mid-June, might not meet flight acceptance standards under this tight schedule.[45] When reviewed by JPL division representatives on June 25, that outcome appeared highly probable. To protect the spacecraft against faults in the hastily prepared instruments, the engineers recommended the electrical fusing of each experiment separately.[46] Despite objections from some of the experimenters, Burke authorized this change in early July, at the same time rejecting the proposal of another group of disgruntled engineers who pressed for a single fuse in front of the power lead to *all* of the sky science experiments. If one of the experiments shorted and went up in smoke on the way to the moon, he agreed, it would not be permitted to take the spacecraft with it.[47] Television pictures of the lunar surface, in the absence of Headquarters directives to the contrary, remained the primary objective.

The electrical power needed for all of these scientific experiments, however, proved to be a far more crucial problem than the potential malfunctioning of any single one. On July 13 Burke learned from Spacecraft System Manager Wolfe that, contrary to initial estimates, Ranger's two trapezoidal solar panels appeared unable to produce sufficient electrical power to operate both the spacecraft *and* all the scientific instruments. The project office faced a choice: an attempt could be made to adapt the larger, rectangular Mariner R solar panels to the new Ranger machines in the time remaining, or the number of secondary experiments would have to be reduced. Given Headquarters' determined commitment to space science, Burke could agree with Wolfe that the latter course had to be viewed as a "drastic option"(Figure 50).[48]

Pending the outcome of more definitive tests, and keenly sensitive to Headquarters feelings at this point, Burke delayed notifying Cunningham of the problem. To suggest now that scientific instruments might have to be removed might be construed as another attempt by JPL to thwart explicit NASA orders. But by the end of the month, a power shortage existed. As Burke explained to JPL Deputy Director Sparks, available power was marginal for Ranger 7, and definitely inadequate for Rangers 8 and 9, which—beyond the scientific experiments—had also to support a 2300-megacycle communications ranging experiment for the Deep Space Network. Under existing schedules, moreover, the Mariner R solar panels could not be adapted in Project Ranger before the flight of Ranger 9. Headquarters had to be informed of this development. If schedules were to be met, the number of secondary experiments had to be drastically reduced.[49]

This time Burke delegated the task. On August 2, Wolfe notified Ranger Program Engineer Walter Jakobowski, his counterpart at Headquarters, of the results of JPL's power review.[50] A few days later Cunningham arrived at JPL to evaluate the situation personally. He promptly concurred in the JPL findings. Without the larger solar panels, electrical power for most of the experiments was simply unavailable. Cunningham agreed to present a full report at the next meeting of the Space Sciences Steering Committee, and also to include

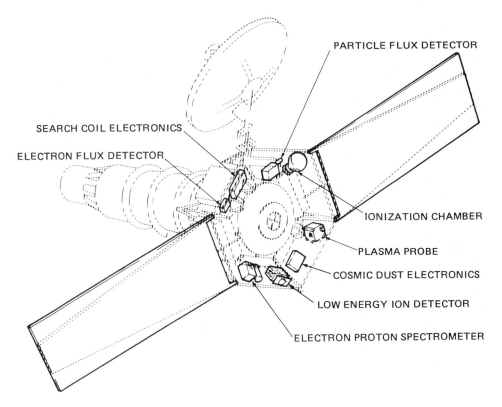

Fig. 50. Placement of Sky Science Experiments on Ranger Block III Spacecraft, Showing Solar Panel Requirements

recommended priorities for the experiments based on their (1) value to Project Apollo, (2) test performance, and (3) suitability for removal at a late date without disrupting the primary mission objective.[51]

On August 23, 1962, Cunningham conveyed the disturbing news to John F. Clark. Now Chief Scientist in Newell's Office of Space Sciences[52] and newly appointed Chairman of the Space Sciences Steering Committee, Clark learned that NASA could delay flight schedules until the Mariner R solar panels could be modified and procured for use on Ranger 7, or, if it wished, the space agency could remove some or possibly all of the secondary experiments. Cunningham reminded the scientist that "acquisition of knowledge of lunar topography sufficient for the determination of gross effects on lunar landing vehicles" was the primary objective of this block of Ranger missions. As though encouraging a choice of the second option, he observed that the fields and particles experiments were "a secondary scientific objective."[53] In the matter of project priorities and objectives, Burke clearly had an advocate at Headquarters.

If the JPL project office had won the support of the Ranger Program Chief in the matter of secondary sky science on Rangers 6 through 9, the entire episode had unquestionably misdirected energies, raised undue expectations, and strained relations all around. It had also cost irretrievable time that might otherwise have been spent on Rangers 3, 4, and 5, plagued as they were with problems of weight, a lack of redundant subsystem components, and heat sterilization with its effects on fabrication and testing.[54] To be sure, the question of the secondary sky science was placed on the agenda for reconsideration by NASA's Space Sciences Steering Committee, but it would eventually be settled out of court by the lunar missions that had already begun rising from Cape Canaveral towards the moon.

Chapter Nine

LUNAR EXPLORATION BEGUN

I N 1962 planetary scientists around the country eagerly awaited the flights of the
Block II vehicles. Rangers 3, 4, and 5 would land seismometers on the moon,
analyze the chemical composition of the surface material, and transmit closeup
pictures. The results could profoundly influence long-held theories concerning the
moon's structure and evolution. At JPL, project engineers expectantly looked
forward to demonstrating the proper functioning of Ranger's ground and space
flight machinery—including the crucial midcourse correction maneuver.[1] And the
upcoming Ranger flights had their national and international ramifications.
Ranger 3 and her sister spacecraft were widely interpreted as "advance scouts for
the manned expeditions…," as vital preliminary links in a still larger chain of
lunar affairs.[2] Ranger offered the nation a chance to advance from challenger to
champion in exploring the moon, the hope of restoring international confidence in
the supremacy of Yankee know-how, and the opportunity to bring to an end a
succession of nine humiliating American moon flight failures.[3] Americans
concerned over the Soviet space lead awaited "The U.S. Moon Shots" of 1962.

PREPARING TO GO

In July 1961, JPL engineers assembled Ranger 3, the first of the Block II
lunar machines. Early in November they completed system testing, calibration, and
checkout of the heat-sterilized, lightweight spacecraft and its scientific instruments.
On November 15, NASA accepted the vehicle. Leaving Pasadena in a specially
fitted, air-conditioned van for southern Florida, Ranger 3 began its journey to the
moon. It arrived at Cape Canaveral on November 20, the day after the Ranger 2
mission ended ignominiously in the earth's upper atmosphere. At the
Aeronutronic Division of the Ford Motor Company in Newport Beach, California,
meantime, other engineers worked double shifts in efforts to confirm the

flightworthiness of the lunar seismometer capsule in time for the scheduled flight in January 1962.

Ranger 3's lunar target had already been selected. Lighting conditions at the lunar surface were suitable for photography from a spacecraft in a near-vertical descent for a few days during the moon's third-quarter phase. In the first month of the new year, the usable days were January 22 through January 26. The seismometer, though unaffected by photometric considerations, had to be directed to a place on the moon's surface where the earth appeared high enough above the lunar horizon to permit radio contact between the tracking stations and the seismometer's transmitter. Certain locations were affected more severely by lunar libration, or rocking of the moon in its orbit, which varied the earth–moon angle, and could cause the loss of as much as half of the data from the seismometer. In a compromise among the experimenters for both instruments, the point of impact selected was just south of the lunar equator and west of the lunar prime meridian, on the eastern rim of Oceanus Procellarum—the Ocean of Storms.

William Kirhofer and Victor Clarke in the JPL Trajectory Group worked out the lunar transfer trajectories from Agena second burn to the completion of Ranger's mission at the selected point of impact. The trajectories strictly limited the permissible time for countdowns. Ranger's launch window on each of the available days extended from 3:30 to 4:45 pm EST. During that period the earth would rotate into a position for the flight to begin. Sixty-six hours after launch, the earth would also be in a position for the Goldstone tracking station to view the spacecraft at lunar impact. By using variable launch azimuths and coasting times between first and second burn of the Agena in earth orbit, and by taking into account the motions and the gravitational attractions of the earth, moon, and sun, flight engineers could employ Ranger's midcourse engine to compensate for minor variations in the flight path, and thus bring the spacecraft to the moon at the specified time and location.[4]

With so many conditions to be satisfied, however, just to achieve the lunar trajectories could be considered a major accomplishment. Ranger 3 was to be directed to a point in space where the moon would be three days later. If the spacecraft were to hit the moon, not to mention the target of interest to the planetary scientists, it would have to be delivered onto its ballistic path in space at an altitude of 193 kilometers (120 miles) above the earth and a velocity of 10,959 ±7 meters per second (24,500 ±16 miles per hour). Only slight deviations in the Agena's altitude, direction of travel, or accumulated velocity would prevent Ranger from being able to correct its flight path sufficiently to hit the moon at all (Figure 51).

Agena B 6003 and Atlas 121D, the launch vehicle components that would place Ranger 3 onto its trajectory to the moon, arrived at Cape Canaveral in mid-December. After inspection and tests conducted by personnel from Lockheed and General Dynamics-Astronautics, these vehicles were erected at Launch Complex 12, preparatory to the performance of integrated tests with the lunar spacecraft. The entire ensemble completed the Joint Flight Acceptance Test on January 5,

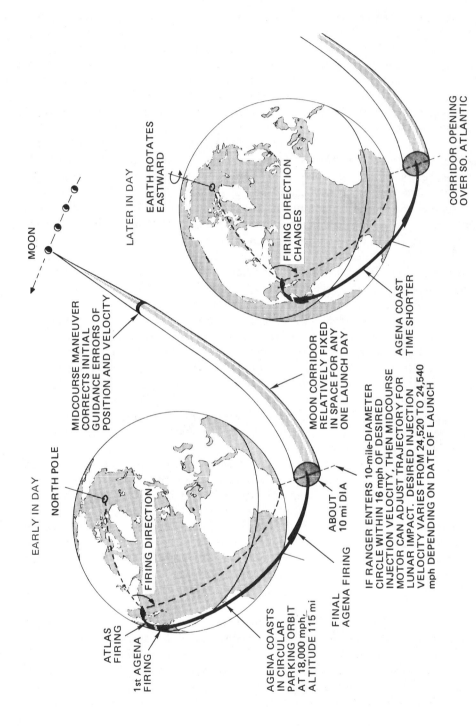

MOON

EARLY IN DAY

NORTH POLE

MIDCOURSE MANEUVER
CORRECTS INITIAL
GUIDANCE ERRORS OF
POSITION AND VELOCITY

LATER IN DAY

EARTH ROTATES
EASTWARD

FIRING DIRECTION
CHANGES

CORRIDOR OPENING
OVER SO. ATLANTIC

AGENA COAST
TIME SHORTER

MOON CORRIDOR
RELATIVELY FIXED
IN SPACE FOR ANY
ONE LAUNCH DAY

FIRING DIRECTION

ATLAS
FIRING

1st AGENA
FIRING

AGENA COASTS
IN CIRCULAR
PARKING ORBIT
AT 18,000 mph,
ALTITUDE 115 mi

ABOUT
10 mi DIA

FINAL
AGENA FIRING

IF RANGER ENTERS 10-mile-DIAMETER
CIRCLE WITHIN 16 mph OF DESIRED
INJECTION VELOCITY, THEN MIDCOURSE
MOTOR CAN ADJUST TRAJECTORY FOR
LUNAR IMPACT. DESIRED INJECTION
VELOCITY VARIES FROM 24,520 TO 24,540
mph DEPENDING ON DATE OF LAUNCH

Fig. 51. Lunar Launch Constraints

whereupon engineers removed the spacecraft to Hangar AE to begin final preparations for launch.[5]

In the two weeks remaining before the first launch attempt, Ranger 3 was disassembled, and its components were inspected for the last time. Reassembled in the newly constructed Explosive Safe Area in Hangar AE in mid-January, the seismometer capsule and retromotor, both balanced and checked out, were installed together with Ranger's fueled midcourse engine, flight battery, and radar altimeter. After successfully completing final systems tests on January 18, engineers returned Ranger 3 to Launch Complex 12 and remounted it on the Agena. They were satisfied that the spacecraft and launch vehicle were flight ready (Figure 52).

Their own plans and procedures established, Ranger's experimenters were also ready. JPL Space Sciences Chief Hibbs had appointed Harold W. Washburn as the Ranger Project Scientist for the Block II missions. Washburn, the conduit between the project engineers and the experimenters, would coordinate their

Fig. 52. Technicians Make Final Adjustments to Ranger 3 at Cape Canaveral

activities during the lunar flights. In meetings at JPL and NASA in preceding weeks, experimenters had acknowledged that closeup pictures of the moon's surface would contain the greatest popular appeal. The television experimenters, "recognizing the great interest of the public," agreed to release representative photographs to the news media immediately upon receipt. Later, after processing was completed, copies of all of the photographs were to be issued to approved lists of scientists and scientific groups.[6] Whatever the popular attraction of any pictorial returns, however, Washburn, Hibbs, and Burke evaluated the potential contribution to knowledge of each individual instrument and determined the standing each should have during flight operations. Wherever priorities could not be avoided in an emergency, they gave first preference to data from the gamma-ray spectrometer, second preference to the television camera, and third preference to the seismometer.[7]

Walter Larkin's Goldstone tracking station in California had installed special equipment to receive the telemetry from the gamma-ray spectrometer and seismometer experiments. A trailer containing RCA equipment to record Ranger's television signals was also at the site and connected to the station. Eberhardt Rechtin and his team of telecommunications specialists in the Deep Space Network began to check out each of the radio tracking stations in the Deep Space Network on January 10, finishing the operational readiness tests involving all of the stations on January 19. The JPL space flight control center in Pasadena transmitted, received, and processed simulated tracking and telemetry data in the proper sequence and format.[8] The Deep Space Network was also prepared to support the project scientists and engineers.

All of Ranger 3's preflight activities continued on schedule through January 19, when launch crews began the task of pumping kerosene fuel aboard Atlas 121D. Upon completing the fueling task, however, they discovered a leak in the bulkhead between the fuel tank and the liquid oxygen tank. While the fuel was pumped back out of the Atlas, project officials hurriedly conferred. It was launch vehicle problems all over again, only this time with Atlas instead of Agena. The situation appeared to preclude the launch of Ranger 3 during January; the Atlas would have to be returned to its hangar for repairs, and the launch postponed until late February.

Air Force and Atlas contractor personnel, however, urged another, more novel course of action. The repair, they suggested, might be made on the pad— from inside the rocket—in time for a launch attempt on January 26. In any case, nothing would be lost, and a month could be saved. Burke approved the proposal. In the next few days engineers disconnected the center engine of the Atlas and lowered it into the flame pit beneath Launch Complex 12. A wooden framework was prefabricated, passed through the engine hole at the base of the Atlas fuel tank, and assembled inside the 3-meter (10-foot) diameter tank by men wearing oxygen packs and masks. Working twenty-four hours a day, the Atlas field crew removed the ruptured bulkhead and restored the launch vehicle to a flight-ready condition for the launch attempt on January 26.[9]

While General Dynamics-Astronautics engineers clambered around inside the Atlas, JPL crews finished last-minute preparations for Ranger 3, still mounted inside the nose fairing high atop the Agena. They performed final sterilization between January 23 and 24, bathing the spacecraft in a toxic concentration of ethylene oxide gas for eleven hours and purging it afterward with dry nitrogen passed through a sterile filter.[10] Whenever launched and whatever the outcome, this United States machine would carry a minimum amount of earth bugs into space.

But after the ingenious repair of the Atlas and the time spent in final preparations, on January 26 only one hour and 15 minutes of the launch window remained. When the countdown began at 10:45 am EST, everyone knew that any further delay would postpone the mission.

A FIRST CHANCE AT THE MOON

The countdown went flawlessly. At 3:30 pm EST, Ranger 3 rose gracefully into a warm, sunlit Florida sky. Spectators gathered there to witness the daytime flight, and bathers on the beaches for miles around joined project engineers in shouting approval. The United States was on its way to the moon (Figure 53).

Yet the vehicle had not left sight before the telemetry monitors knew that something was awry. Commands generated by the Cape computer and radioed to Atlas 121D went unacknowledged—the rocket's airborne radio guidance system had failed. The Atlas continued to ascend under control of its autopilot, using internal program information to establish the sequence of flight events. Unable to command shutdown of the engines at the precise times, NASA and JPL officials could expect a deviation in the planned lunar trajectory. The project officers drove across the Cape to the JPL command post at Hangar AE to await further word from the tracking stations downrange, including the Department of Defense ship between Antigua and Ascension Islands. Within twenty minutes data from these stations indicated that the Agena B had separated properly and its engine had burned once, placing the vehicle in an earth parking orbit. Unable to respond to commands from the ground, however, the Atlas had flown higher and faster than planned. Like its predecessors, Ranger 3 was being accelerated into an unplanned orbit.

The crew of the mobile tracking antenna in South Africa waited to receive the values for the actual trajectory from Cape Canaveral. But communications equipment in Florida malfunctioned, preventing the transmission. Using the time of launch and planned trajectory figures, the station acquired Ranger 3 as it rose over the horizon five minutes ahead of schedule, at 3:55 pm EST. The Agena B had completed its second burn, although a further variation in the trajectory, later traced to an error in the Agena's flight program, had also occurred.

Within an hour, the radio tracking station at Woomera confirmed these facts. Fired into space with excessive speed, Ranger would pass ahead of and below the moon at a distance of some 32,000 kilometers (20,000 miles). This kind of

Fig. 53. Launch of Ranger 3

miss-distance exceeded the corrective capabilities of Ranger's midcourse engine; Ranger 3 would continue into orbit around the sun. Yet Burke and other project engineers had some consolations. The spacecraft separated properly from the Agena, acquired the sun and earth, deployed its high-gain antenna, switched to solar power, and turned on the gamma-ray spectrometer experiment. Despite problems with the launch vehicle, Ranger 3 was working. Except for an unexplained drop in the signal strength of the low-gain antenna, NASA and JPL controlled an operating Ranger on a deep space trajectory.[11]

Confident in this knowledge, Burke left the Cape for the West Coast. That same evening, tired and disheveled, he assumed command of the space flight operations at the JPL control center. Together with his assistant Gordon Kautz and Flight Test Director Marshall Johnson, Burke considered what they might do with a spacecraft destined to hurtle wide of the moon. The deep space trajectory of Ranger 3 might have supported the earlier experiments of sky scientists handsomely, but it was of little value to the planetary experiments actually aboard the spacecraft. Since the lunar seismometer could not be deposited on the moon, it could never listen for a moonquake. The gamma-ray spectrometer, busily establishing the radiation background generated by the spacecraft, would be limited to measuring for sky science the gamma-ray flux in interplanetary space. Among the lunar experiments, only the television camera might be used to good effect. If Ranger 3 could be positioned properly using a terminal maneuver, pictures of the moon, including portions of the backside hidden to view from the earth, would be a distinct possibility.

Evaluating these and other matters, Burke, Kautz, and Wolfe decided that Ranger 3 would be used to test all of the spacecraft functions, including the midcourse and terminal maneuvers, and to obtain television pictures of the moon. Although they would have liked to launch the lunar seismometer capsule and ignite its retrorocket as a test, that course of action was not possible. The capsule system had been intentionally designed to separate from Ranger only on command from the radar altimeter. And since Ranger 3 would pass the moon at a great distance, the altimeter could not detect that celestial object and trigger the ejection.[12]

A few hours later, in the early morning of January 27, the Spacecraft Command Group finished preparing the three stored commands defining Ranger's midcourse correction maneuver. Sent to Goldstone, the commands were radioed to Ranger 3 by the tracking station. The spacecraft confirmed their receipt and stored them in the register of its electronic brain, the central computer and sequencer. A final command to initiate the maneuver was transmitted at 2:00 am PST. Ranger turned through the desired angles in roll and pitch, and the midcourse rocket engine fired at the appointed moment and for the proper duration. The attitude control system then reacquired the sun and the earth, returning the spacecraft to its former attitude, and normal cruise operations resumed. Cheers and handshakes were the order of the day in the JPL control center: for the first time an unmanned spacecraft had altered its course in flight upon commands from the earth (Figure 54).

But as the sky lightened in the east above the San Gabriel Mountains, postmaneuver tracking information relayed from the Woomera and Johannesburg stations revealed Ranger 3 to be moving in a direction that was the mirror image of the one anticipated. Members of the Spacecraft Data Analysis Team pored over the available data in attempts to determine the cause of the unorthodox course change. During the period of about thirty hours remaining between the midcourse and terminal maneuvers, they analyzed the fault. A sign had been inverted between the digital maneuver code used in the ground computer at JPL and the spacecraft computer. While preflight tests had checked the magnitude and polarity of the commands, they had not checked their meaning. In any case, the flight path had been changed in such a way that Ranger would fly nearest the moon at the end of the second period during which the spacecraft was in view of the Goldstone tracking station, instead of earlier in the period as planned. That change limited the time for picture taking.

Even though headed on an erroneous course, Ranger 3 continued otherwise to perform flawlessly. Shortly after the midcourse maneuver, the gamma-ray spectrometer was extended 1.8 meters (6 feet) by its gas-actuated telescoping boom. To the delight of experimenters James Arnold and Albert Metzger, the

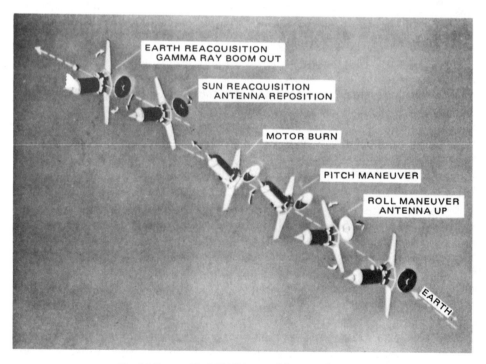

Fig. 54. Ranger Block II Midcourse Maneuver Sequence

instrument began returning data on the gamma-rays in interplanetary space. The next day, on January 28, as Ranger neared the moon, in the period termed "lunar encounter," additional commands caused the protective cover over the television camera lens to swing aside, and that instrument began to warm up.

At JPL, members of the Spacecraft Command Group prepared new coded commands for the terminal maneuver—this time with the erroneous digital sign reversed (Figure 55). At least the spacecraft could be turned so that the television camera pointed directly at the moon. Beginning at 8:29 am PST, Goldstone transmitted the commands to the spacecraft. The command to initiate the terminal maneuver followed one hour later. Ranger 3 began to pitch in the proper direction, confirming the diagnosis of a sign inversion. Midway through its prescribed turns, however, Ranger's signal strength began to waver, momentarily dropping below the threshold of the sensitive receivers at Goldstone. Ranger's high-gain antenna, which pivoted and continued to operate during this maneuver, was no longer pointing directly at the earth as it should. The spacecraft was out of control.

Ranger's central computer and sequencer had failed, immobilizing the earth and sun sensors. Under the control of its gyroscopes alone, the spacecraft continued to turn. Goldstone intermittently reacquired the radio signal, but its strength fluctuated greatly as Ranger 3 drifted aimlessly about its axis in space. The television camera, nevertheless, began sending back pictures as it already had been programmed to do. Because the high-gain antenna remained pointed away from the earth, the video signal was extremely weak and obscured by noise. When developed, the pictures showed that the camera system had operated; reference crosses on the lens were clearly observable, illuminated by light apparently reflected from the spacecraft superstructure. But the moon was nowhere to be seen.

Unaware at first of the true cause of Ranger's situation in space, the controllers transmitted urgent backup commands to the spacecraft. Now brainless, Ranger 3 could not process them. The unresponsive spacecraft swept past the moon six hours later at a distance of 37,000 kilometers (23,000 miles), on its way into solar orbit. Intermittent tracking continued for three more days, until January 31, when the attitude control gas ran out and the spacecraft began to tumble. The mission was over.[13]

REFLECTIONS ON A NEAR MISS

There were hardly enough hours in the succeeding weeks for engineers to accomplish all that needed doing before the launch of Ranger 4. Final system tests were completed on schedule at JPL, and on February 26 the spacecraft arrived at its Cape Canaveral hangar. There it was readied for the series of preflight tests to meet the launch period from April 21 through 26. Other members of the Project Office in Pasadena, meanwhile, undertook an investigation of the flight of Ranger 3, since the success of Ranger 4 obviously hinged on identifying and resolving the technical failures of its predecessor.

Fig. 55. Ranger Block II Terminal Maneuver Sequence

On February 8 Burke sent the preliminary findings of the postflight investigation to Oran Nicks at Headquarters. Burke itemized the malfunctions and the corrective measures to be taken for future Ranger flights. The Atlas bulkhead rupture during fueling, as well as the airborne guidance failure in flight, the breakdown in communications between Cape Canaveral and the deep space tracking stations, and the reversal of signs in the digital command codes prepared for Ranger spacecraft would be precluded by new tests and revised procedures. The exact causes of the drop in signal strength in Ranger 3's low-gain antenna and the failure of its central computer and sequencer during the terminal maneuver remained question marks. Attempts to duplicate these malfunctions with the Ranger proof test model in simulated flight situations proved inconclusive. Minor revisions were made to both pieces of equipment on the "basis of possible causes." New components added to Ranger's electronic brain, Nicks learned, would not be heat sterilized.[14] Heat sterilization was suspected to have contributed to the failure of the central computer and sequencer.

One further difficulty, mentioned but briefly in the preliminary findings, caused special concern to Burke and the project office. Ranger 3 had provided the first opportunity for ground controllers to compute a deep space orbit, generate commands for a midcourse maneuver, and direct the flight of a spacecraft. Although JPL personnel who served on the control teams reported administratively to Johnson and Burke for the flight operations, they worked "on loan" from the technical divisions at the Laboratory. During the flight a number of the division chiefs, anxious to know how the respective spacecraft subsystems were performing, had entered the control areas. Too often they had interfered with the flight operations by demanding time from and issuing orders to their functional subordinates—orders that conflicted with the planned flight routine. Several of the experimenters had also bypassed Ranger Project Scientist Harold Washburn to deal directly with Flight Test Director Marshall Johnson, further complicating the operations.[15]

Burke and Johnson pressed for changes in procedure and in the arrangement of space inside the temporary control center in Building 125 at JPL. With the support of Laboratory leaders in the weeks preceding the launch of Ranger 4, they added glass partitions in the control center, insulating control team members from others nearby, and prepared lists of personnel authorized to be in the various control areas, with entry to be monitored by security guards. Finally, they redefined the authority and responsibilities of the various participants. The Flight Test Director, not the division chiefs, would direct the control teams *and* the spacecraft during a mission.

Because the spacecraft had performed well until the central computer and sequencer failed at lunar encounter, the project engineers considered the capabilities of the machine demonstrated. The next flight, they were sure, would remove all remaining doubts. Even the experimenters who had yet to realize their scientific objectives shared in that confidence. Having nearly succeeded, "we were sure," James Arnold observed, that one or both of the remaining capsule missions "would give us good data."[16] Their optimism found its way into the news media. The U.S. would shortly have pictures of the moon and more. Scientists and engineers in Pasadena "were less than disconsolate. They were pleased that most of the complex mechanisms...had functioned perfectly in response to radio signals over distances of hundreds of thousands of miles."[17] Ranger 3 after all was the most sophisticated spacecraft yet launched by the United States, "as crammed with electronic tricks as a barrel of transistor radios," *Time* magazine enthused.[18] Another flight, or maybe two, would see America on the moon.

But not everyone shared the confidence evident among project engineers and scientists. Dissent surfaced within the engineering fraternity, particularly that segment affiliated with the aerospace industry that preferred Air Force contracting methods. The aerospace trade journal *Aviation Week* suggested that the Ranger system, developed for NASA "in-house" by a university-affiliated laboratory, was too ambitious and not designed conservatively enough. The flight of Ranger 3 had "greatly aggravated the continuing controversy over reliability vs performance." The NASA-JPL pursuit of "very high performance" from Ranger had been at the expense of its reliable operation. Improving the reliability of the spacecraft—by adding backup systems, for example—called for increased weight. And more weight on a spacecraft as crammed with electronic tricks as Ranger was surely impossible.[19] Ranger, the journal implied, would have been handled differently for NASA by American industry.

The implications of prior launch vehicle failures, however, and the miscalculated weight of the Ranger spacecraft that precluded redundant engineering features, were nowhere evident in the reporting. Addressing himself to these and other questions raised by the flight of Ranger 3, on February 7, JPL Director Pickering publicly summarized the results of the postflight investigation at a news conference. He explained the thinking behind the original Ranger design and sequential pattern of flights: "Because of the great complexity of the project— necessarily great if we are to advance in this space flight area—three Ranger [lunar] shots were planned to achieve a mission success." Furthermore, while the most recent mission was not a complete "success, or even a moon impact, I can safely say we face the next shot...with a great deal more knowledge and confidence because of the facts learned from Ranger 3."[20] The project, Pickering assured its budding critics, was in good hands, and making excellent progress.

ANOTHER CHANCE

At Cape Canaveral in March, project engineers subjected Ranger 4 to the prescribed series of system tests. The launch vehicle, Atlas 133D/Agena B 6004, began preflight tests at the same time; all of the Ranger elements moved steadily toward a flight on April 21, the first day of the prescribed lunar launch period. With minor adjustments in the lunar trajectory, the experimenters targeted Ranger 4 to rough-land its seismometer on the moon at nearly the same point as that selected for Ranger 3: just south of the lunar equator on the eastern rim of the Ocean of Storms.[21]

By April 20, all of the preflight inspections and tests had been finished.[22] When Ranger 4 was placed atop the Atlas-Agena launch vehicle at Launch Complex 12, the precautions and changes adopted after the flight of Ranger 3 had been followed to the letter. The seismometer capsule sported a new saw-tooth paint pattern designed better to maintain thermal balance during its flight to the moon. Terminal sterilization of the spacecraft with ethylene-oxide gas had been completed. The tracking stations in the deep space network were checked out, and for the first time the 25-meter (84-foot) antenna at Johannesburg was equipped with a transmitter making it possible to command as well as track Ranger 4 from South Africa. Four tracking ships instead of one had been detailed to support the flight of Ranger 4, deployed from the Caribbean to Ascension Island. Once again everything was ready (Figure 56).

Fig. 56. Technicians Prepare Ranger 4 for Launch at Cape Canaveral

Air Force and NASA officials postponed the countdown on April 21 one day to permit another launch at Cape Canaveral; with fair weather forecast for April 22, Ranger's participants were eager to proceed. Around the nation, Ranger interest was again at a high pitch. In widely quoted testimony before Congress just a few weeks earlier, NASA Administrator James Webb had informed the country that the Soviet Union would continue for the time being to dominate space near the earth with its manned orbiting missions. "But for [manned] landing on the moon and return," he stated, "we are ahead of them." And "if a thousand things go right" with Ranger 4, the *New York Sunday News* summarily proclaimed, "the U.S. will hit the paschal moon with a spacecraft that will blaze a trail for the first American—and, hopefully, the first earthling—to follow to the moon."[23]

On Saturday, April 2, Ranger officials from NASA, JPL, and the Air Force agreed to schedule the first launch attempt for Monday, April 23, having rejected the preceding day, Easter Sunday, out of regard for people's religious sensibilities. At Cape Canaveral on Monday, clear weather and mild winds made an ideal day on which to undertake or watch a lunar flight. The countdown began in midmorning and proceeded without interruption. Under new arrangements, Ranger Program Chief Cunningham joined Burke in the JPL command post at Hangar AE. The two men remained in constant voice communication with the Air Force personnel conducting launch operations in the blockhouse at Launch Complex 12. With all checks completed, the launch sequence began at 3:50 pm EST, with the familiar puff of white smoke signaling start of the Atlas engines. Ranger 4 rose, on its way to the moon.

Tracking antennas at Cape Canaveral monitored the performance of the Atlas, the Agena, and Ranger until they passed out of sight over the horizon eight minutes after liftoff. By that time the Atlas had staged and separated from the Agena, and Agena first burn had begun. As the Agena coasted into low earth orbit over the South Atlantic, the tracking stations downrange received a steady flow of data. First returns looked good. They confirmed the start of Agena's second burn, and the normality of Ranger's vital signs. The Cape transmitted trajectory data properly to the Johannesburg tracking station in South Africa. As the Agena swung to the southeast and out of view of the antenna on Ascension Island, the picture was very bright indeed.

The mobile tracking antenna promptly located Ranger 4 as it rose over the South African horizon twenty-three minutes after launch. The Agena had completed its second burn; the spacecraft signal, however, indicated immediate, serious difficulties. Though Ranger's transponder was radiating at the 960-megacycle tracking frequency, all telemetry commutation was absent. The state of health of the spacecraft could not be ascertained. Two minutes later Ranger 4 separated from the Agena and, as designed, tumbled very slowly on its lunar trajectory. The moment passed when the machine was to extend its solar panels and high-gain antenna and begin stabilizing itself. But without telemetry, flight controllers lacked confirmation that these events had taken place. The fluctuating strength of the transponder signal strongly suggested that the spacecraft was still tumbling, its solar panels tucked up firmly against the superstructure and its high-gain antenna still stowed over the midcourse engine beneath the spacecraft. The mission was in desperate trouble.

The station at Woomera acquired Ranger's transponder and began tracking the spacecraft as it moved out and away from the earth. These data, relayed to the JPL control center and fed into a computer, permitted an initial trajectory to be calculated. This time the near-perfect performance of the Atlas-Agena B had put Ranger 4 on a collision course with the moon even without a midcourse correction maneuver. But the spacecraft remained inoperative. Burke, who had arrived in Miami aboard an Air Force plane on his way home, approved a series of trouble-shooting commands to be sent from the Johannesburg station in an effort to fathom the spacecraft's true status. Commands to advance the telemetry, to switch the transponder signal from the low-gain antenna to the high-gain antenna, to change the high-gain antenna hinge angle, and to override the spacecraft roll control system were all quickly transmitted. All of them proved futile. At JPL the Spacecraft Data Analysis Team issued a grim prognosis: the master clock in Ranger's central computer and sequencer had stopped. Without that timer, the telemetry decommutator had ceased operating, all timed functions had failed to take place, and the vehicle could not accept and act on commands from earth. For all intents and purposes, Ranger 4 was dead.[24] At the Cape a NASA official lamented: "All we've got is an idiot with a radio signal."[25]

The flawless performance of the Atlas-Agena launch vehicle made Burke's own disappointment all the more bitter. Hitting the moon with an idiot Ranger was no consolation. Burke knew there would be "nothing" for science, and little for project engineers. Even the reason for the timer failure could elude detection. Antennas at the downrange stations south of Antigua tracked only the launch vehicle; their receivers were never equipped to monitor the 960-megacycle spacecraft frequency. A six-minute and 17-second gap in coverage of spacecraft telemetry existed between the time Ranger passed out of view of the stations in the Caribbean and the time it was first picked up in South Africa. And the timer had stopped some time during that period. Every single preceding Ranger spacecraft, he recollected, had functioned in space given any opportunity at all—but not Ranger 4.

As Ranger 4 moved inexorably toward its lunar rendezvous, NASA Administrator Webb arrived in Los Angeles for a speaking engagement. Early in the morning of April 26 he flew to the Goldstone tracking station in the Mojave desert, where he joined Nicks, Pickering, Cummings, and Burke. There, the space officials awaited word of a lunar impact and discussed the Ranger 4 mission with representatives of the press. Electrical power from the spacecraft battery had run out hours before, and Ranger's transponder had ceased to operate. Stations in the Deep Space Network continued their radio tracking nonetheless, homing on the 50-milliwatt signal produced by the tiny battery-powered transmitter in the seismometer capsule. Helping to pass the time the Goldstone station staff supplied their guests with the amplified sound of the seismometer's transmitter and closed-circuit television pictures of the moon taken from earth. The transmitter's beep droned off into an inaudible hum as the spacecraft approached the moon and accelerated under the pull of lunar gravity. The lunar pictures, provided by a camera and telephoto lens aligned through the center of one of the large dish antennas trained on Ranger from Goldstone, flickered silently on the monitors. During the final moments the cross-sight was centered on the moon's leading edge. At 4:47 am PST, sixty-four hours after launch, Ranger 4 skimmed over the leading edge and, two minutes later, crashed out of sight of the earth on the far side of the moon.

Success had eluded the project once more. Webb told the newsmen that the mission still had contributed to the "long strides forward in space" made recently by the United States. And, he added, this spacecraft was far more sophisticated than Lunik 2, which deposited a Soviet pennant on the moon in 1959. Pickering, putting the best face possible on an otherwise dismal outcome, observed that the accuracy of Ranger's launch vehicle had been fully demonstrated. Ranger 5, he was encouraged to believe, would bring this original phase of the project to a successful conclusion before the end of the year.[26] Later that day, at a formal press conference in Los Angeles, James Webb was even less restrained.

Taken in context, he explained, the flight of Ranger 4 had to be considered "an outstanding American achievement." For the first time an American spacecraft had reached the moon. The first American astronaut, Colonel John Glenn, had orbited the earth two months earlier. Just the day before, on April 25, moreover, the second Saturn rocket, the kind that would eventually convey man to the moon, had been launched in a perfect test flight from Cape Canaveral. The nation's manned lunar program, Webb insisted, remained right on schedule.[27] The press and the public seemed to agree. "Ranger 4 Hits Moon, Scores U.S. Space Feat," was a typical lead in the newspaper columns.[28] Though disappointed at the failure of Ranger's electronic brain, *The New York Times* said that in this flight "the fact that the moon was reached tells much about the increased power and improved accuracy the United States has achieved in rocket technology."[29] Summing up the week's events in space, a major West Coast paper concluded: "At last the cheers are drowning out the fears that we are 'losing' the space race" (Figure 57).[30]

But Soviet Premier Nikita Khrushchev could not resist the temptation to needle the American space agency just a little. "The Americans," he observed in a press conference of his own, "have tried several times to hit the moon with their rockets. They have proclaimed for all the world to hear that they launched rockets to the moon, but they missed every time..." The Soviet pennant already deposited there, he quipped, was getting lonesome waiting for an American companion.[31] The Soviet Premier's well-targeted remarks were like salt in an open wound at JPL. Pickering wasted no time reacting. "On April 26, at 4:47.50 am Pacific Standard Time," he declared speaking for the Laboratory as well as for NASA, "Ranger 4 was tracked by the Goldstone receiver as it passed the leading edge of the moon. At 4:49.53 am it crashed on the moon at a lunar longitude of 229.5 degrees East and a lunar latitude of 15.5 degrees South."[32] If the Russians wished

Fig. 57. Red, White, and Blue Cross (Copyright, *Los Angeles Times*;
Reprinted With Permission)

to confirm that fact, they could dispatch one of their own astronauts to the spot and investigate it first hand.

But however much space officials or the press praised Ranger at home, there were growing doubts whether the project was living up to American expectations for the space race or, for that matter, even up to expectations held for it by the project engineers and scientists themselves.

Chapter Ten

WHICH WAY RANGER?

O NLY Ranger 5, the last of the Block II machines, remained to secure the primary objectives of lunar science. Beyond that, attention focused on the objectives for Rangers 6 through 9. Simmering at NASA Headquarters and at JPL during 1962 was the question: Would these missions indeed fly for science first and support Apollo second, or the other way around?

FAMILY RELATIONS

Having reorganized NASA in late 1961, James Webb possessed a management structure with which to prosecute both the nation's program of landing a man on the moon and the other varied activities of the space agency. Dividing the lunar program into manned and unmanned segments had made it necessary to strike arrangements to coordinate the dichotomous enterprise. But in January 1962 these arrangements had not yet been made, and they would not be made before the end of the year. In the interim, Homer Newell and Brainerd Holmes, the directors of the two responsible NASA offices, managed affairs as best they could—in the interest of NASA and the country as each perceived it.

In March 1962, after the pro-science modification of the Ranger Block III objectives, Newell appeared before a Congressional Committee to speak in favor of NASA's budget authorization for the coming year. His Congressional questioners, less concerned than ever with any information that automatic spacecraft might gather to shed light on the origin of the solar system, were considerably interested in accelerating Project Apollo. They peppered him with queries about how much support the unmanned lunar program was providing to the manned. Disturbed by the reasoning that inspired these questions, Newell took time out to write a personal letter to Representative Joseph E. Karth, the Farmer-Labor Democrat from Minnesota's Fourth Congressional District in Saint Paul-Minneapolis, and the firm, personable Chairman of the Subcommittee on Space Sciences of the House Committee on Science and Astronautics.

"Space science, in addition to laying the groundwork for future activity in space," Newell told the Subcommittee Chairman, "is one area in the national space program where we [already] hold a clear lead over the Soviets, and in which we can continue to hold the lead if we maintain the vigor and breadth of our effort." It was imperative, he averred, that Congress "support space science for the sake of science, maintaining our faith that the practical benefits thereof will assuredly accrue." Newell added that a broad rather than narrow science program would best serve the admittedly crucial goal of a manned lunar landing. In any case, he insisted, space science was the ultimate raison d'être for all unmanned *and* manned lunar flights. Aside from competition with the Soviet Union, "the single greatest foreseeable reason right now for sending man to the moon and planets is to explore and investigate them..."[1]

A month later, on June 7, at the first Senior Council Meeting of the Office of Space Sciences in NASA Headquarters,[2] Newell told his assembled field directors, including Pickering, that he intended to incorporate quality science on every flight. But the field directors wondered whether Newell's office or the Office of Manned Space Flight would determine and order the requirements for unmanned support of Project Apollo. The Office of Space Sciences might set these requirements, Charles Donlan of Langley Research Center observed, but it still seemed intent on pursuing scientific objectives selected before landing a man on the moon had been established as a national goal. "Space Sciences," he continued, "was rather unbending in not getting scientific data which would assist the manned program." Donlan had touched the nub of the entire problem: What science would best serve Apollo and the overall NASA program? Apollo desperately needed information on the hardness of the lunar surface and the mechanics of soil erosion. A soil penetrometer experiment proposed by Langley for future Ranger missions had been designed to meet these needs, Donlan asserted, but it had already been rejected in favor of additional seismometer experiments. While the information requirements of Project Apollo "may not be quite as elegant in a scientific sense," he concluded, the Office of Space Sciences "should support the entire NASA program."

Was it not possible, Pickering inquired, to use the available funds of the Office of Space Sciences to support the ends and objectives of the manned program and at the same time advance the scientific program? If not, then Donlan was right: the Office of Space Sciences remained too heavily committed to pure science in the face of the national goal to land a man on the moon. In Pickering's view the objectives for Ranger Block III should be restored to those specified by Silverstein in 1961, with support of the manned program coming first, at least until its requirements were met. Still, Pickering continued, Apollo seemed too devoted to engineering without regard to space science. Some "melding" and accommodation between the two offices was called for.

Though resolving some of the management problems between field centers and Headquarters, the day-long meeting achieved no consensus on scientific objectives. Unfortunate as the lack of unanimity might be, Newell preferred it to altering the Headquarters scheme for unmanned lunar exploration. Newell and his colleagues also sensed that any melding of the manned and unmanned lunar programs in NASA would likely result in the total subordination of the Office of Space Sciences to Project Apollo. So long as the Office of Space Sciences planned a program and paid the bills, it intended to establish the scientific objectives for unmanned lunar flights. "Pure science experiments," Nicks reiterated, "will provide the engineering answers for Project Apollo." Besides, more immediately important than any melding with Apollo was the poor reliability so far demonstrated by the unmanned missions already budgeted. Ranger's overall track record, 0 for 4, was worse even than Nicks' estimate of one success in every three flights. Clearly, Newell and his deputy insisted to the Senior Council meeting, the project required more effort.[3]

Two miles away, in a rented building on 19th Street, Brainerd Holmes and his colleagues in the Office of Manned Space Flight were busy deciding issues of their own. Project Apollo was the cynosure of NASA and the nation,[4] where everything—the Saturn rockets, the facilities that would house and launch them, and the funding—was of Brobdingnagian proportions. And the most compelling of the decisions faced by Apollo's managers in 1962, perhaps, pivoted on the method of staging to reach the moon: Would the astronauts rendezvous in earth orbit or in a lunar orbit? Until answered in July, that single question claimed the lion's share of attention of Holmes as well as NASA's top leaders.[5] The scientific priorities of Newell's unmanned lunar projects, even if each of them did cost several hundred million dollars, quite simply did not rate with these affairs in importance.

Holmes and deputies Joseph Shea and George Low pursued the objective of landing a man on the moon with a singlemindedness of purpose. Every resource at NASA's command had to be bent to this task, and that included the unmanned lunar program. When Project Apollo needed data on lunar surface conditions from the unmanned flights, Project Apollo officials expected to specify these data. Bespeaking that need was the document Holmes issued on June 15, *Requirements for Data in Support of Project Apollo,* which declared that "the basic requirement for maximum United States accomplishment in the Apollo Project time period dictates that all space activity provide maximum support..." The document called for the unmanned lunar projects to furnish Apollo with three classes of information during the next three years. First, environmental data on fields and particles in space near the moon to assist in the design of spacecraft and ensure the safety of the Apollo crew in flight and on the moon. Second, information on the physical characteristics of the lunar surface to confirm spacecraft landing gear designs. Third, photoreconnaissance and topographic data to permit early selection of Apollo landing sites and aid in surface operations. Nowhere in the text was the Office of Space Sciences mentioned. Rather, all NASA groups engaged in the

unmanned lunar program were asked to "obtain the technological data specified...
on a priority basis."[6]

The message was straightforward enough, but the peremptory manner of its
release and wording almost guaranteed ruffled feathers in Newell's shop, not to
mention the Office of Administrator Webb.[7] Not only had Holmes crossed well-
defined jurisdictional boundaries and demanded support, but he expected the
Office of Space Sciences to perform the bulk of this work and pick up the tab
without so much as a thank you. The repercussions rattled offices in Pasadena,
where JPL, unlike most of the other field centers, was affected directly. James
Burke was scarcely eager to repeat the discomposing experience of March, when he
had been caught between the factions contesting over Ranger for science or
Ranger for Apollo, between his manager-superiors at Headquarters on the one side
and the project engineers at JPL on the other. But the *Requirements for Data in
Support of Project Apollo* appeared to rekindle the issue of Ranger's primary
purpose. Could Project Ranger serve—and please—two masters at once?

Pickering, Cummings, Burke, and Surveyor Project Manager Eugene
Giberson considered that issue during the next week at JPL. If the unmanned
lunar projects were to proceed in an orderly manner, they agreed, the respective
Offices of Space Sciences and Manned Space Flight had to agree upon their
objectives and priorities. In their opinion, moreover, considering the national
commitment to a manned lunar landing, serving the immediate needs of Apollo
seemed the only realistic approach to take.[8] In the following days, that view was
reinforced by other pressing events. Despite the best of intentions, the secondary
sky science experiments crammed onto Rangers 6 through 9 threatened the
reliability of the spacecraft.[9] Plans for another block of five Ranger spacecraft
seemed close to approval at NASA Headquarters, and fabrication of these vehicles
would have to be contracted to an industrial firm if JPL was to avoid a manpower
crisis. The experiments for them had to be selected with clear objectives in mind.
In this connection, the Aeronutronic firm, which had all but completed work on
the original lunar seismometer capsules, wanted to know if it should disband its
team of personnel or prepare to make more of the hard-landing capsules for future
Rangers.

On June 29, 1962, Cummings flew East to discuss these questions with his
counterparts in the Office of Space Sciences. Project Ranger, Cummings said,
needed to be streamlined technically and in terms of objectives. Future spacecraft
should be contracted to industry, and the secondary experiments at least limited to
those supporting Apollo requirements. As a first step, he strongly urged the Office
of Space Sciences to take the lead and engage the active participation of the Office
of Manned Space Flight in the unmanned lunar program, if possible with financial
support, but at least in written ground rules for the conduct of Project Ranger.[10]
Nicks and Cortright listened intently, but they ultimately rejected any role in their
theater for the Office of Manned Space Flight.

Newell's people knew that Project Apollo already needed more funds than
NASA had budgeted for it. Financial support of the unmanned program by the

Office of Manned Space Flight thus seemed highly unlikely. Furthermore, the active participation of that office was liable to squeeze pure science out of all the lunar flights preceding Apollo—leaving as the final product only an engineering residue. Evidently annoyed by JPL's willingness to risk that outcome, Newell and his staff resented the Laboratory's resistance to their direction as well as its enthusiasm for Apollo requirements. Besides, Ranger's distressing flight record to date hardly seemed to match the reputation for excellence in science and engineering of JPL-Caltech. If Pickering, Cummings, and Burke still doubted the wisdom of those responsible for establishing Ranger priorities, Newell, Cortright, and Nicks shared growing reservations over JPL's management of Ranger.

NASA's LUNAR OBJECTIVES RECONSIDERED

JPL's difficulties were compounded in July 1962, when its first Mariner R flight to Venus ended in disaster because a "hyphen" had been omitted from the launch vehicle guidance equations. Shortly before the launch, Texas Democrat Albert Thomas had questioned NASA plans to renew its contract with Caltech to operate the Jet Propulsion Laboratory for another three years. This nonprofit operation, he charged, allowed NASA to "end run" Civil Service pay scales for nongovernment research talent. This method of government contracting for research and development with an institution of higher learning, moreover, did not seem to be producing the results expected of it. It might be better, he suggested, to separate the Laboratory from Caltech and operate the installation as a NASA field center under Civil Service regulations. Caltech President Lee A. DuBridge promptly rebutted the suggestion in a letter to Webb, but the letter did not completely satisfy Thomas.[11] On July 24, two days after the Mariner disaster, Republican James Fulton admonished D. D. Wyatt, NASA Director of the Office of Programs, on the poor record of unmanned space flight. The first Mariner R alone cost the taxpayers close to $14 million. "We should," Fulton declared, "be beyond this stage."[12]

At 2:53 am EDT, August 27, 1962, another JPL Mariner R—it was designated Mariner 2—rose from Launch Complex 12 at Cape Canaveral. Another Atlas guidance malfunction occurred, momentarily appeared to doom the mission, but was miraculously overcome just in time for the separation of the Agena second stage (Figure 58). Soon Mariner 2 was on its way to Venus. In the following hours, then days, the spacecraft obediently extended its solar panels and high-gain antenna, acquired the sun and earth, stabilized in space, and flawlessly executed a necessary midcourse correction maneuver. Postmidcourse trajectory computations indicated a projected "miss distance" of approximately 41,000 km (25,476 miles), well within the operating limits of two onboard experiments designed to measure conditions in Venus' atmosphere and on its surface during the flyby encounter on December 14, 1962.

The expected mission success blunted Congressional criticism, restored confidence in and at JPL, and ratified the soundness of the basic Ranger design. At

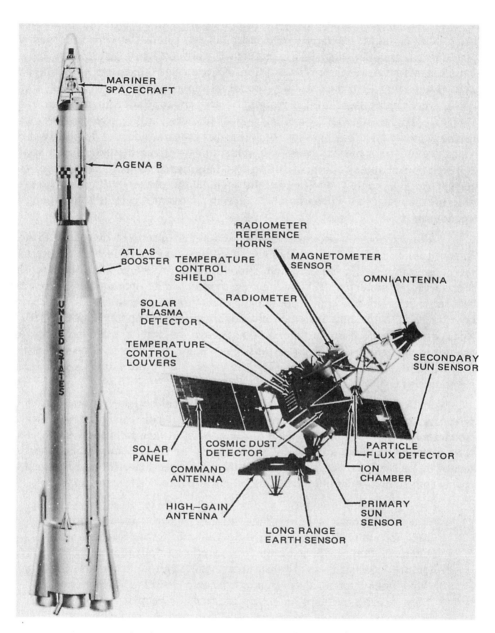

Fig. 58. The Mariner R Spacecraft and Launch Vehicle

that moment, however, the international implications seemed most important. Though the Soviets had orbited the first satellite and the first man around the earth, now Americans led them in planetary exploration—at least on the way to

Venus.[13] But NASA insiders knew that the objectives and priorities for the lunar Rangers 6 through 9 remained as clouded as ever. Both the Office of Manned Space Flight and the Office of Space Sciences continued to insist that their own requirements be served first. The position of Pickering, Burke, and Cummings at JPL—engineering first, then science—received support on August 10, 1962, when the Space Science Board of the National Academy concluded in a report: "As [Apollo's] engineering tasks are accomplished..., scientific investigations and missions will also be phased into the program; and, as flexibility and sophistication are achieved, scientific investigations will become the primary goals. Appreciation of these concepts is of critical importance to the acceptance of the current Apollo program by scientists throughout the country..."[14] But however welcome the Board's conclusion, it did not advance the JPL position at Headquarters.

Although NASA Headquarters had tentatively approved extending Project Ranger to include five more vehicles (Rangers 10 through 14), Newell's office had them programmed for science first, Apollo support second. And despite the shortage of personnel at JPL and an even more acute shortage of space, the vehicles, Nicks informed the Laboratory, were "to be assembled and tested in-house [at JPL], with some increased divisional contractor support allowed."[15] JPL faced a manpower crisis in the months ahead as well as the likelihood of prolonged contention at Headquarters over Ranger's objectives between the manned and unmanned factions. The Laboratory's position in NASA's lunar program was fast becoming untenable.

On August 15, 1962, Pickering, acting to force the issue, wrote a letter to NASA Associate Administrator Robert C. Seamans, with copies to Newell and Holmes, alluding to the differences at Headquarters and their deleterious effects on JPL. All of the questions raised by the extension of Project Ranger, he asserted, related to a more basic one: "How should the unmanned lunar efforts be shaped to produce maximum benefit to the manned flight program?" He went on:

> I do not believe it appropriate for JPL to seek to exploit the differing points of view within the Office of Space Sciences and Office of Manned Space Flight in an effort to advance our own point of view as to what constitutes a viable and useful program. I would greatly prefer that you and I, Homer Newell, and Holmes or Shea reach agreement and then instruct our people accordingly. We at JPL are now in favor of a faster-paced and technically simpler program than the ones that we have in the past advocated or at least accepted..., and we strongly support the ... [Space Science Board] recommendation that lunar environmental and engineering data for Apollo design be sought with urgency and even, if necessary, at the expense of data having greater intrinsic scientific value.[16]

If Seamans or Webb had been unaware of any differences before, they knew about them now.[17]

Over a month later, Seamans called a meeting of Pickering, Newell, and Holmes for October 11.[18] At the meeting itself, which included Nicks, Seamans stressed that Newell and Holmes were to coordinate their respective unmanned and manned lunar programs. Both sides agreed that the extension of Project Ranger could contribute toward the goal of landing a man on the moon. At Seamans' request, Nicks explained how Holmes' needs might be met through the future Ranger television and capsule missions. Holmes asked that plans for these missions be coordinated with his deputy, Joseph Shea. Finally, Seamans instructed Nicks to act as the coordinator between the two programs, and to report on the "probable contribution" that the unmanned lunar program would make to Apollo.[19] Having been asked by the NASA Associate Administrator to cooperate in this area with the Office of Manned Space Flight, officials of the Office of Space Sciences had no choice but to proceed as directed. Pickering, pleased, considered the meeting a breakthrough. In Project Ranger, so everyone seemed to have agreed, Apollo's requirements shared a priority at least equal to those of space science.[20]

Holmes and Newell added further to this impression a few days later, when they announced formation of a Joint Working Group of representatives from each of their offices. The group was to be responsible for "recommending a detailed program of scientific exploration to the Office of Manned Space Flight, recommending to the Office of Space Sciences a program of data acquisition to assure a timely flow of environmental information into the planning for manned projects, and establishing and maintaining close liaison with field centers, government agencies, and universities in the development of an integrated scientific program for manned space flight."[21] Named Chairman of the Joint Working Group: Eugene Shoemaker, coexperimenter on the Ranger television system and a geologist with the U.S. Geological Survey. Agreement among all of the affected organizations on the objectives for future Ranger missions now appeared possible. Ranger 5 might also aid the manned lunar program immediately; the scientific information it was supposed to transmit, especially the television pictures and data from the radar altimeter, could help confirm the design of the Apollo lunar lander.

ONE MORE TIME

Ranger 5's launch had been scheduled in June 1962, but the interposition of the Mariner R mission to Venus had forced a delay until October,[22] which allowed project engineers more time to solve the technical malfunctions of the previous flight. Nevertheless, the reason for the failure of the central computer and sequencer, as Burke had guessed, could not be determined with certainty because of the missing spacecraft telemetry during the crucial period over the South Atlantic. Adding to the puzzlement, the electronic brain had been operated for some 700 hours in tests at Cape Canaveral and JPL; this vital component had been continually cycled as the spacecraft passed each of its tests without experiencing a catastrophic failure.[23] Engineers who examined the available Agena and Ranger telemetry tapes ascertained the approximate time of failure: it had occurred during the few seconds between electrical separation of the spacecraft and Agena, when the Agena umbilical plug was detached and the "power-up" command issued to the spacecraft transmitter by the central computer and sequencer. The subsequent performance of Ranger 4 pointed to the inverter or transformer units in the electronic brain as the sources of the failure; the units had most likely short-circuited.[24] The short-circuiting, in turn, was suspected to have been caused by a tiny, undetected flake of aluminum or gold floating in the zero gravity and touching two "hot" pins on the exposed umbilical separation connector.[25]

As a result of these findings, a number of changes were made in Ranger 5. Diodes and fuses were added to isolate the "hot" lines in the connector. With weight limitations no longer a problem, engineers added a backup clock to ensure that the telemetry would be provided with synchronizing pulses even in the event of a malfunction of the electronic brain. Another nitrogen gas bottle was added to the attitude control system to decrease pressures in the cold gas system, and backup pyrotechnics were added to the propulsion system to guarantee ignition of the midcourse engine.[26] All of these changes were, of course, for the single purpose that Ranger 5 would not fail.

The project office also changed the temporary space flight control facility at JPL. The communications control room received new teletype and voice conference systems designed to reduce operator errors. Closed-circuit television consoles were installed in adjoining rooms to permit flight controllers to view incoming data. At Goldstone and the deep space radio tracking stations overseas, the voice conference circuits were modified to allow controllers to talk on several nets at once. Patrick Rygh replaced Marshall Johnson, who had been assigned full time to manage the design and construction of the permanent Space Flight Operations Facility at the Laboratory, as Ranger Space Flight Test Director. Rygh, an aeronautical engineer and graduate of Caltech, would work directly with Burke during future Ranger flights (Figure 59).

One change was not made. Like its predecessors, Ranger 5 had been heat sterilized, and that deeply disturbed the man charged with the sterilization program, Rolph Hastrup. To his mind, there was no longer any question that heat sterilization seriously degraded the performance and life expectancy of equipment in the Block II machines. On July 25, 1962, he expressed these fears to George Hobby, who had originally developed the JPL lunar sterilization procedures. "We are...paying a substantial price...to sterilize all of our lunar spacecraft," Hastrup asserted. "The application of sterilization procedures such as the dry heat cycle presents a serious risk to reliability..." Whether instituted to placate scientists or to perfect techniques to be used on planetary landers, Ranger's sterilization requirements had to be relaxed, and the public "informed beforehand" so it "would not be 'shocked' by the situation."[27]

Fig. 59. JPL Space Flight Control Center Readied for Ranger 5 (Patrick Rygh is Seated at the Center Console.)

Despite his initial advocacy of the procedure, heat sterilization was now a sore topic with Burke, who was also increasingly concerned over its effects on the performance of electronic components. But the die had been cast for Ranger 5. In the project office, what appeared most important now was ensuring the performance of critical equipment on Rangers 6 through 9. Burke and Ranger Spacecraft System Manager Wolfe had a new spacecraft test model assembled electrically equivalent to Ranger 6, and asked for more exceptions to heat sterilization of crucial components. Pickering agreed. JPL requested and NASA Headquarters granted more waivers, exempting from heat sterilization all of the important electrical cable harnesses in these spacecraft, the six RCA television cameras, and the critical electronic brains as well. Pickering also advised Newell that the heat sterilization procedure was believed seriously to reduce the reliability of the central computer and sequencer. He pointed to the possible significance of the failure of Ranger 4 having been traced to the computer-sequencer array or its related power modules. But any waiver of sterilization for the computer and sequencer would cover only Ranger 8 and later spacecraft, since earlier spacecraft modules had already been sterilized.[28]

Piecemeal waivers on important equipment, however, were not Hastrup's idea of how to proceed. On August 30 he dispatched a nine-page memo to

Cummings, expatiating further on the situation and recommending a more definite course of action. Experience to date, he observed, warrants "complete reappraisal of [U.S.] lunar sterilization policy." Many component failures detected in tests had positively been caused by heat sterilization. Tight schedules and a minimum of spare units often combined with component failures due to sterilization to force testing of the spacecraft without some of its flight hardware. In addition, in many instances the repairs required by the failures had been difficult to accomplish and had resulted in subsystems "basically less reliable than one which had not been 'dug' into."

The problems encountered in sterilizing Rangers 3 through 5, Hastrup warned, were being repeated on Rangers 6 through 9 on all items not already exempt from the dry heat cycle. The practices of "clean room" assembly in a dust-free environment and terminal sterilization of the spacecraft with ethylene oxide gas inside the Agena nose fairing should be continued, he concluded; however, heat sterilization had to be abandoned to assure the performance of lunar spacecraft. JPL should request NASA's permission to redetermine "the extent, if any, to which the lunar program would ...| carry on the development of sterilization procedures" for planetary exploration.[29] There could be no mistaking Hastrup's alarm, but Cummings would not recommend so radical a change to Pickering or to NASA Headquarters, at least not until the flight of Ranger 5 offered more evidence one way or another.[30]

The Ranger 5 spacecraft left for Florida on August 20, arriving at Cape Canaveral on the day of Mariner 2's launching toward Venus. Atlas 215D and Agena 6005 arrived shortly thereafter, and all three machines began the cycle of tests leading to a launch on October 16.[31] At the Cape, the attention of project officials fixed on the Atlas and Agena. Except for the flight of Ranger 4, one or both of them had malfunctioned on every single Ranger and Mariner mission. The Atlas, most agreed, now represented the principal problem. "Testing and checkout [at Cape Canaveral] have revealed that none of the six Atlases delivered to NASA were flight worthy," Hans Hueter's staff asserted. "The problem is further magnified by the fact that the last Atlas delivered was no better than the first."[32] Atlas equipment discrepancies had delayed Ranger launches before, and all of the project personnel worked hard to ensure that none remained in Atlas 215D before the scheduled flight of Ranger 5.

Ranger's experimenters, the Deep Space Network, and the Space Flight Operation Teams, meantime, finished their preflight preparations. Despite the stated needs of Project Apollo, the objectives of space science still shaped this last original Ranger mission, and NASA's experimenters planned their operating priorities accordingly. In the event that the spacecraft missed the moon, obtaining data from the gamma-ray spectrometer would take precedence over taking television pictures of the surface. Further gamma-ray experiments, they recognized, were not scheduled for some time to come; an improved television system, on the other hand, would fly in 1963 on Rangers 6 through 9.[33]

Equipment checkout and net integration tests at the deep space stations occurred while tracking of Mariner 2 continued.[34] After the launch of Ranger 5, these stations would track for both missions simultaneously. On October 11, JPL Deputy Director Brian Sparks issued guidelines for the space flight operations. In any situation where the deep space stations could support only one of the two missions, Ranger 5 would command priority. With one exception, all JPL computing facilities would also be placed at the disposal of Ranger 5, and the Control Area in the JPL Space Flight Operations Facility was to be used exclusively by Ranger 5 flight controllers during that three-day mission.[35]

Officials terminated the launch countdown of Ranger 5 on October 16 when technicians discovered an electrical short circuit in the spacecraft radio transponder. Detached from the Atlas-Agena, Ranger 5 was returned to Hangar AE for a spare unit, in time for another launch attempt on October 17. It too was canceled, this time by high winds from Hurricane Ella, which loitered nearby in the Caribbean. The next day, October 18, was overcast. Surface winds, however, had decreased sufficiently for another launch attempt to begin. The vehicle passed all checks. The Atlas engines ignited at 12:59 pm EST, and Ranger 5 lifted from launch Complex 12 into a leaden sky (Figure 60). The roar of the Atlas engines subsided as the vehicle soared out of sight. Inside the JPL command post, Burke learned that the rate beacon in the Atlas guidance system had malfunctioned. The rocket, nevertheless, continued on course, responding to backup commands radioed to it from the Cape computer. The same Atlas failure had caused the loss of Mariner 1, but the proper "hyphen" in the guidance equations saved the day for Ranger.

The spirits of project officials rose as all information received from the downrange stations proved positive. The launch vehicles separated, the Agena's engine ignited, and Ranger 5 coasted into its planned orbital path over the South Atlantic. Tracking ships, now equipped and located to preclude losing any Ranger telemetry downrange, confirmed the normal functioning of the machine in space.

The mobile tracking antenna acquired the spacecraft as it rose above the South African horizon a half-hour later, at 1:30 pm EST. Second burn of the Agena engine had occurred successfully, injecting Ranger 5 into its lunar transfer trajectory. A few minutes later the spacecraft was observed to separate from the Agena, and at 1:46 pm the large dish antenna in Woomera contacted Ranger 5 as it moved out and away from the earth. All the data continued excellent: Ranger had extended its solar panels, acquired the sun, begun generating solar power, and stabilized in orbit. In a short time Ranger's electronic brain would command the craft to extend its high-gain antenna and roll in search of the earth, and normal cruise operations would begin. Burke, Albert, and Cunningham conveyed the encouraging news to waiting newsmen. "All of us," Burke said, "are keeping our fingers crossed."[36]

At 2:12 pm EST, the pleasure turned sour. At that moment the temperature in the power switching and logic module of the central computer and sequencer rose sharply; electrical power from Ranger 5's solar panels was lost. Though the spacecraft switched to battery power, it performed erratically. The programmed

Fig. 60. Ranger 5 Ignition

command to turn on the gamma-ray experiment occurred, but Ranger's electronic brain did not issue the command to "acquire earth." To make matters worse, spacecraft telemetry now being received in the United States was garbled, making detailed analysis impossible. The telemetry-to-teletype encoders at the deep space stations in South Africa and Australia had both malfunctioned.

During the afternoon Burke, stunned, was on the phone with Space Flight Test Director Rygh at JPL (Figure 61). The Spacecraft Data Analysis Team reported that short-circuiting must have caused the loss of solar power. Even if that were true, however, knowing it now was of little help. The battery furnishing Ranger's electrical power would be exhausted within hours, ending operations of any kind. Together Burke and Rygh determined on an immediate midcourse maneuver to achieve lunar impact as an engineering test. Without earth reference, however, no one knew the true roll position of Ranger 5, and insufficient time remained to obtain it before the electrical power ran out. A midcourse maneuver,

Fig. 61. Officials Assembled for the Ranger 5 Postlaunch Press Conference at Cape Canaveral. Left to Right: Friedrich Duerr, Major J. Mulladay, Lt. Col. Jack Albert, Kurt Debus, William Cunningham, and James Burke

they decided therefore, would be conducted using only the sunline reference, and certain turns were excluded.

Override commands extending the high-gain antenna away from its stowed position over the midcourse motor were radioed to Ranger 5 at 8:00 pm EDT from the deep space station in Johannesburg. The spacecraft reacted obediently. The maneuver execute command followed at 8:29 pm. The crippled machine acknowledged the message and began the novel thirty-minute maneuver sequence. Minutes later the craft experienced further electrical shorting, and telemetry ceased for 13 seconds, then came back on again. The midcourse maneuver could not be completed. The battery was drained. Half-way through the midcourse maneuver Ranger's radio transmitter fell silent, and, as its gyroscopes ran down, the spacecraft began to tumble.

Two days later, on October 21, the deaf and blind Ranger 5 cartwheeled past the trailing edge of the moon at an altitude of 720 kilometers (450 miles) and accelerated into solar orbit. The stations in the Deep Space Network continued to track the tiny battery-powered transmitter inside the seismometer capsule for a few days longer, until that minute signal dropped below the threshold of the sensitive receivers on earth. This time, the tense and shaken engineers and scientists at NASA and JPL refrained even from announcing a near-success.[37]

Ranger 5 was a disaster. So too, it seemed, was the project.

Chapter Eleven

IN THE COLD LIGHT OF DAWN

THE mushrooming Cuban missile crisis quickly diverted public attention from the failure of Ranger 5. Ranger project participants, nevertheless, felt keenly the blow of the abrupt and unexpected ending of the original Ranger flights. Project scientists had learned from the gamma-ray measurements on these missions of the existence of a cosmic gamma-ray flux, but nothing about the moon,[1] which remained just as remote an object of scientific inquiry as ever. Project engineers had failed in five flights to demonstrate the capabilities of the JPL Ranger spacecraft.

These distressing results increased tension and provoked conjecture at NASA and JPL. Had too many experiments been heaped on Ranger, forcing the scientific cart before the engineering horse? Or did this automatic machine and its associated systems possibly represent too ambitious and rapid an advance in technology? Or had the spacecraft system been designed and tested improperly? Or was it truly a victim of the sterilization program? Whatever the answers, one thing was clear: a reappraisal of the entire effort had to be made, and so, on October 20, Homer Newell told Pickering.

THE RANGER INQUIRY AT JPL

On October 22, his eye on the evaluation of Ranger by Headquarters, Pickering formed his own special assessment committee at JPL. The members, senior personnel not associated with Ranger, were to investigate the project, contrast its operation with that of the Mariner R project, and submit to him any recommendations for change by November 12. The JPL committee conducted extensive interviews with Ranger and Mariner R as well as technical division personnel, and submitted a seven-page, sharply critical report to Pickering.

The report declared that the weight-limited Ranger Block II spacecraft, like the Mariner R spacecraft, did not possess redundant engineering features sufficient to guarantee flight operations after a partial failure. Successful operation had depended upon the proper function in series of numerous components, and the failure of any one among them would have disabled the machine. The detailed excellence of subsystem engineering and hardware found on Mariner R, they asserted, had not been achieved on Project Ranger. With adverse effects for both Ranger and Mariner R, the report continued, engineers assigned to flight projects from the technical divisions often lacked the required experience; in many instances, their section chiefs were unresponsive to the needs and directives of the project offices. Moreover, the committee investigators judged the section chiefs to be "quite unfamiliar with either the project management or the details of the subsystems developed and furnished by their sections." The whole arrangement left the integrity of JPL spacecraft "heavily dependent on the inspiration, skill, and attention of individual cognizant engineers. Little if any evidence of review of the work of these engineers by responsible section chiefs was discovered."

The report also called attention to errors of detailed design and hardware assembly common throughout the project, errors compounded by personnel turnover, by attempts of the undermanned project office to delegate direction of the technical sections to representatives of the Systems Division, and by inadequate procedures for design verification, inspection, and failed-parts reporting. They considered Project Manager Burke, moreover, to have spent excessive amounts of time negotiating matters of space science, the launch vehicles, and launch operations with NASA Headquarters and Air Force personnel when he should have been attending to the JPL spacecraft. He had, furthermore, consistently stressed meeting schedules, and the hurry-up atmosphere engendered in beating the Soviets to the moon had reduced further the "motivation to pursue details which could lead to engineering excellence." Curiously, the JPL committee paid little attention to heat sterilization, its implications for component reliability, or to the differences in performance between the sterilized Ranger and the unsterilized Mariner R spacecraft.[2]

On the basis of its findings, the committee concluded that Rangers 6 through 9 could not be expected to perform any more successfully than Rangers 3 through 5 unless JPL took "specific and forceful action":

1. Further flights should be delayed until the Laboratory named a new manager for Project Ranger, completed a thorough spacecraft design review, introduced new inspection, testing, and project management procedures, and provided a revised plan and schedule to top management.

2. The Ranger 6 flight should be abandoned, and that spacecraft converted to a Design Evaluation Vehicle to be used for testing purposes.

3. Plans for the Ranger 10 through 14 spacecraft should be modified to reflect the changes made in the design of Rangers 6 through 9.

After the committee submitted its report on November 14, Pickering learned that one member had refused to endorse the recommendation to replace the project manager, arguing that Burke could not be held entirely responsible for the conditions prevailing at JPL and NASA. Still, Burke's job was clearly in question.[3]

THE KELLEY BOARD INVESTIGATION

On postlaunch day, October 19, in the NASA Office of Space Sciences, Homer Newell had reacted to the loss of Ranger 5 by instructing Oran Nicks to compose a board of inquiry of individuals not presently associated with Ranger. Besides determining the cause for the failure of Ranger 5, the board was to review all operations of the project and the potential reliability of future spacecraft.[4] The next day Newell explained his decision to Pickering and asked for "JPL's full cooperation." The board, he continued, would report in time to decide whether Ranger 6 should be launched as scheduled in January 1963 or delayed "to introduce necessary changes."[5] Some changes, the wording implied, were indeed necessary.

On October 29, 1962, Newell named and established the Ranger Board of Inquiry. Its Chairman, Albert J. Kelley, was to return by November 30 recommendations "necessary to achieve successful Ranger operation," including but not limited to changes in management procedure, systems, components, testing, quality control, reliability assurance, and operational procedures.[6] In a separate memo to Kelley, Lunar and Planetary Programs Director Nicks stressed the importance for science, Apollo, and the nation's prestige of Project Ranger and hence, the serious nature of the investigative assignment.[7]

Kelley, a former commander in the Navy with an ScD in instrumentation, directed the Electronics and Control Division in NASA's Office of Advanced Research and Technology and had chaired the Agena B Coordination Board during the latter half of 1960. He knew Project Ranger and was widely respected at Headquarters and JPL. The Board members appointed with Kelley represented a number of important NASA Headquarters offices and field installations, as well as one business firm.* During the next three weeks the Kelley Board, as it came to be called, set about an exhausting schedule of conferences and interviews with Ranger participants from coast to coast. The first stop on October 30 was the Jet Propulsion Laboratory.

---

* Other members: John M. Walker, also of the Office of Advanced Research and Technology; Herman Lagow, Goddard Space Flight Center; John Foster, Ames Research Center; Francis B. Smith, Langley Research Center; Frederick J. Bailey, Manned Spacecraft Center; John Hornbeck, President, Bellcomm Incorporated; Arthur H. Rudolph, NASA Office of Manned Space Flight; and James Koppenhaver, NASA Office of Programs. NASA Ranger Program Chief and Program Engineer, Bill Cunningham and Walter Jakobowski, served ex officio to assist the Board.

The JPL staff that met the Board hardly displayed its usual self-confidence, its accustomed readiness to consider no job too big or too complex to accomplish. One long-time JPL hand recollected the common laboratory attitude: "We were good and we knew it, and we did not hesitate to request permission [of NASA] to take on another child while still nine months pregnant." Previously, in fact, the easy self-assurance and esprit exhibited by employees of the Laboratory had often irritated NASA personnel, who saw themselves patronized as tyros in the field of rocketry and astronautics.[8] Now, for the first time, JPL had failed to deliver in a key applied research and development program. And now it had to suffer the further humiliation of an investigation by a NASA board of inquiry.

On November 30 the Kelley Board submitted its report to the Office of Space Sciences. Kelley himself presented an oral summary of findings and recommendations—they were supported unanimously—to Newell and his staff on December 5, 1962.[9] While the investigation had not been confined to JPL operations, the principal focus of the assessment was the JPL spacecraft. With attachments and appendices, the Board's final report was ten times longer, and, although similar in many respects, even more critical than JPL's own investigation.

Clearing the scientific experiments of any direct culpability in Ranger's failures, the Board fastened on the Ranger spacecraft itself, especially the extent to which it was based on the design of a planetary machine. The members considered such a design not "optimum for lunar missions," but rather "more flexible and complicated... Ranger is sufficiently complex, and its mission sufficiently demanding, that a high order of engineering skill and fabrication technology is required in its design and construction to make it work successfully." That high order, they observed, had not been achieved. Consequently, "the present hardware comprising Rangers 6 through 9, as was the case of Rangers 3 through 5, is unlikely to perform successfully." Going further, the Board indicted JPL's entire approach to the project. The fast-paced, high-risk Ranger undertaking smacked of the military method of research and development, where actual shakedown flights substituted for data on a vehicle's performance that should be obtained in elaborate ground tests. Proving the spacecraft technology was crucial, but the tests should be conducted on the ground and not in flight. Board members labeled this practice as a "shoot and hope" approach.[10]

The Kelley Board also judged Ranger's heat sterilization program to have contributed measurably to the poor performance of the spacecraft. "Even on a system especially designed to withstand sterilization, this practice would no doubt degrade performance... On marginally designed equipment, as in the case with some parts of the present Rangers, the performance degradation due to sterilization can be extremely serious. Under these circumstances the Board doubts whether a program of testing and inspection to find weak or failed parts, and then replacing them, will even lead to a high confidence that the spacecraft [Rangers 6 through 9] will be capable of completing the lunar photographic mission successfully."[11]

Besides the sterilization problem and the deficiencies in spacecraft design, construction, system test, and checkout, the Board also determined that:

1. JPL had improperly organized spacecraft system management for large, long-term projects such as Ranger.

2. The spacecraft did not possess sufficient redundant engineering features to ensure reliable operation. Success hinged "entirely on the perfect execution of every event in a long sequential chain."

3. Network drills conducted prior to Ranger flights were inadequate, and the effectiveness of the deep space tracking stations was impaired by poor voice communication channels.

4. The Atlas-Agena launch vehicle had demonstrated such a poor record of reliability that "current estimates place the probability of a successful launch no higher than 0.5," or fifty percent.

5. Poorly stated and improperly understood, mission objectives added "complexity damaging to mission success." At any one time the project was supposed to (a) deliver a payload to the lunar surface in support of Project Apollo; (b) carry out extensive scientific experiments; (c) be essentially nonmagnetic so as not to interfere with the scientific experiments; (d) be biologically sterile to avoid contaminating the moon with earth organisms; and (e) develop technology useful to planetary missions.[12]

The Board's primary recommendations correspond to these findings:

1. Project management should be strengthened to provide adequate staffing in the JPL project office and clear-cut lines of responsibility and authority at JPL. Extraneous duties of the Project Manager, such as serving as Deputy Director of the Lunar Program, should be eliminated, and new procedures for design review, design change control, testing, and quality assurance introduced.

2. Heat sterilization of all Ranger components should be abandoned immediately. NASA should also reexamine all other methods of sterilization to determine the need for them and their effectiveness.

3. To ease the JPL workload, an industrial contractor should be assigned to fabricate Ranger 10 and all subsequent Ranger machines.

4. The assignment of further new flight projects to JPL, they suggested, also could be withheld until the Laboratory resolved the Ranger situation to NASA's satisfaction.

5. The job of procuring and launching Atlas-Agena vehicles for the space agency should be assigned entirely either to NASA or to the Air Force.

Whatever the choice, NASA had to institute more and improved prelaunch inspection and monitoring of these launch vehicles.

6. The objectives for Project Ranger should be restated, and all other activities not directly related to these objectives (e.g., sterilization and the development of technology for planetary missions) should be eliminated.[13]

Within a few days of its submission at Headquarters, JPL received the classified final report. The disquisition was much farther-reaching and more severe than officials at JPL—or even NASA—had expected it to be. The JPL staff resented the implication in the "shoot and hope" indictment that the Ranger spacecraft were substandard at the time of delivery and that NASA had accepted them on the outside chance that they might work when launched. They also found unfair that the report ignored the facts that a race to the moon had required fast-paced schedules, or that large environmental facilities for ground testing did not exist, or that in 1960 NASA itself had accepted Project Ranger as a high-risk enterprise. Still, notwithstanding the wide latitude for action reserved to JPL under the terms of the NASA contract with Caltech, Headquarters did control the JPL purse strings and could enforce any of the recommendations by withholding further space flight projects from the Laboratory. Certainly there would have to be a prolonged delay in the flight of Ranger 6, and major changes in the conduct of unmanned lunar projects. On every side in Project Ranger, past sins of omission and commission had come home to roost. And most of them, it seemed, were to be found perched above the door of the JPL Project Office.[14]

NEW MANAGEMENT AND NEW OBJECTIVES

On Friday afternoon, December 7, 1962, a secretary hurriedly left the Lunar Program Office, her face a mask of grief. "The Lunar Program..." she wept, "Cliff is out. Jim Burke too."

Personnel attached to JPL's Project Ranger and Lunar Program Office had discussed and speculated on the pending changes. Burke, on his own, had already ordered an immediate halt to the heat sterilizing of all Ranger flight hardware.[15] The flight of Ranger 6 had been postponed indefinitely. Revised procedures for testing, design control, and quality assurance were in the offing. So were a design review of the complete Ranger spacecraft and changes in the management relationships between JPL's Project Office and the technical divisions that supported it. But few had guessed that Burke *and* Cummings would be relieved of command.

Five days later, Deputy Director Brian Sparks and a delegation of JPL officials met with Newell's representatives at NASA Headquarters to review the findings of the Kelley Board and settle upon a course of action. Systems Division Chief Schurmeier and Spacecraft System Manager Wolfe represented the Project Office for JPL, while Burke waited outside the room. Heat sterilization and secondary scientific experiments had become controversial in NASA's unmanned lunar program. Hammering out new project objectives and guidelines in this and a final meeting on December 17, the participants agreed to (1) eliminate all secondary sky science and communications experiments approved by the Space Sciences Steering Committee for flight on Rangers 6 through 9; (2) discontinue heat sterilization of Ranger components and terminal sterilization of the spacecraft with ethylene oxide gas, and replace Ranger hardware previously heat sterilized; (3) displace the development of technologies for the planetary program to a byproduct and not an objective of Project Ranger; (4) postpone the flight of Ranger 6 until JPL and NASA were satisfied that the launch vehicle and the spacecraft would succeed.

JPL, on its part, would accord Ranger the highest priority of any flight project underway at the Laboratory. It would conduct a complete design review of the Ranger spacecraft, inventory sterilized and unsterilized hardware, introduce changes in project management, testing, and quality assurance practices, and prepare detailed project organization charts, a project recovery plan, and recommendations for phasing in an industrial contractor for subsequent Ranger spacecraft. NASA would fund all of these changes, which would obviously cause a major increase in the project's final cost. Finally, and most important to the future of the project, it was agreed that Rangers 6 through 9 would have a single objective: to acquire pictures of the lunar surface "significantly better" than those obtained from the earth, and of the greatest possible benefit to Project Apollo and the scientific community. Although a definition of "significantly better" remained to be worked out, in the event of any dispute in which "the scientific objectives are found to compromise the information needed for the manned program, every consideration will be given to meeting the Office of Manned Space Flight needs."[16]

While Newell, Cortright, Sparks, and Schurmeier put the finishing touches on the Ranger reprogramming at NASA Headquarters, Mariner 2 flashed past the planet Venus. During December 14 the craft radioed to earth scientific findings of the temperature and properties of the Venus atmosphere and surface. With all of Mariner's objectives attained, the world-wide acclaim boosted sagging morale at the Laboratory and provided Pickering a welcome if brief respite from the repercussions of the Ranger investigation. A few days later, on December 18, Pickering announced the replacements for Burke and Cummings. Robert Parks would direct the Lunar and Planetary Programs at the Laboratory (Figure 62). Reporting to him as the new Ranger Project Manager would be Harris M. Schurmeier.[17]

It was with mixed emotions that Schurmeier, universally known as "Bud," succeeded his Caltech classmate and friend as Manager of Project Ranger. The two men had been undergraduates, naval aviators, and graduate aeronautical engineers together. Both had come up to the Laboratory from Caltech in 1949, lived a few houses from one another, and shared skiing, sailing, and soaring in Southern California. But whatever Schurmeier's personal reservations about replacing Burke, he rapidly established himself as the Project Manager on the job.

Fig. 62. JPL Lunar and Planetary Program Director Robert Parks

Inquisitive and determined, he possessed a hard streak of common sense, a fine feeling for making decisions at the proper moment, and an ability to use the talents of those who worked with him to the best advantage. As the former Chief of JPL's Systems Division, where the diverse elements of Ranger were reconciled, Schurmeier was familiar with the task at hand (Figure 63).[18]

The task itself was formidable, with morale a shambles and the project in disarray. The sterilized and unsterilized components of Rangers 7 through 9, in various stages of assembly, had to be separated and identified. Rangers 10 through 14 were now to be assigned to an industrial contractor. Prior flight plans and firm schedules had evaporated. Burke elected to remain with JPL and on the staff in the Ranger Project Office. "That eased the transition in command considerably," Schurmeier recalled, "and much of what we accomplished in succeeding months I owe to Jim. He was absolutely dedicated to making Ranger work, and he had what

Fig. 63. JPL Ranger Project Manager Harris Schurmeier

it took to stay and help when another would have turned his back and walked away.''

Early in 1963, shortly after William Pickering and a Mariner 2 float had lead the Rose Parade through the streets of Pasadena, Schurmeier, his assistant Gordon Kautz, Burke, and Spacecraft System Manager Wolfe took the first steps to right the overturned project. They composed a Ranger System Design Review Board of JPL engineering Section Chiefs to evaluate the spacecraft, find any weak points in the system and subsystem designs, and recommend changes to eliminate them and to improve reliability. Burke could applaud the ground rules Schurmeier issued to guide members of the board. Chief among them: "A few TV pictures of the moon, better than those taken from earth, is the only mission objective. No advanced development experiments or additional scientific instruments will be carried!"[19]

The elimination of the eight sky science experiments from Rangers 6 through 9 had released 22.5 precious spacecraft kilograms (50 pounds) which the project engineers could use to incorporate redundant engineering features. Schurmeier intended that every possible measure would be taken to increase the reliable performance of all Ranger components. No longer to be geared for success on razor-thin margins of weight and power, the spacecraft was to be able to achieve its picture-taking objective even in the event of a partial failure in flight.

Schurmeier wanted the design review and other measures completed and evaluated by the end of January. He also instructed Patrick Rygh, Ranger's Flight Test Director, to identify the steps necessary to improve space flight operations. Similar orders were dispatched to Hans Hueter's launch vehicle team in Huntsville, and a detailed evaluation of the Atlas-Agena booster began. In New Jersey, Bernard P. Miller, RCA's Ranger manager, evaluated the television payload and proposed improvements in those areas where testing indicated weaknesses, or where the design had been compromised to meet the extremely tight Ranger schedules. Ranger 6, converted to a design evaluation vehicle, started to undergo extensive testing. JPL engineers microscopically reinspected all Ranger flight hardware for possible weaknesses in workmanship and packaging.[20] Once more, in the project and program offices around the country, things began to hum.

Changes in management and operating practices at JPL assisted the work of Schurmeier's Project Office. Pickering sharply curtailed the feudal independence of JPL's technical divisions. Section Chiefs could no longer transfer their personnel from one assignment to another without the consent of the affected project manager. The divisions also lost their prerogative to set and maintain separate standards for quality assurance and flight acceptance testing of components; new, Laboratory-wide standards were established. To ensure that these standards were met, Pickering combined the functions of quality assurance and reliability in a single office under the direction of Brooks T. Morris. With these actions and control over design changes strengthened, JPL's Section Chiefs found themselves involved directly in project activity and held accountable for any poorly designed or malfunctioning equipment produced by their engineers. Finally, Pickering announced that Ranger enjoyed the highest priority of any flight project at the Laboratory. Hereafter, Schurmeier could be sure of the authority he needed to command action at JPL.[21]

At NASA Headquarters, Newell and his deputy Cortright completed their own planning for the new Ranger and, on December 19, publicly announced the changes.[22] They also furnished Congressman Joseph E. Karth, now Chairman of the House Subcommittee on Space Sciences and Advanced Research and Technology that approved authorizations for Newell's office, more of the detail on the findings and recommendations of the Kelley Board. And they sent final instructions on the plans for Ranger to Pickering at the end of the month.[23] Most of the trauma for project engineers, at least, was over.

SPACE SCIENCE AGAINST THE WALL

But many scientists affiliated with the project were angry. Rangers 6 through 14 were all programmed by NASA to fly television cameras to take pictures of the moon's surface. None would carry nonvisual planetary experiments such as gamma-ray spectrometers and seismometers that might yield data on the chemical composition and internal structure of the moon. Prominent among the discontented planetary scientists was the Nobel Laureate Harold Urey, a Ranger television experimenter, and prime instigator of NASA's lunar exploration program. In late October 1962, Urey wrote to Newell to complain about the "enormous emphasis on TV photography... I understand the importance of TV photographs from the engineering standpoint, but I do think they have small scientific value." So did a number of other planetary scientists, who wrote the White House, the NASA administrator, and the Ranger Program Office to say so.[24]

Wherever the protests were sent, they eventually made their way to the beleaguered director of Space Sciences, Homer Newell. For all Newell's past determination to maintain a program for space science, he recognized that space science was in no bargaining position now, not after the persistent flight failures on the one side and, on the other, the hard engineering requirements of the manned lunar landing. Newell responded to the planetary scientists by patiently explaining the situation in Project Ranger. The best designed experiments, he told Harold Urey, were useless if the spacecraft failed to deliver them to their target and transmit the results back to earth. Besides, Newell added, putting the best face he could on the matter, a close-up look at the moon would be of "real scientific consequence." In the future, Newell told Urey along with all the other scientific dissenters from the pro-Apollo emphasis, he hoped—and expected—that NASA would reestablish a program better balanced in the interests of planetary science.[25]

Since Project Ranger's inception in late 1959, both engineers and space scientists had acted forcefully to see that their views prevailed. Lacking a consensus on what aspects of the project needed to be accomplished first, they had sought to do everything at once. Engineers had compromised reliability by attempting to fashion simultaneously a new-generation launch vehicle as well as a sterilized planetary spacecraft, by conducting test and flight operations without adequate facilities, and by compressing everything into an extremely tight schedule to "beat the Russians." Burke, moreover, had not insisted on accountability and uniform procedures of quality control from the JPL technical divisions, nor had JPL Director Pickering supported him enough in his project dealings with the technical chiefs or with NASA Headquarters. In Washington, Homer Newell and his lieutenants had added mightily to the difficulties by insisting that ambitious scientific objectives be met while the technology was under development, by equating Burke's objections essentially with attempts to thwart policy decisions, and by vigorously adding experiments over the project manager's valid technical protests.

Both groups had lost. Ranger 5 proved an agonizing personal defeat for James Burke and his engineers, and a decisive programmatic defeat for Homer Newell and space science. The aftershock of that failure, in fact, would reverberate inside the space agency for years to come. Without exception, every single unmanned lunar spacecraft launched thereafter in the 1960s would fly first in support of Apollo, and second in the interests of planetary science.[26]

Part II

THE NEW RANGER

Chapter Twelve

HOMESTRETCH ENGINEERING

I N the months following the Kelley Board investigation, all four Ranger system components—the spacecraft, space flight operations, the Deep Space Network, and the Atlas-Agena launch vehicle—were modified and requalified. Mission success dominated project thinking, as every conceivable measure was taken to improve the reliability of the entire system before launching Ranger 6 to the moon.

REDESIGNING FOR IMPROVED RELIABILITY

The principal item under fire was JPL's Ranger spacecraft, and first on the agenda in Pasadena was a reevaluation of its design. On February 5, 1963, Project Manager Schurmeier, his assistant Gordon Kautz, Spacecraft System Manager Wolfe, and members of the JPL Ranger Design Review Board and Senior Staff met to consider the Board's recommendations. While sticking to the basic design of the Ranger spacecraft, they agreed to use the newly available weight for ten major alterations to improve and assure the reliable performance of Rangers 6 through 9:

1. Change the spacecraft bus structure from magnesium to aluminum alloys to improve thermal efficiency.*

2. Substitute the electrically segmented, rectangular solar panels used on Mariner 2 for the smaller, trapezoidal panels previously used on Ranger.

---

* The darker magnesium alloy had been adopted during the Block II weight reduction efforts in 1961 (see Chapter 4). Although lighter in weight, it absorbed solar energy, and the exposed members had to be wrapped in aluminum foil. The brighter though heavier aluminum reflected and radiated heat and sunlight much more efficiently.

3. Add a second 1200-watt/hour battery, doubling the backup power capacity and assuring time to conduct a midcourse maneuver even if the solar panels were completely disabled.

4. Add a second 1/4-watt transponder to the radio frequency equipment, permitting both the high-gain and low-gain antennas to be supported by a separate unit.

5. Fabricate new central computer and sequencer units, without heat-sterilized components, together with a backup timer capable of initiating certain vital commands if the electronic brain malfunctioned.

6. Retain the backup clock, first flown in Ranger 5, to ensure telemetry commutation in the event the normal synchronizing pulses generated by the electronic brain were lost.

7. Add a duplicate attitude control gas system, including gas bottles, piping, and jets.

8. Modify the squib firing assembly (used to unlatch the solar panels and trigger some other crucial functions) to permit existing redundant ignition wiring to be driven separately by the two spacecraft batteries.

9. Add a larger midcourse rocket motor, able to compensate for larger trajectory errors caused by the launch vehicles.

10. Make mandatory "conformal coating," that is, complete coating with a plastic substance of all exposed terminals and points in the spacecraft electronic modules, to preclude any electrical shorting caused by undetected flakes or debris floating in the zero-gravity environment of outer space.[1]

Bernard Miller and the RCA contract team at that firm's Astro-Electronic Division in Hightstown, New Jersey, already had submitted a proposal for changes in the design to the bank of six television cameras that were to return pictures of the lunar surface.[2] RCA had urged that the two full-scan and four partial-scan cameras and their related electronic subassemblies be separated into two independent electrical chains, each operating on separate circuits; thus, some pictures would be taken even in the event of a malfunction in one chain. In addition, to protect against the failure of Ranger's electronic brain to command operations of the television subsystem at the proper moment, a backup timer to turn on the subsystem might also be added. If the spacecraft attitude control system were still properly pointing the cameras toward the moon, the transmission of pictures would thereby be assured.[3] Collectively these recommended changes and those pertaining to the spacecraft had the effect of separating and duplicating nearly all key spacecraft functions. Schurmeier and Kautz heartily endorsed them. If they were put into effect immediately, and if time were allowed to complete retesting all Ranger's system components, the slip in the project schedule might be held to eleven months. Ranger 6 would be ready for launch in early December 1963 (Figures 64 and 65).

In a meeting on February 12 and 13, JPL's plans for modifying the spacecraft and the procedures for testing and requalifying it were presented to NASA Headquarters. Schurmeier and Kautz received approval for all of the design changes and for a first launch in December as well. But there was some opposition to the swift pace of the plans among Cortright, Nicks, and Cunningham. What concerned them was the potential risk of using the Atlas-Agena B launch vehicle on such short schedules. Eleven months seemed hardly enough time to guarantee the performance of that machine *and* iron out the operating differences still remaining between NASA and Air Force launch offices. The early Ranger and Mariner flights had painfully demonstrated, after all, that without a dependable launch vehicle the best of spacecraft and scientific instruments meant nothing.[4]

LAUNCH VEHICLES REVISITED

The concern voiced by Cortright, Nicks, and Cunningham was widely shared at Headquarters and among the JPL engineers who had followed launch vehicle developments in the past. In fact, since NASA had authorized use of the Atlas-Agena B in January 1960, the difficulties encountered in procuring and

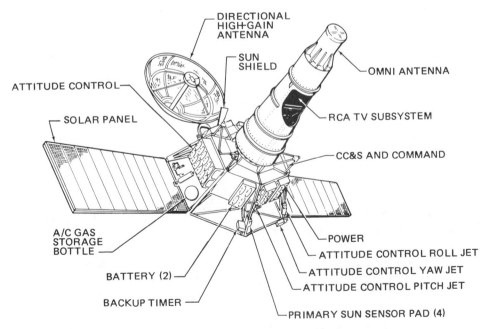

Fig. 64. The Ranger Block III Spacecraft as Viewed From Above

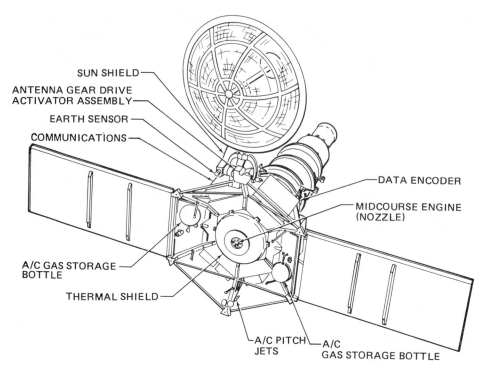

SUN SHIELD

ANTENNA GEAR DRIVE
ACTIVATOR ASSEMBLY

EARTH SENSOR

COMMUNICATIONS

DATA ENCODER

MIDCOURSE ENGINE
(NOZZLE)

A/C GAS STORAGE
BOTTLE

THERMAL SHIELD

A/C PITCH
JETS

A/C
GAS STORAGE BOTTLE

Fig. 65. The Ranger Block III Spacecraft as Viewed From Below

launching them had proved a constant source of contention. It remained the one
Ranger system component that JPL did not directly control. And though Friedrich
Duerr and Hans Hueter in the NASA Light and Medium Vehicles Office at the
Marshall Space Flight Center in Huntsville had been given charge of this
procurement, the Atlas and Agena remained Air Force, not NASA, products. At
Huntsville, the interest of Marshall's Director Wernher von Braun continued to be
riveted on developing the nation's manned lunar launch vehicle, the Saturn.[5]
Lacking support in the years that followed, the strenuous efforts of Duerr and
Hueter had been proved insufficient for the task at hand. Frequent confusion over
roles and the tension created between the NASA and Air Force offices and
committees responsible for procurement on the West Coast and launch operations
at Cape Canaveral had further contributed to delays, technical mixups, and hard
feelings all around. Caught up in an interagency dispute over launch vehicles much
larger than Ranger, neither the vigorous complaints of Duerr and Hueter nor those
of Burke and Cummings had ever had any remedial effect.[6]

By the time Ranger 5 was launched in October 1962, the performance of
the Atlas-Agena B in flight, as well as the methods used to procure and launch that
combination, had welled into problems of major proportions for the space agency.[7]
After the Kelley board recommendations in December 1962, NASA Administrator

James Webb acted to settle the nagging launch vehicle issue once and for all. First, he removed responsibility for monitoring the Atlas- and Thor-Agenas from Marshall Space Flight Center, and reassigned that function to the Lewis Research Center in Cleveland. Second, he pressed the Secretary of Defense to allow NASA to perform all of its own launch operations at Cape Canaveral, and to procure Atlas and Agena launch vehicles directly from the contractors without the Air Force's interceding.

Webb and his deputies Dryden and Seamans recognized that moving the management of these Air Force-produced launch vehicles from Marshall to Lewis would involve introducing new personnel into Ranger and other flight project organizations at midstream. The new people would need to learn on the job. But a change could be avoided no longer, and the Lewis Research Center was responsible for only one other NASA launch vehicle project, the Atlas-Centaur, transferred earlier in 1962. Webb believed that Abe Silverstein, now the Director of the Center, and his staff would devote to this task the kind of attention and resources it desperately needed.[8] By the same token, additional personnel would be freed at Marshall to help develop the important Saturn rockets needed for Apollo.

Silverstein appointed a long-time Lewis hand and trusted colleague, Seymour C. Himmel, as Agena Systems Manager. Himmel, a mechanical engineer and long-time employee of the Lewis Research Center, was a stickler for fine detail. With Silverstein's active support, he moved quickly to learn all there was to know about the Atlas-Agena and to make his presence felt among project participants. Enjoying a well-staffed office in Cleveland, Himmel increased the numbers in the Field Representatives Branch at Lockheed in Sunnyvale and at a similar office at the General Dynamics-Astronautics plant in San Diego to monitor all work on the launch vehicle. Within six months—by the end of June 1963—his staff in Cleveland and on the West Coast swelled from 32 to 67.

The JPL engineers had no difficulty establishing a close and important working relationship with the eager engineers from Cleveland. Although Himmel at Lewis monitored the procurement and launch of all Atlas-Agenas for NASA, in Project Ranger he reported to its manager, Schurmeier. On March 11, Schurmeier informed Himmel of JPL's recommendations for improving the Atlas-Agena B. Nearly all of them focused on the Atlas booster, in particular the trouble-plagued General Electric "Mod III G" airborne guidance package which controlled Atlas' ascent. New, definitive specifications and requirements for building and testing the Atlas, "more stringent than the original USAF Weapon System requirements of several years ago," were among the most important changes to be made. The G.E. guidance equipment, Schurmeier advised, had to be modified and tested to withstand a higher vibration environment in flight, and, if the cooperation of the Air Force could be secured, it would be advantageous to fabricate electrical wiring harnesses for the Atlas using the high standards employed on the Atlas-Mercury manned flight project.[9]

In a three-day Launch Vehicle System Review held at the Lewis Research Center between June 3 and 5, Himmel and his staff recommended final changes to the Atlas-Agena. To improve confidence in the Atlas-Agenas in protective storage and those yet to be built, they recommended that improved flight equipment be substituted for items with known weaknesses, all of the vehicles modified to one standard configuration, and the procedures used to qualify them for flight tightened. NASA Ranger Chief Cunningham and JPL Project Manager Schurmeier approved the Lewis program. Members of JPL's Launch Vehicle Review Board, nevertheless, still shared the reservations expressed earlier at Headquarters by Cortright and Nicks. Could these improvements, though necessary, be accomplished in time for the scheduled launch of Ranger 6 in December?

Harry Margraf expatiated on this subject in his summary report to Schurmeier. The launch vehicle improvement program, he observed, had to march in cadence with the Air Force Space Systems Division. If the Air Force dragged its feet in Inglewood, the effort would be delayed. The program would also face uncertain hazards from the possible deleterious effects on the launch vehicles of prolonged storage, from the organizational changes already made or anticipated in the management of NASA launch vehicles and launch operations, and because Ranger was to act once again as groundbreaker—using the modified configuration before anyone else. No other Air Force or NASA flight projects would share "any of the potential risks involved in flying 'first time' improvements or the burden of obtaining additional flight environmental data for [Atlas D-Agena B] vehicle configurations." It was the judgment of the JPL review board, he concluded, that the Ranger Project Manager "consider the *proper* and *complete* implementation of the Lewis plan to be a *minimum* prerequisite for the launch of Ranger 6, and if at some later date the need for more time becomes apparent, the Board would recommend that Ranger 6 be rescheduled accordingly."[10]

Even as Schurmeier considered these risks and decided to proceed with the launch vehicle improvement program for Project Ranger, NASA Headquarters had already acted on the question of launch operations. In meetings with representatives of the Department of Defense and the appropriate Air Force commands, NASA Administrator Webb and his associates successfully negotiated an *Agreement Between the Department of Defense and the National Aeronautics and Space Administration Regarding Management of the Atlantic Missile Range of DOD and the Merritt Island Launch Area of NASA.*[11] This bilateral treaty between the United States military and civilian space agencies called for major changes in the conduct of launch operations at Cape Canaveral. It was a breakthrough for NASA.

The Air Force would continue as the single manager of Cape Canaveral under the terms of the agreement, but complete responsibility for logistics and administration in the Merritt Island launch areas would be assigned to NASA. Most important, however, all of the prelaunch activity, the launchings, and the postflight evaluations were to be performed solely by NASA for NASA missions regardless of whether they took place at Cape Canaveral proper or the adjacent Merritt Island complex. Hereafter NASA, not the Air Force, would command its launch operations. The 6555th Aerospace Test Wing would disappear from NASA organization charts, and so would most of the interagency offices, panels, and committees that had so often hamstrung orderly launch operations for Ranger and other NASA flight projects. Himmel's Agena Office would select the NASA organization to manage future Atlas- and Thor-Agena launches from Cape Canaveral. The Field Projects Branch of NASA's Goddard Space Flight Center, reporting to Himmel, would supervise all launch operations; the Lockheed and General Dynamics-Astronautics personnel who prepared the launch vehicles and conducted the actual operations, would now report to the Field Projects Branch instead of to the Air Force.

In the spring and fall, Himmel's hand was further strengthened. On June 14, 1963, the Air Force contract with Lockheed for procurement and modification of Ranger Block III Agena vehicles—those remaining to be built—was transferred to NASA. Then, on August 9, the USAF and NASA executed a new Memorandum of Understanding. Superseding the one of February 14, 1961, this agreement gave NASA full responsibility to procure and direct modifications to its Atlas and Agena vehicles.[12] Himmel's Agena Office would no longer "monitor" procurement by the Air Force Space Systems Division, but directly supervise this activity at Lockheed and General Dynamics.

Himmel now had the authority to match his responsibility for procuring, modifying, and launching the Atlas-Agena. Secure in the knowledge that direct control over all four Ranger system components, for the first time, would rest in the JPL Project Office, Schurmeier could proceed with requalifying the flight machinery.

REQUALIFYING RANGER: PROGRESS AND PROBLEMS

However encouraging to Schurmeier the new agreements between NASA and the Air Force, the eleven-month delay in Project Ranger was a setback for Manned Space Flight Director Brainerd Holmes and his colleagues hard at work on Project Apollo. They had counted on closeup photographs of the moon early in 1963 to confirm a landing gear design for Apollo's lunar module.[13] Now the best they could hope for was some pictorial return at the end of the year. Adding to the pressures, on April 2, 1962, the Soviet Union launched the 1350-kilogram Luna 4 toward the moon. Russian news releases hinted strongly that the robot spacecraft was a precursor to manned flight and might land. A Russian astronomer proclaimed that Luna 4, two years ahead of NASA's Surveyor soft lander, "would send back detailed reports on the most topical issue—what the moon's surface is like." Actually, on April 5, the spacecraft missed the moon and continued into solar orbit without returning pictures of the surface. "Radio Moscow cancelled a special program entitled 'Hitting the Moon' and broadcast music instead." That old devil moon, *Newsweek* reported, "appeared to be plaguing Russia as it had the United States' Ranger program . . . "[14]

In the Office of Manned Space Flight, meantime, Holmes authorized the Grumman Aircraft Corporation in Bethpage, Long Island, the contractor for Apollo's landing module, to proceed with the design of the landing gear in accord with the best available estimates of the moon's surface slopes and soil composition. In June, a deputy, George Low, explained the situation to an inquiring Senator Clifford Case of New Jersey. Apollo, Low insisted, would continue to rely on data furnished by the unmanned lunar program to confirm or modify existing designs. In a short time, he declared, "we hope to obtain much more information about the surface of the moon from the unmanned Ranger surveillance program." If the United States was to be confident that it could land the first man on the moon, Ranger had to make good, soon.[15]

Pressured or not by external considerations, Schurmeier, Kautz, and Wolfe had learned well the lessons of the Burke era, and they were not about to repeat them. Ranger 6 was proceeding on an extremely tight schedule, Wolfe warned JPL Ranger personnel. "It is intended that this will *only* be reflected in a lowered confidence in meeting the existing launch date, and *not* in lowered reliability of the spacecraft."[16] Until the launch vehicle, the spacecraft, the tracking network, and the spaceflight operations were all requalified to the satisfaction of NASA and JPL project officials, Ranger 6 would not fly.

To assist in this work, earlier in the year the staff in JPL's Ranger Project Office had been increased from three to five. Marjorie J. Boyle, an experienced statistician and Stanford graduate with a high tolerance to Ranger's pressure-cooker environment, assumed the duties of controlling and budgeting financial resources for the project. Donald R. Schienle, a business administration graduate student at the University of Southern California, took over the details of scheduling the multifarious activities in Ranger's four system components.

JPL engineers evaluated, modified, and checked out two more of the principal system components—the deep space tracking stations and space flight operations supporting Project Ranger—in the latter half of 1963. Schurmeier formed another JPL Design Review Board in May to examine their configuration, capabilities, plans, and procedures, and recommend solutions to any deficiencies that were found. The Board completed its review in early August and issued two reports that outlined and guided the remedial action taken at the Laboratory. The bulk of the changes made to the tracking net and spaceflight operations involved more stringent operating procedures, improved reporting of failures uncovered during tests, and a greatly expanded series of preflight tests. All of these network tests involving the JPL Control Center and deep space tracking stations overseas, moreover, were now to be run for nonstandard missions—in which a partial failure of the launch vehicle or spacecraft was simulated[17]—as well as normal missions.

Beyond new documentation and test procedures, changes were also made to the telecommunications equipment in the tracking and command network. Data phone lines were installed between Cape Canaveral and JPL. Ranger's composite analog telemetry signal would no longer need to be recorded and sent by plane to Pasadena for scrutiny hours later, but could be received by the Spacecraft Data Analysis Team during the countdown and the early phases of flight. A third teletype circuit was added between JPL and each of the tracking stations, which were equipped with new telemetry decommutators and teletype encoders, replacing those that had malfunctioned during the flight of Ranger 5. Finally, with the installation of a 200-watt transmitter at the tracking station in Woomera, all three deep space stations could send commands to the spacecraft.

While Ranger Space Flight Operations Director Rygh certified these alterations in extensive tests, Schurmeier and his staff busied themselves requalifying the spacecraft using JPL's new reliability and quality assurance procedures. The Ranger test models and flight spacecraft, besides using preferred parts, or similarly screened commercial parts, now benefited from more frequent design reviews, uniform standards of testing, and improved documentation to guide the work.[18] The detailed reporting of any anomalies or problems encountered in tests, in particular, operated to good effect. These "problem/failure reports," known as PFR's, were circulated widely at the Laboratory and helped generate competition among engineers and section chiefs who sought to prove the quality of their technical judgment, analyses, and corrective action. Collectively, they had determined to erase prior bad marks by adopting the attitude: "By God, *my* piece of equipment will not fail."[19]

Early test results bespoke the improvements. By mid-June a Life Test Vehicle completed thirteen sixty-six-hour lunar missions inside JPL's new 7.6-meter (25-foot) diameter space simulator, with but one subsystem failure, and validated the integrity of the basic design. A proof test model incorporating all of the changes in the spacecraft bus and television subsystem made after the flight of Ranger 5 then began its round of system and environmental tests, demonstrating that the performance of the new Ranger spacecraft met or exceeded specifications.[20] When assembly of the Ranger 6 flight spacecraft began on July 1, Schurmeier was confident that problems of design, fabrication, and testing at JPL were truly things of the past (Figure 66).

Troublesome problems, nevertheless, dogged work on Ranger's television subsystem at RCA's plant in New Jersey. JPL Television Project Engineer Donald Kindt and two colleagues, detailed to RCA to oversee development of the payload, kept Schurmeier and Kautz apprised of these potential threats to Ranger's schedule and performance. Most perplexing among them, perhaps, was the uneven resolution and erase characteristics exhibited by the vidicon tubes produced for Ranger's television subsystem. Running short of time to complete the flight package for Ranger 6, RCA issued subcontracts to three other tube manufacturers for additional support. Specialists from these firms found that the external masks on the sensitive surface of the vidicon tubes caused shadowing and affected resolution.

In the spring of 1963 RCA's Ranger team settled on another tube design employing a laminated vidicon faceplate with the mask sandwiched between two

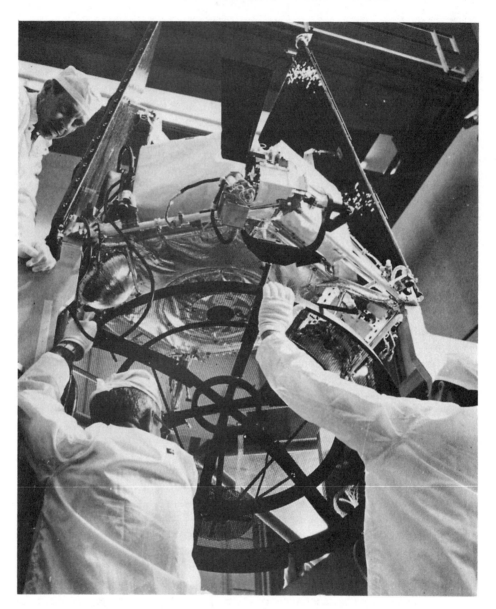

Fig. 66. Assembly of Ranger 6: Installation of the Midcourse Motor

layers. Although this helped to eliminate shadowing and improve resolution, it still did not end all of the differences in performance between individual, seemingly identical tubes. Other disturbing differences, RCA engineers now learned, resulted from system interactions that often could not be detected before a complete television package—six cameras—was assembled and tested. Standard quality control procedures in the fabricating process did not measurably improve the puzzling situation. As summer turned to fall, the production of high-performance vidicon tubes with uniform quality still seemed to be as much an art as it was a science.[21]

Electrical arcing and short-circuiting in the television subsystem was another, potentially more serious problem. Among Ranger's engineers, arcing in electronics was the single most feared agent of flight failure. Thermal-vacuum tests at RCA in February revealed arcing in the cables and connector associated with the redesigned television package's 60-watt power amplifiers. Early investigations pinned the cause on changes in the design of the amplifiers that "increased the power-output of this assembly." Plans were made to modify the affected cables and test them to a 90-watt power level before installation. Despite these and other modifications, however, arcing persisted.[22] At the Ranger Television Quarterly Review meeting on May 2, RCA's Ranger television manager Bernard Miller pledged to purchase high-power test equipment and devote maximum attention to the problem. The procurement and test cycle, nevertheless, would require 45 days, which closely approached the scheduled delivery of the flight package for Ranger 6 at the end of July.[23] Schurmeier agreed to slip the delivery date to August 15, to provide RCA the time needed to complete the test program and qualify the television subsystem.

As if these development problems were not enough to occupy the hard-pressed television engineers, especially severe vibrations and noise at launch, Miller speculated, might step a relay and turn on one or both channels of the television subsystem. If not turned off, the batteries powering the television cameras would be depleted long before reaching the moon.[24] Donald Kindt and others at JPL analyzed Miller's theory in succeeding weeks. The vibration and noise necessary to activate the television channels, they concluded, were above the levels so far encountered or expected during launch. Nevertheless, the JPL project office made precautionary plans to transmit a series of real-time commands to the spacecraft from the Johannesburg station, shutting off the television subsystem in the unlikely event that it had been turned on early in the flight.[25]

The flight package with its six cameras, meantime, completed the extra high-power tests at RCA and was accepted for shipment to JPL in mid-August. The arcing difficulties had been traced to the "four-port hybrid," an electrical device that combined the radio frequency signals from both TV transmitters and routed them to a common antenna. The observed corona effects had been eliminated by redesign. JPL's TV subsystem engineer, Kindt, and his RCA associates were satisfied that the arcing problem had been solved and that the basic design of the split TV system was a sound one. Ranger Spacecraft System Manager Wolfe agreed, and, in the interests of time, waived a final design review of the RCA TV subsystem (Figure 67).[26]

But no answer to the erratic picture yield of the vidicon tubes had proved altogether satisfactory to JPL or RCA. This nagging difficulty had been skirted by fabricating additional tubes and then selecting the best and most uniform of the lot

Fig. 67. The Television Subsystem Is Readied for Tests at JPL

for Ranger 6. NASA's Ranger Program Chief Cunningham, however, remained concerned over the chances of procuring enough acceptable tubes in time for the flights to follow. After conferring with Schurmeier at JPL, he recommended that NASA delay the schedules of Rangers 7, 8, and 9 to ensure "that these missions have at least as good a TV system as Ranger 6."[27] Nicks and Newell agreed; all

Ranger pictures for Project Apollo and for science were to be comparable in resolution and detail. On September 10 they amended the Ranger Block III launch schedule to allow time to fabricate more vidicon tubes: [28]

Ranger 6	Early December 1963 (no change)
Ranger 7	Postponed from January to February 1964
Ranger 8	Postponed from March to May 1964
Ranger 9	Postponed from April to July 1964

At JPL, Ranger 6 and the flight television subsystem began the final round of qualifying tests—on schedule. But a new problem, uncovered by Himmel's Agena team in Cleveland, now derailed all launch plans. JPL personnel first learned of the problem on September 14, during a review of the Atlas guidance improvement program at Lewis Research Center. Two serious test failures in the General Electric guidance components had resulted from short circuits caused by loose gold flakes in certain diodes. The diodes in question, each the size of a grain of rice, acted as tiny electrical switches, permitting current to flow only in one direction. If shorted, they permitted current to move in either direction with potentially disastrous consequences. Hundreds of these same diodes, unhappy JPL engineers realized, were already installed in Rangers 6 and 7.

Schurmeier was incredulous. How, he asked, could flight equipment, almost certainly containing many contaminated diodes, have passed all JPL tests without failure? The event touched off a major investigation to answer that question, assess the extent of the damage to Ranger 6, and establish plans to remedy the situation. This investigation, completed in early October, showed the incidence of gold-flake contamination in these diodes to be so high that most equipment containing them was unfit for flight. The flaking originated from poorly bonded excess gold cement at the attachment of a silicon wafer that supported the post inside the diode. The condition had escaped detection, Schurmeier learned, because system and environmental tests did not—in fact could not—continously measure the reverse-current resistance of every diode, and if a temporary short occurred while forward voltage was applied, there was no measurable effect. Finally, no tests could simulate the zero gravity of outer space, where the gold flakes would float inside the diodes.[29]

Schurmeier had no alternative. All of the suspect diodes purchased from Continental Devices and contained in critical locations in the Ranger spacecraft had to be replaced. He directed that this work begin, and that procedures for ordering, inspecting, and sample-testing diodes and other electronic parts used in the Ranger spacecraft be revised to prevent any recurrence. On October 18, Pickering, Parks, and Schurmeier told NASA officials in a meeting at Headquarters that to assure success of the lunar missions, the flight dates of Rangers 6 and 7 had to be postponed until all of the questionable diodes were replaced in these spacecraft and in the General Electric guidance components used on the Atlas launch vehicles.

Homer Newell and his Ranger associates in the Office of Space Sciences agreed to the proposal. Three days later they postponed the launch of Ranger 6 two lunar months (from early December 1963 to late January 1964), and that of Ranger 7 one month (from February to March 1964). The flight dates of Rangers 8 and 9 remained as posted on September 10. Newell instructed JPL to use the additional time to replace the diodes, procure additional spare flight equipment, and finish corrective action on all outstanding problem/failure reports.[30]

Project Apollo and Ranger's scientists would simply have to wait a few months longer for closeup pictures of the moon.

Chapter Thirteen

SPACE SCIENCE: A NEW ERA AND HARD TIMES

THE reprogramming decisions of December 1962, along with shifting the scientific emphasis of the Block III missions once more toward support of Project Apollo, saddled NASA Space Sciences Director Homer Newell with a number of pressing tasks. There was the unpleasant duty of notifying sky science experimenters that their instruments were eliminated from Rangers 6 through 9. There was the requirement of completing, then implementing, the remaining experiment—the RCA television cameras. And finally, there was the necessity of setting, if the revised project was to mesh neatly with the complete NASA lunar program, the number and scientific content of the Ranger spacecraft planned to follow these flights.

GUARANTEEING SUPPORT TO PROJECT APOLLO

At the Senior Council Meeting of the Office of Space Sciences on January 3 and 4, 1963, Newell acquainted all of his field center directors with the changes in policy made in December. Six months before, he had assured them that the requirements of space science would take precedence over those of Project Apollo in all of the unmanned missions for which he was responsible.[1] Now, because of intervening events, he said, the Office of Space Sciences would further support the manned lunar missions by removing from the unmanned flights whatever secondary experiments might interfere with this goal. Ultimately, NASA intended to rely upon manned technology to explore the moon for science.[2]

On January 21 Ranger's experimenters were notified of the scientific cancellations on the Block III missions. Ranger Program Chief Cunningham handled the task as diplomatically as possible in a meeting at JPL, announcing that although sky science experiments were to be eliminated from Rangers 6 through 9, the experimenters and their instruments would remain a part of the project. NASA Headquarters intended to reschedule them as soon as possible in later Ranger missions. Although "the majority of the experimenters were understandably disappointed," Cunningham informed Newell, "they appeared to appreciate the program problems" entailing the cancellations.[3]

The imaging objective of Rangers 6 through 9 and the tentative requirements for Project Apollo were defined soon afterwards. On January 28 Ranger television experimenters Harold Urey, Gerard Kuiper, Eugene Shoemaker, and others convened at JPL to consider these matters in view of the preceding changes. The best photographs of the moon taken through telescopes on earth afforded a resolution at the lunar surface on the order of 300 meters (1000 feet). "Significantly better" pictures of the surface taken from Ranger spacecraft, the group decided, demanded an improvement by at least an order of magnitude, or by a factor of ten. But pictures with a resolution of 30 meters, though of value to planetary science, would not afford the closeup detail of surface slopes needed to confirm the landing gear design of the Apollo lunar module. For that purpose the experimenters settled on a best resolution at the surface of 1.8 to 3.0 meters (6 to 10 feet) in the final frames. That much resolving power was claimed for the telephoto cameras in RCA's television subsystem, and that was to be the goal. The experimenters advised NASA and JPL project officials that the trajectories prepared for the first day of each Ranger launch period should permit achieving this resolution without any terminal maneuver of the Ranger spacecraft.[4] If the needs of Project Apollo differed greatly from this anticipated imaging goal, it was up to Apollo's managers to specify them.

On March 1 the Office of Manned Space Flight issued a completely revised *Requirements for Data from Unmanned Spacecraft in Support of Project Apollo*, this time based on prior consultation and cooperative planning with the Office of Space Sciences.[5] The data needed from Newell's unmanned lunar missions were carefully defined. Ranger's television resolution at the moon's surface was deemed sufficient to confirm the design of the landing gear. Beyond considerations of design, however, Apollo required other "critical" information about the intensity of radiation and the distribution of micrometeoroids in cislunar space, and on lunar surface features and contours, to determine actual probabilities of failure. NASA, the report noted, could only proceed confidently if there was "less than one percent probability . . . of disaster" involving the loss of a crew.

Because design of Apollo's command, service, and lunar landing modules already had begun but could be altered to some extent later, the time to provide this information was extended until the fall of 1965. And, rather than ordering compliance, the Office of Manned Space Flight tactfully "requested that the Office of Space Sciences review the document and state its plans for obtaining the data..." If all of the information could not be obtained in time, the Office of Space Sciences could state "the limits of expected capabilities...so that new means of obtaining data may be generated or, as a last resort, the requirements for support from the unmanned program may be changed."[6]

Newell and his Lunar and Planetary Program Director Oran Nicks were reasonably confident that the unmanned projects could provide Project Apollo the specified data. However, the slow progress of planning for Apollo's scientific experiments and objectives, particularly at the Manned Spacecraft Center in Houston, the field center responsible for the manned lunar mission, increasingly concerned them.[7] Having committed the Office of Space Sciences to support Apollo engineering, at least for the near future, they expected a quid pro quo: Apollo's astronauts to explore the moon for science. In Washington, the Office of Manned Space Flight might agree, but its manned Spacecraft Center had yet officially to endorse that objective.

At the Manned Spacecraft Center there was little evidence of any scientific planning. Holmes and Low at Headquarters set down the policies and plans for Project Apollo, Center Director Robert Gilruth and his associates in Houston were responsible for allocating and supervising the enormous engineering tasks of manned lunar flight and completing the designs of Apollo machines. As specified in the Requirements Document, engineering for Apollo allowed no tolerance for failure. The colossal repercussions for NASA and the nation that would follow the loss of a crew were simply unacceptable; guaranteeing the safety of the astronauts dominated all considerations. Consequently, Gilruth seemed to be saying to space scientists: "Wait until the Apollo machine is perfected and the hazards to man are calculated before discussing any program of experiments."

The continued neglect of science in Project Apollo offended others who counted themselves among the nation's scientific spokesmen. Increasingly in 1963 some influential scientists began to criticize Apollo's pervasive engineering flavor. Caltech President Lee A. DuBridge, speaking before a Panel on Science and Technology of the House Committee on Science and Astronautics in January, declared it imprudent to send jet pilots to the moon just to "look around." If they were to return without any scientific information, the program would not be "worth 1 percent or even a tenth of a percent of our national budget..." Space science was the proper goal. NASA and the Congress, he observed, had to have "a broader and deeper view toward the space program than only to beat the Russians...to the moon."[8] The former president of the American Association for the Advancement of Science Warren Weaver and the chemistry Nobelist Linus Pauling joined with DuBridge to berate the competitive aspects of the program and the low station accorded science, and to discuss other uses that could be made of Apollo funding.

Philip Abelson, Director of the Geophysical Laboratory at the Carnegie Institution and editor of the influential *Science,* the journal of the American Association for the Advancement of Science, published a mordant editorial on the subject. Surveying the rationale advanced to justify Apollo, he pronounced the propaganda value of placing a man on the moon vastly overrated. "Interest in lunar exploration will be sustained only if there are important military implications, exciting scientific accomplishments, or technological fallout." The advantages of a military base on the moon or new technology for civilian uses, he asserted, did not justify the immense costs of the program. Scientific exploration of the moon, though certainly justified, had "been accorded a secondary priority in the lunar program. This has been indicated in the attitude surrounding presentation of the new budget to Congress and underlined by the decision not to have a scientist in the first lunar astronaut crew." Unmanned vehicles, on the other hand, could explore the moon at a cost "on the order of 1 percent of the cost of the manned variety... A reexamination of priorities," he insisted, "is in order." Warming to the task in a speech at the University of Maryland a few days later, Abelson declared that science had been "used" by NASA "as a 'front' for technological leaf raking" in Project Apollo.[9]

Of course, in the spring of 1963, Apollo did have its influential scientific friends. Physicist James Van Allen, astronomer Lyman Spitzer, chemist Willard Libby, and geneticist Joshua Lederberg, among other luminaries who had worked with Newell's team,[10] supplied a testimonial statement in support of the manned lunar landing. But their colleagues merely pronounced pro-Apollo scientists "captives" of the space agency. Worse, the Office of Space Sciences commanded neither the funds nor the flight opportunities necessary to satisfy both sides in the escalating scientific contention over Project Apollo.[11] Even Holmes' declaration that Apollo's mission objectives included "manned scientific exploration of the moon"[12] failed to allay the suspicions of the unmanned enthusiasts. After two years of Apollo planning, scientists like Urey expected action, not words, from the space agency. NASA Administrator James Webb, hoping to discourage what he judged to be intemperate attacks, fueled them instead by publicly criticizing, among other scientists, Caltech President DuBridge and Warren Weaver, Vice President of the Alfred P. Sloan Foundation. These gentlemen, he said, set single motives for exploring space above all other considerations; their misgivings were thus "magnified out of all proportion to reality."[13]

Deprived of a major role or voice in Apollo, many scientists had begun to perceive its awesome budget as a malignant tumor, consuming such other, more favored scientific endeavors as unmanned lunar exploration. Future Ranger missions were still programmed for scientific purposes. If some measure of scientific favor and healthful balance could be restored, it remained for Homer Newell and the Office of Space Sciences to achieve it.

PLANNING IN THE FACE OF CHANGE

Plans made at NASA Headquarters in the summer and fall of 1962 called for five additional RCA television missions, Rangers 10 through 14, to be launched in 1964, followed in 1965 by at least four more hard-landing capsule missions, Rangers 15 through 18. Designated Blocks IV and V, respectively, these JPL-built spacecraft were scheduled to be flown to points on the moon of interest to Ranger's scientific clients. On October 3, 1962, the Office of Space Sciences authorized JPL to begin the design of flight hardware for the Block IV vehicles,[14] but a week later, on October 11, the objectives were modified.

Meeting at Headquarters on October 11, Seamans, Newell, Holmes, and Pickering placed more emphasis on Apollo's requirements for lunar data.[15] When Newell publicly announced the extension of Project Ranger a few days later, he observed that the additional television missions would "increase the probability of obtaining lunar surface detail information that could be used in the manned landing system design."[16] The proposed hard-landing capsule missions still under study remained committed to scientific objectives. The Office of Space Sciences instructed the Aeronutronic Division of the Ford Motor Company, having already developed the seismometer capsule for the original Rangers, to commence work on a prototype capsule that could hard-land a small photo-reconnaissance television camera on the moon's surface.[17]

More Ranger missions might furnish lunar data for Apollo's engineers as well as planetary scientists, but neither of these groups was expected to pay for them. Homer Newell and the Office of Space Sciences had to arrange for the funding. Prospects for obtaining that money in Congress unquestionably dimmed on October 18, 1962, when the flight of Ranger 5 brought the entire project into question. The subsequent delay and requalification of the Block III Rangers 6 through 9, furthermore, increased the costs of this portion of the project from an anticipated $64 million to $118 million, and cut heavily into existing funds that might otherwise have been programmed for the follow-on missions. Without doubt, if the needed funds were to be had, Congressman Karth and his colleagues in the House Subcommittee on Space Sciences and Advanced Research and Technology would have to be assured that Ranger's problems were solved.

Newell, Cortright, and Nicks considered these and other matters during the Ranger reprogramming meeting at NASA Headquarters on February 12 and 13, 1963, when they approved of the changes and plans proposed by JPL for prosecuting Rangers 6 through 9. They were unquestionably anxious for the success of these and other flights to follow. But the Ranger spacecraft was no longer the novel design it had been in 1959 and 1960. And as the Kelley Board had pointed out, the serial production of some eight or ten more Rangers was a task best suited to an industrial firm rather than a university-affiliated research and development laboratory. The Jet Propulsion Laboratory, results thus far seemed to indicate, had had a hard enough time designing and building the first five machines.

At JPL, Sparks and Schurmeier agreed with their NASA colleagues to engage an industrial firm both to furnish engineering support on the Block III and IV Rangers and to fabricate all of the Block V spacecraft. Although the large number of temporary trailers located at JPL would have to be increased even further to house the necessary contractor personnel, industrial capabilities could be expected to alleviate the manpower shortage at JPL, permit Schurmeier's project office to devote maximum attention to rebuilding the Block III machines, and meet the demands of producing and qualifying large numbers of Ranger spacecraft.

At the Ranger Quarterly Review in Pasadena on February 19, more of the details were defined, and the Block IV schedule of activities was established:

mission objectives were to be available on March 4, spacecraft preliminary design on April 15, and final drawings were to be finished and the procurement of hardware initiated on July 1, 1963. The rationale for so large a number of lunar impacting television missions planned for Ranger, already a sore point with many planetary scientists, was also reviewed—and found wanting. NASA and JPL officials decided to reduce Block IV from five to three television flights (Rangers 10 through 12), to be launched in July, August, and September 1964. The number of Block V hard-landing capsule spacecraft, to be built by a contractor, was increased from four to six (Rangers 13 through 18), all to be launched in 1965.[18]

In keeping with the planning to date, Pickering furnished Headquarters with the Laboratory's proposal "for bringing an industrial contractor into the Ranger Program." It called for the competitive selection of a qualified firm, and, in the first phase of a two-phase contract, the assignment of their personnel at JPL to provide engineering support in Project Ranger during the fabrication and testing of the Block III and IV vehicles. If the work was judged acceptable, the firm would become the spacecraft system contractor in Phase 2, responsible for the assembly, test, and launch of the six Block V vehicles. JPL would continue to provide technical direction through Schurmeier's project office. Given the proposed launch dates, Pickering advised NASA, a Source Evaluation Board made up of NASA and JPL personnel had to be formed "as soon as possible" to select the contractor.[19]

Time was indeed crucial, and not only for the reason of Ranger's schedule. Newell and Cortright had to appear before Congress in two weeks to testify in favor of the fiscal 1964 authorization for the Office of Space Sciences. The naming of a contractor to assist JPL would complete action on the Kelley Board recommendations for Project Ranger and buttress the case for proceeding with Ranger Blocks IV and V, but soliciting bids and awarding the contract on a competitive basis as urged by JPL could not possibly be accomplished before the hearings began. Newell and Cortright consulted NASA Deputy Administrator Robert Seamans. Time could be money in this instance, and together they decided on a sole-source award to the Northrop Space Laboratories in Hawthorne, California.

Northrop was presently building Ranger's attitude control system and the central computer and sequencer for the Block III spacecraft; in the opinion of NASA's leaders, it was also a qualified firm, and was conveniently located near JPL. Moreover, a number of Northrop engineers recently released when a missile contract had been cancelled by the Air Force were available for immediate assignment, and Northrop Corporation's Senior Vice President Richard E. Horner, responsible for the division in question, had once been NASA's Associate Administrator, and knew JPL.[20] Of course, more disinterested observers might have seen the sole-source contract as a dubious practice in the public interest, but this consideration was overlooked as NASA officials sought immediately to complete the actions called for by the Kelley Board.

On March 5, Pickering was summoned to NASA Headquarters in Washington. There Newell and Seamans informed him of NASA's decision to proceed with a sole-source award. Surprised and dismayed, Pickering protested. JPL was not opposed to contracting the Ranger work—indeed it had proposed such a move a year earlier and had no special objection to Northrop—but this kind of Headquarters fiat, he asserted, had never before been exercised under the terms of the NASA–Caltech contract and operating agreement. It left the Laboratory with no voice whatever in choosing the firm with which it must work, and it put NASA at some disadvantage in pricing the contract. Seamans and Newell were unmoved; Pickering's objections were overruled. NASA intended to direct the unmanned lunar program, and Northrop had been selected as the firm to participate in Project Ranger.

Upon returning to Pasadena, Pickering hesitantly considered appealing the decision.[21] After talking further by phone with Cortright and with his own advisors, however, he dropped the idea. It could only strain relations even more between the space agency and its one contract-operated laboratory. Aware of the reluctance to proceed on the West Coast, Newell dispatched a terse letter to Pickering on March 8: Confirming the decision reached in the meeting at Headquarters on March 5, "the Jet Propulsion Laboratory is directed to immediately issue to the Northrop Corporation...a Letter Contract," to be made final "as expeditiously as possible, establishing the Northrop Corporation as Ranger spacecraft system contractor" for the follow-on missions.[22]

That afternoon NASA publicly announced the selection of the Northrop Corporation. The estimated $2-million Phase 1 contract, the statement declared, assured JPL industrial assistance in the "work related to the Rangers to be launched through the end of 1964." Eventually, NASA expected Northrop to be "assigned complete spacecraft system responsibility for the Ranger program in a Phase 2 effort, beginning with Rangers to be launched in 1965."[23] In Pasadena, the Jet Propulsion Laboratory issued Letter Contract 95059 to the Northrop Corporation.

MAKING A CASE FOR MORE RANGERS

In the spring of 1963, when the space agency was asking for a budgetary increase from its current $3.7 billion to a new $5.7 billion, Newell and Cortright faced tough questions about Project Ranger in the appropriations hearings before Joseph Karth's subcommittee. The questions focused on the performance of the Jet Propulsion Laboratory. The problems uncovered in Pasadena appeared to be "probably as much management as technical" in nature, Chairman Karth observed. Exactly what management problems, he inquired "were brought to your attention by the Kelley report,...and what has been done about it?"[24] Cortright reviewed the findings and recommendations of the Kelley Board, and recent changes at JPL, and did his best to defend the overall performance of the

Laboratory. The award of a contract to the Northrop Corporation to assist JPL in Project Ranger, Cortright affirmed, completed action on these recommendations.

Committee members, nevertheless, remained skeptical. Newell and Cortright were faulted for the delay in engaging industrial support for Project Ranger until four months after the Kelley Board report became available.[25] When they left the hearing they could not be certain that the $90 million requested for Project Ranger in fiscal year 1964 would be approved. Indeed, NASA officials appearing before other committees and subcommittees in the House of Representatives found themselves on the defensive as well, and by the end of the week reports were circulating that the agency might face a reduction in its budget for the first time.[26]

Reservations about Ranger and its objectives were not confined to Washington. In the spring of 1963, some project engineers in Pasadena had also begun to join experimenters who openly questioned the wisdom of more Ranger television missions. Deloyce Alcorn, recently appointed JPL Block IV Spacecraft System Engineer, raised the issue with Ranger Spacecraft System Manager Wolfe. Considering the proposed objectives and the design of the Block IV spacecraft, Alcorn mused, "the apparent reason for the missions is becoming less and less clear. The scheduled launch dates dictate [that these] spacecraft incorporate only slight improvements and the addition of three or four experiments. If there are one or two successful Block III flights, the value of Block IV is extremely questionable. If there are no successful Block III flights, does it make sense to launch more of the same machine?"[27]

Ranger's objectives had often been a sore point with Burke, and they were fast becoming a source of concern to his successor, Schurmeier. On March 18 Program Chief Cunningham sent JPL the wording of the objectives preferred at NASA Headquarters. The mission objective of Ranger Block III was to return "television pictures of the lunar surface which will be of benefit to both the scientific program and the U.S. manned lunar program. These pictures should be at least an order of magnitude better in resolution than any available earth-based photography. Should the requirements of the manned lunar program conflict with the scientific requirements, every consideration will be given to meeting the manned lunar program needs." The fundamental objective of the Block IV Ranger missions was "to acquire significant new knowledge of the moon. Specifically, the objective is to obtain by means of high quality television photography and by means of non-visual techniques, new and useful data concerning the lunar topography, environment, and composition."[28]

Schurmeier was satisfied with the Block III objectives, which hewed closely to the wording agreed upon at NASA Headquarters in December 1962. But he was disturbed that the objectives for the follow-on Block IV did not specify support of Project Apollo, as expressed in the public announcement of these flights.[29] Without such a specification, these missions might also become a center of friction between Ranger's scientists and Apollo's engineers. On April 1, 1963, Schurmeier submitted to NASA Headquarters his own attempt at wording the objectives. The television pictures returned to earth by the Block IV Rangers 10 through 12, he insisted, should be recognized as benefitting "both the scientific program and manned lunar program."[30]

During the following week Cunningham and Nicks considered JPL's recommendation. Although space science had been the motive for extending Project Ranger and proposals for secondary experiments already had been solicited, the original announcement of the Block IV missions could not be denied. On April 8 Schurmeier welcomed the final wording approved by Nicks:

> The primary objective of the Ranger Block IV flights is to obtain television pictures of the lunar surface which will be of scientific significance, and which will contribute to the United States manned lunar landing program. These pictures should be of higher quality than those of Block III and should provide coverage of as many sites as possible within the Ranger spacecraft capability.
>
> The secondary objective is to obtain by means of non-visual techniques significant new and useful data concerning the lunar topography, environment, and composition.[31]

A few days later, in choosing the nonvisual secondary experiments for the Ranger Block IV spacecraft, NASA's Space Science Committee passed over sky scientists in favor of their planetary colleagues. The Committee selected James Arnold's gamma-ray spectrometer and Walter Brown's pulse radar as the scientific passengers to complement the RCA television cameras. In making known their choice to Homer Newell, however, Committee members expressed misgivings over "the subordinate role in which the non-visual experiments have been placed" in the "objectives established for these flights" by the Office of Space Sciences.[32]

As though responding to these misgivings, Newell observed a few days later before the American Society of Newspaper Editors that opinion was divided over the merits and motives of NASA's space science program. The program was now viewed variously as advancing knowledge, sowing the seeds for future applications, or supporting either the manned flight program or military efforts. Newell, consistently in character, asserted: "Such discussions [imply] that the science program would be different if the motivation was different. However, it is our conviction that the kind of research that best supports any of these objectives is a good, sound, scientific program."[33]

On April 25 Newell added to the fretful Ranger experimenter Harold Urey that it was too soon to tell conclusively whether pictures of the moon's surface would prove of greater scientific value than nonvisual experiments. "We have asked the Space Science Board [of the National Academy of Sciences] a number of times for its recommendation in this area, and each time the answer has been that first priority should go to obtaining pictures." Undoubtedly, there would be "many questions about the lunar surface that will not be answered by the taking of pictures, and that, indeed, in the long run will be more important scientifically than the early pictures are likely to be. These measurements," Newell pledged, "will most certainly be made in the course of the [Ranger] program, and they will be made just as soon as we have the opportunity to do so."

Newell also explained to the Ranger scientists that they were to be given greater opportunity at JPL to influence the design and use of their experiments. That included Arnold and Brown on the nonvisual gamma-ray experiment and the radar surface properties experiment, as well as the experimenter team associated with the television cameras on Ranger Block IV. "In this block of Rangers, then, we will be placing greater emphasis on scientific measurements, and it is our understanding that the experiments selected are those that you and our other advisors would like to see done."[34]

The next day Newell notified JPL Director Pickering of the additional experiments selected by the Office of Space Sciences for Ranger Block IV, and of his desire to see that these instruments not be subordinated to television photography in the conduct of the missions. "While I concur in the importance of the television system," he said, "it should be recognized that we have not de-emphasized the importance of obtaining scientific measurements which can be provided by non-visual experiments. In light of these reservations, we will want to discuss possible modifications of the existing [Block IV] objectives with your Ranger Project management."[35]

Schurmeier was truly disturbed. Inconstant project objectives modified at the insistence of Ranger's scientists, he was certain, had contributed mightily to Burke's woes a year before. Jack James, the Mariner R manager, could unreservedly agree with the sentiments that Schurmeier expressed in an immediate telegram to his NASA counterpart, Cunningham:

> The mission objectives are the guiding statement for all elements of the project. They should be clear, concise, and unambiguous. If they are unclear or fail to resolve known issues, then various people or organizations will necessarily interpret them differently and as a result, confusion, inefficiency, misunderstanding, and often hard feelings will reign.
>
> If there are multiple objectives they should either be stated in the proper priority order or if they are equal it should be so stated...
>
> As problems arise throughout the course of a project...the mission objectives are referred to as an aid—in fact the controlling criteria—for the proper tradeoff in specific decisions.[36]

If the Office of Space Sciences failed to appreciate this rationale and intended to change the Block IV objectives, then, Schurmeier was also certain, Project Ranger could expect serious trouble once again. But if Schurmeier hoped for help from Cunningham, he hoped in vain; by late spring of 1963 Ranger Block IV was beyond all help from NASA or JPL.

LUNAR ORBITER AND CONGRESS INTERVENE

For some time Edgar Cortright, Newell's deputy at Headquarters, had been keenly interested in a lunar orbiter—a spacecraft placed in orbit about the moon, capable of mapping its surface and conducting other scientific experiments. Original plans had called for JPL to modify the automatic Surveyor lunar soft-landing machine to meet the requirements of an orbiting vehicle. However, these plans had been scrapped during the reprogramming meetings in December 1962 when JPL reported that the technical demands of such a conversion militated against Surveyor orbiter, and that the Laboratory, furthermore, was in no position to provide the manpower needed for another flight project.[37]

On April 25, 1963, considering the prospects of future unmanned flight projects, Cortright argued to Newell that the time was ripe for beginning a lunar orbiter. The Langley Research Center in Hampton, Virginia, he observed, could provide the management "and is highly motivated to do so." Also, the Apollo project might plan a lunar photoreconnaissance mission if the Office of Space Sciences did not,[38] and, as far as surface coverage was concerned, one successful orbiter would be worth "dozens of successful Ranger TV impacters." Under the circumstances, Cortright recommended that Ranger Block IV be canceled and that Block V be continued at a reduced level of effort, as a backup in case the Surveyor Project faltered and it became "necessary to press Ranger into service at some later date." Among other advantages, this course would save $50 million in fiscal year 1964, permit the start of a lunar orbiter, and cushion the budget shock in the event of a "Congressional cut."[39]

In Pasadena, where manpower was still very much at a premium, Schurmeier faced difficulties of his own in meeting commitments on the approved Ranger Block III, let alone Block IV. Even with 63 of an eventual 125-man contingent of engineers from the Northrop Corporation already in residence by the end of May, he had barely enough men to requalify Rangers 6 through 9 at JPL and RCA. These Rangers commanded priority; Schurmeier and JPL Lunar and Planetary Program Director Robert Parks agreed that if they were to be completed on schedule, Ranger Block IV had to be delayed.[40]

On June 17 they made that recommendation to Newell, Cortright, and Nicks during a budget review meeting in Washington. Though unaware of the thinking at Headquarters, Schurmeier and Parks' findings coincided neatly with Cortright's suggestions. A few days later Schurmeier announced the resulting decision to all project personnel. "As an outcome of the current budgeting and reprogramming activities being conducted...with Headquarters," he said, "it has been mutually agreed that the Ranger 10 schedule will be slipped six months. The timing of flights subsequent to Ranger 10 has not been established... A moratorium should be placed on Block IV work. All possible effective effort should be applied to Block III."[41] With or without industrial support, Block IV was barely alive.

On June 27 the ax fell on the quivering remnants. Karth's subcommittee sent the budgetary authorization for the Office of Space Sciences on to the House Committee on Science and Astronautics. The report recommended that Ranger's fiscal 1964 budget be slashed from the $90-million requested to a sum of $65-million. "Of the five Ranger spacecraft already launched," Representative Karth stated in announcing the disallowance, "none has been a...success. The so-called Kelley report on the Ranger Project casts grave doubts upon the adequacy of the management of this project, both by NASA Headquarters and the Jet Propulsion Laboratory... The subcommittee feels that in view of the poor record of Ranger to date, Congress should be given reasonable assurance of success before going forward full speed with more spacecraft."[42]

Newell and Cortright knew that the full Committee and the House itself could be expected to endorse Karth's report. They also believed that the Senate was unlikely to restore the funds—a belief soon substantiated when the Senate concurred in the House authorization. On July 2 and 3, JPL officials again flew to Washington to present their recommendations for the future of Project Ranger. Given the increased costs, the anticipated budget cut, scheduling difficulties, and the similarity between the two blocks of television missions, Schurmeier and Parks advised canceling Ranger Block IV. However, they urged continuing the Block V hard-landing capsule missions with launches in 1965 and 1966. Cortright, Nicks, and Cunningham presented that same recommendation to NASA Associate Administrator Robert Seamans a few days later, on July 9, advising as well that a lunar orbiter project be authorized. Seamans agreed. Indeed it was most desirable, he said, that NASA "have both a Ranger Block V and an Orbiter Program."[43]

On Monday, July 15, Newell formally notified Pickering. The extremely tight fiscal situation and the potential of the Surveyor and Orbiter programs, he explained, caused NASA to direct the Laboratory to "terminate all efforts [on] the impacting TV missions beyond Block III." Maximum effort would be devoted to Rangers 6 through 9, with the intent of achieving a flight success by early 1964. Work on Ranger Block V could proceed; however, he also instructed JPL to "evaluate the implications of terminating the Ranger Project after Block III,...and to prepare a plan to handle this eventuality..."[44]

Shortly thereafter, NASA announced commencement of the Lunar Orbiter Project, to be managed by the Langley Research Center. Designed to photomap the moon's surface at a resolution equivalent to the best Ranger pictures, Lunar Orbiter would both benefit science and permit Apollo's managers to select appropriate landing sites. One successful orbiter would furnish "1/2-million times the area coverage of a Ranger TV impacter." Orbiters rather than Rangers were to "take over [moon] flight photography in 1965 and beyond."[45]

Planetary scientists expressed no overt dismay at the passing of the Ranger follow-on television missions. James Arnold and Walter Brown both anticipated selection of their nonvisual experiments for flight in Ranger Block V. But the mood in Congress offered small comfort for the future, and Ranger's pathetic record still engendered recrimination among scientists and engineers. *Science* summarized: "The House Committee's rough handling of Ranger certainly reflects a new militancy in judging NASA's plans and performance,...five straight failures not only embarrassed the space agency and frustrated the scientists who had worked hard on the Ranger experiments, but also disappointed the scientific community at large."[46]

Chapter Fourteen

MORE MISSIONS FOR SCIENCE?

W HETHER planet or sky, the long-promised return for the nonvisual
experimenters in Project Ranger now had to be met by Block V or not be
met at all. "It was clearly anticipated," to be sure, that these flights would conduct
"most of the scientific program of Ranger."[1] But with Apollo demanding an ever
larger share of the NASA budget, earmarking the funds for more scientific
Rangers was becoming an increasingly worrisome problem.

BLOCK V UNDERWAY

On May 6, 1963, Schurmeier's office issued the *Project Description,
Spacecraft System Contract, Ranger Block V* that described the tasks and
requirements expected of the spacecraft system contractor, Northrop Space
Laboratories. At the same time, the project office asked the parent corporation to
prepare a proposal for the second phase of work covering design, fabrication,
assembly, test, and launch of the Block V machines. On June 10 Northrop
submitted the proposal and program plan, estimating the total cost for six
spacecraft to be $72.4 million, including a 6.8-percent fee for the company. JPL
officials judged the tentative plan acceptable and, pending definition of some
spacecraft technical details and negotiation of a precise cost, authorized Northrop
to begin Phase 2.[2] With this approval, Ranger became the first major spacecraft
contract Northrop held with NASA—a solid foot in the door of the growing
civilian space market.[3] That the firm's performance undoubtedly would affect
future business with the space agency went without saying. The Northrop
management, determined to take every measure to assure the satisfactory
completion of this project, selected as its head V. William Howard, a former
associate professor of mechanical engineering at MIT who had successfully
managed a number of missile and aeronautical projects for the company's Ventura
Division (Figure 68).

Fig. 68. Northrop Manager of Project Ranger William Howard

After the cancellation of Block IV on July 12, Northrop engineers at the Jet Propulsion Laboratory assisted in requalifying Rangers 6 through 9. Although reporting administratively to Howard, they acted as functional members of the project team, receiving day-to-day direction from JPL supervisors in the Systems, Telecommunications, Engineering Mechanics, and Quality Assurance and Reliability Divisions. And while becoming familiar with the project in Pasadena, they also prepared to assume responsibility for the complete Block V spacecraft system. Howard and other Northrop engineers, meantime, busied themselves with this important second phase: planning and implementing Ranger Block V.

In his project office in Hawthorne, Howard began a Ranger spacecraft system design evaluation. This work involved analyzing the Ranger spacecraft, determining its potential for successfully performing the Block V mission of hard-landing a capsule on the moon's surface, and recommending changes to the design.

The final product, a set of spacecraft "Design Evaluation Recommendations," was to be returned to JPL in November. Schurmeier provided guidelines that limited what could be submitted; time did not permit the creation of a new spacecraft. The existing design would be modified only as required to improve reliability and complete the mission.[4] With a final design approved by JPL, a firm program plan and performance specifications could be issued, and a definitive cost negotiated. All of these details, Howard knew, had somehow to be wrapped up by the end of the year if Northrop hoped to meet a first launch date in 1965. Before any design could be made final, however, Howard also had to know exactly what scientific experiments would be carried and the specific objectives for these missions—matters controlled by NASA and JPL rather than by Northrop.

A few months before, when space science and Apollo support momentarily appeared on a collision course once again in Project Ranger, Schurmeier and Burke had hurriedly prepared mission objectives in consonance with their concern for engineering. The primary objective of the Block V Rangers 10 through 15* they informed the JPL division managers, should be "to obtain [lunar] data which will be useful to both the scientific and manned lunar landing programs... The secondary objective is to obtain...lunar composition and environmental data which will support the capsule experiments and subsequent missions."[5] Premature and quite unacceptable to scientists, these objectives could not hope for Newell's sanction, for he had already pledged Block V as a scientific venture. In fact JPL Space Sciences Division Manager Meghreblian immediately rejected them. The question of proper objectives, he counseled Schurmeier, could safely wait until the experiments had been submitted, evaluated, and selected.[6]

On August 5, 1963, after an initial evaluation at JPL, Meghreblian returned to Schurmeier with his own recommendation. Penetrometer or photofacsimile capsules might provide information on the bearing strength of the surface or pictures in situ, but developing reliable capsules in the time available he judged technically risky and unwise. The primary mission objective for Ranger Block V, he suggested, should "be the successful landing and operation of a seismometer capsule" on the moon. The secondary objective should "be acquisition of data concerning lunar surface composition by means of gamma-ray spectrometry and surface topography by means of television photography and radar probing."[7] Schurmeier quickly approved this choice of experiments and objectives recommended by JPL's science manager, and transmitted the message virtually word for word to Ranger Program Chief Bill Cunningham in Washington. Ranger Block V would become an updated, unsterilized copy of the Block II missions—for planetary science.[8]

By the fall in Pasadena, Ranger Block V, with its six capsule missions, had begun to consume efforts equivalent to if not greater than the four visual imaging

---

*When NASA cancelled the Ranger Block IV missions, the six Block V vehicles were renumbered sequentially in their place, beginning with Ranger 10.

missions. Schurmeier and Kautz continued to concentrate upon the first-order problems of requalifying Rangers 6 through 9. Kenneth C. Coon, as electrical engineer and graduate of the University of Minnesota, was appointed Block V Spacecraft System Manager to oversee developments at Northrop. But even with Coon's able assistance, the growing demands of the capsule flights threatened to overwhelm the small staff in Schurmeier's project office.[9] Burke's attention had been spread among three different blocks of nine spacecraft in 1961, a situation that Schurmeier now sought to avoid. Pickering and Parks accepted Schurmeier's proposed changes, and on August 9 they divided Project Ranger in two: Schurmeier remained in charge of the Block III visual imaging missions; a new project manager was named for the Block V capsule missions—Geoffrey Robillard, a solid-propellant engineer and manager of the JPL Propulsion Division (Figure 69). With Ken Coon, he was to furnish technical direction to and monitor the performance of the Ranger team at Northrop Space Laboratories. Both Howard and Robillard were to continue to report functionally to Cunningham at NASA Headquarters in Washington.[10]

At NASA Headquarters meantime, Cunningham and Nicks considered the Meghreblian—Schurmeier recommended scientific experiments for Ranger Block V. Using the seismometer and the other instruments first flown on the Block II machines made sense; they were planetary experiments designed to investigate the composition of the moon's surface and internal structure. Little time would be required to requalify them for flight, and efforts could be devoted to improving their reliability. The cost to NASA of using equipment already developed was negligible compared to starting something new. Both agreed on the advantages and, following JPL's lead, they rejected twelve proposed sky science instruments in favor of the planetary experiments already tried.[11]

On August 26 Nicks carried this message to the Space Sciences Steering Committee's Planetology Subcommittee. Academic members of the Subcommittee serving as consultants to NASA also concurred in the choice: "A lunar spacecraft of the Ranger configuration should be primarily concerned with the many questions about the moon itself and not attempt to make sweeping measurements of the space between the earth and the moon."[12] A week later Nicks presented what was by now the Subcommittee's recommendation to the full Steering Committee. Information about the moon from the proposed planetary experiments, Nicks assured the commmittee members, could be obtained in two successful flights. Two more flights could be held in reserve as backup missions and, if desired to save funds, the last two of the six planned capsule flights could be canceled. The Space Sciences Steering Committee ratified the Ranger Block V experiments recommended by NASA's Director of Lunar and Planetary Programs, by its Planetology Subcommittee, and by JPL (Table V).[13]

When Newell confirmed this recommendation, Howard and Robillard on the West Coast could proceed to finish the spacecraft design and propose firm

Fig. 69. JPL Ranger Block V Project Manager Geoffrey Robillard

Table V. Ranger Block V Planetary Science

Experiment	Principal Investigator
Single-axis passive seismometer[a]	Frank Press, Caltech Seismological Laboratory
Gamma-ray spectrometer[b]	James Arnold, U.C. San Diego
Surface scanning radar[b]	Walter Brown Jr., JPL
Approach television[b]	Gerard Kuiper, University of Arizona

[a] Capsule to survive hard landing.
[b] Spacecraft bus – destroyed on impact.

mission objectives. But Newell's Office of Space Sciences was also engaged in a detailed program and budget evaluation; one week passed, then two weeks, without formal notification from NASA Headquarters. On September 18 Newell approved the experiments selected by the Space Sciences Steering Committee he chaired,[14] and conveyed the news to JPL. These experiments, he wrote Pickering, were to be flown "on Rangers 10 and 11, with Rangers 12 and 13 to be planned as backup" missions. JPL and Northrop still had time to complete the payload design and integration to meet "the desired flight schedule." Nevertheless, in view of the recent funding cuts and the ongoing evaluation of NASA's space science commitments, "we are withholding letters to the Principal Investigators notifying them of their selection pending a review of the revised FY 1964 program..."[15] Ensuring adequate funds for Ranger Block V, the letter implied, was still a source of real concern at Headquarters.

Adding to the uncertainty, the prospects of more missions for science were darkened by general difficulties cropping up for NASA in 1963. The Air Force was demanding a manned role in space of its own, and various Congressmen, including Senator Barry Goldwater of Arizona, were declaring loudly that on grounds of national defense it ought to have one. Congressmen of both parties were also beginning to seriously question the legitimacy of the rapidly burgeoning NASA budget.[16] In April President Kennedy asked Vice President Johnson to examine the space program and determine the ways and to what extent it could be reduced without affecting Project Apollo. Responding to that request for Newell's Office of Space Sciences, Nicks claimed that, as Apollo support missions, Ranger and Surveyor were exempt from interference. Funds planned for the Pioneer and Mariner planetary projects could be reprogrammed for Apollo, Nicks observed, but such a course would have serious repercussions for unmanned space exploration and the U.S. scientific community.[17]

Although Webb and his deputies Dryden and Seamans had strongly defended the President's manned space flight program in Congress and before its detractors on the President's Science Advisory Committee, they also agreed that the NASA program as a whole should not be driven to the single purpose of landing U.S. astronauts on the moon. They viewed the President's suggested trimming of other NASA work in favor of Apollo as ill-advised.[18] A few weeks later in a meeting with Kennedy to discuss the agency's plans and goals, Webb told the President that NASA's triumvirate believed strongly in continuing an active and diversified program in aeronautics and astronautics. He personally was unwilling to remain and "run a truncated program." Furthermore, Webb explained, NASA's leadership did not agree with Brainerd Holmes, who also wanted to obtain more money for Apollo by returning to Congress for a supplemental appropriation as recommended by the White House.[19] "We wanted to establish the principle that we went out and fought for the President's Budget, and then we lived within it," Webb later recalled. "We had to do that once a year, and we couldn't do it two or three times a year.''[20] President Kennedy listened attentively to his space administrator, and in the end agreed with him. Apollo might have to be cut back, but Webb would continue to run a diverse NASA program.

In October 1963 Webb tightened his control of the space agency through a reorganization. Although Homer Newell became Associate Administrator of the Office of Space Science and Applications, and his responsibilities measurably expanded, this change had little effect on the makeup or duties of his staff in space sciences. Newell and his lieutenants counted on the reworked Ranger 6 nearing completion at JPL, for launch in January 1964, as the next increment in the steadily improving record of flight success. NASA issued the schedule for the Block V Rangers built by Northrop Space Laboratories on October 18; it called for Rangers 10, 11, and 12 to be launched in the second, third, and fourth quarters of 1965.[21] At Headquarters, Nicks and Cunningham had now to guarantee financing for them. At JPL and Northrop on the West Coast, Robillard and Howard continued their efforts to finish the Block V design, flight objectives, and a host of other details before the end of the year.

NONVISUAL SCIENCE: ALL OR NOTHING AT ALL

On October 23, 1963, the Aeronutronic Division of Philco-Ford at Newport Beach received the contract to build four seismometer capsules for the Block V machines. Issued by the JPL Space Sciences Division, the contract called for Pasadena-based scientists and engineers to monitor this work. Under a separate contract with Caltech, the Pasadena group would also monitor the development of an improved single-axis seismometer. These scientific components were to be integrated and tested at JPL, then supplied as government-furnished equipment to Northrop for installation on Rangers 10 through 13. Northrop would directly supervise the development of all other spacecraft components, including the capsule support structure and retrorocket system at Aeronutronic, and would assemble and test the complete spacecraft. At the end of the month, Howard's office submitted the design evaluation recommendations for these spacecraft and the performance specifications for the subsystems to Robillard at JPL.[22] A final design awaited only review and approval of the Northrop recommendations for the spacecraft and the scientific payload.

But by the end of October, it was not at all clear to Robillard and Coon how the NASA-selected Block V experiments could be fitted onto the new spacecraft. These experiments had been maintained on the light-weight Block II machines by removing all redundant spacecraft engineering features and increasing the risk of flight failures.[23] Now, Headquarters expected the Laboratory to allocate some 158 kilograms (350 pounds) to carry an improved set of the same experiments, while Northrop was proposing numerous design changes, also expected by Headquarters, to increase the backup engineering features on Block V. Although these machines would be heavier than their Block II predecessors, the weight, power, and space requested separately by Ranger's engineers and scientists were already in conflict.

At the beginning of November Robillard explained the impending collision of demands between science and engineering to Gerard Kuiper, principal investigator for the approach television. Since the Block III flights each carried six television cameras for visual imaging, Robillard observed, perhaps the television cameras might be the items eliminated on the Block V capsule missions with the least discomfort to all concerned. Long an outspoken champion of the importance of visual imaging in planetary exploration, the Dutch-born astronomer from the University of Arizona was immediately aroused. Not one picture of the moon had as yet been produced in the Ranger project, he asserted. If the Block V project manager insisted on making television a candidate for removal after it had been approved by the Space Sciences Steering Committee, Kuiper warned, he would take the matter to higher authorities at NASA Headquarters. But, wherever the issue might be taken, Robillard knew something had to be eliminated before he and Howard froze the spacecraft design. Another showdown between scientists and engineers in Project Ranger appeared inevitable.[24]

Robillard had already issued the mission objectives for Block V. Based on discussions with Bill Cunningham, the objectives were divided between two mission assignments, leaving Headquarters the option of replacing some of the planetary experiments with sky science experiments on later flights:[25]

> The first mission of the Block V Ranger flights is to land an operating package containing a single-axis seismometer on the surface of the moon, to obtain information regarding seismic activity on the lunar surface, and to obtain gamma-ray measurements, lunar radar reflectivity measurements, and photographs of the lunar surface from lunar altitudes greater than about 20 km.

> The second mission objective of the Block V Ranger flights is to obtain other information from the surface of the moon and in the vicinity of the moon.

> The six flights will be divided between the first and second missions. The first two flights are assigned to the first mission. The third and fourth flights are tentatively assigned to the first mission, and the remaining two flights to the second mission.

If funds could be found for all six flights, nonvisual sky science experiments might also be accommodated.

Funds—that was the ultimate rub. Since work had begun on the new capsule missions, finding the money to pay for them had been the outstanding problem confronting Nicks and Cunningham. The problem became more acute on October

24, when Newell advised his staff to consider the possibility of canceling these flights. Such a move, he said, would free some $30 million in fiscal 1964 funds urgently needed to begin the Lunar Orbiter project, retrofit the hundreds of contaminated diodes on Rangers 6 and 7, and, it was hoped, bail out the hard-pressed Surveyor lander project where costs were skyrocketing. Moreover, the funding levels for all NASA projects in fiscal 1965 had to be decided upon before the end of 1963. The Office of Space Science and Applications could not postpone a decision on these scientific missions much longer.[26]

In Washington on December 2, Nicks and his associates presented a plan to expand NASA's unmanned scientific lunar and planetary programs to Webb, Dryden, and Seamans. By reprogramming some funds, Nicks said, NASA could "increase the probability of scientific returns [in the lunar program] without requiring the initiation of new projects."[27] NASA's ability to provide these returns in the present unmanned lunar program, Planetology Program Chief J. R. Allenby said, had been impaired by "two major constraints": first, the obligation to certify landing sites for Apollo, "which, if the lunar surface is extremely complicated, could require that nearly all of our currently planned vehicles be used in mare areas to outline a safe site;" and second, the limited weight and power of contemporary spacecraft, which made it impossible "to satisfy for the foreseeable future the scientific needs of a variety of specialities..." Moreover, rather than immediately answering questions, each successful flight, "especially in the early stages, may only raise puzzling problems that will require increased data from points covering the entire lunar surface"—beyond the mare regions of prime interest to Apollo. Thus, if a few more lunar flights in the existing unmanned projects were to be approved, the value of the program to science could be greatly increased.[28]

Each of NASA's lunar program chiefs urged that more flights be added to the Surveyor lander and Lunar Orbiter projects. More flights in Project Ranger, Cunningham asserted, were not requested, but the funding for those already planned was certainly needed. For a very modest increase in the planned investment in Block V, more experiments could be added, and the wishes of more scientists granted. "If this money is to do the most good," Cunningham advised Webb and his deputies, "the contracts [for the additional scientific experiments] would have to be let by February 1," 1964.[29]

Money to meet the needs of Ranger Block V might be nearly invisible in the space agency's budget, but all of Newell's unmanned lunar and planetary project proposals combined meant that his segment of NASA's program would about double in size and demand a share "on the order of 10 percent of the NASA budget" in the years ahead. If NASA hoped to land a man on the moon by 1970, retain a diverse program, and meet the fiscal guidelines issued by Budget Director David Bell, the agency simply could not afford these laudable scientific objectives. Moreover, the chances of obtaining more funds were poor, in part because of the growing Congressional dissatisfaction with the size of the NASA budget, in part because the new President, Lyndon B. Johnson, was publicly promising a special emphasis on economy in goverment. Before the week was out, James Webb informed Newell that the proposals for an expansion in the unmanned lunar and planetary program could not be considered.

About the same time, on December 9, Howard's office at Northrop Space Laboratories submitted to JPL its proposal for the development program of the Block V spacecraft system.[30] The next day, on December 10, Secretary of Defense Robert McNamara announced that the Air Force would begin work on a Manned Orbiting Laboratory to carry two Air Force astronauts in late 1967 or early 1968 to explore the problems and potential of manned military operations in near-earth space.[31] McNamara's announcement, the liberal and influential *New York Times* observed editorially, "makes it more imperative than ever that a new hard look be taken at what we are trying to do [in space], the wisdom of the choices made, and the adequacy of our resources to the total effort."[32]

Newell, Cortright, and Nicks had already taken a hard look at the resources available to the Office of Space Science and Applications and decided: existing fiscal conditions dictated that part of that program be trimmed immediately. On Friday the 13th, Newell wired Pickering:

> Since we are now reasonably certain of the final FY 1964 NASA appropriation by Congress, as well as the maximum probable level of support in FY 1965, it has become necessary to conduct an extensive review and evaluation of all programs in order to arrive at a budget which will meet our program requirements and still remain within the funds which will be available to us. One of the decisions resulting from this evaluation was to terminate immediately all activities associated with the Ranger Block V program...[33]

In the final analysis, Cortright observed, "it was purely funds that cancelled the work."[34]

The decision eased the tight money situation in the Office of Space Science and Applications and the manpower shortage at JPL, where the engineers returned to their respective technical divisions or were quickly reassigned to other NASA deep space flight projects. But shortly before Christmas, 200 unlucky Northrop engineers received layoff and termination notices. And the nonvisual planetary scientists who had worked and waited and hoped on the assurances of Headquarters could hope for nothing more at all from Project Ranger.

"The first thing sacrificed is *always* science," Harold Urey snapped irately during an interview at the annual meeting of the American Association for the Advancement of Science in Cleveland, Ohio. Though some lunar pictures might yet be returned by Ranger, scientists possessed "no possibility now of finding out anything about the composition of the moon's surface until about 1966 or 1968.'' As to the popular purpose of NASA's Apollo, "namely, putting someone on the moon," the Nobel Laureate added tartly, "any man or woman with an attractive personality would do."[35]

Chapter Fifteen

SIX TO THE MOON

A LL that remained of Ranger was flights 6 through 9. All that remained of the original scientific equipment were the television cameras—the visual imaging "experiment." Some months before, in early 1963, the RCA television system for the Block III Rangers had been modified to improve its prospects for taking pictures of the moon's surface. And about the same time, for the flight of Ranger 6, certain organizational changes had been introduced in the management of the crucial visual imaging effort.

Organizing Visual Science for Ranger Block III

JPL space sciences manager Meghreblian appointed Charles Campen, an engineer and physicist who had helped establish the Aeronautical Chart and Information Center for the Air Force, as the Block III space sciences project representative in Schurmeier's Ranger Office. Thomas Vrebalovich, an aeronautical engineer with his PhD from Caltech, who succeeded Harold Washburn as project scientist, assisted and represented the NASA-appointed TV experimenters full-time at JPL (Figure 70). Raymond Heacock continued to serve as the space sciences project engineer, making certain that the television system met the requirements specified by the experimenters.[1] All three provided increased support at JPL for Ranger science activities.

But establishing an easy working relationship among project engineers and scientists, the three men quickly learned, was not so simple. Ranger's television experimenters remained in a novel and, in the view of some, not altogether desirable position. Collectively, they were a "team" of experimenters rather than sole proprietors of individual instruments.[2] Although chartered to support them, Campen, Vrebalovich, and Heacock acted as intermediaries, buffering relations between the experimenters and Schurmeier's project office. RCA designed, built, and tested the "experiment" itself, which was of course the television system,

Fig. 70. JPL Ranger Project Scientist Thomas Vrebalovich

under the supervision of JPL, thereby removing its development from the direct control of the experimenters. Project engineers even calibrated the optics. Thus, unlike their colleagues who designed and supervised the fabrication of spectrometers or magnetometers under contract to NASA, Ranger's visual experimenters were left with few responsibilities save those of specifying attainable imaging characteristics at the outset, selecting the targets of interest on the moon before each flight, and—they hoped— analyzing the pictorial returns afterwards.

The often-tenuous connection between NASA-appointed experimenters and the engineers who ran the space agency's project offices was well known to Homer Newell. He fully intended to rectify the real isolation of the visual experimenters from the Ranger Project Office, which was a source of particular concern. On April 23, 1963, Newell took up the matter in a letter to Harold Urey. With Ranger 5 having concluded the Block II missions, Newell said, NASA Headquarters expected to rename the television experimenters for Rangers 6 through 9. In these flights "it is our intention to eliminate situations...in which experimenters are for some reason insulated from the actual work going on." Those scientists appointed "shall have the opportunity to influence the conduct of the experiment. Indeed," he concluded, "in our view it is an inescapable responsibility of the experimenter to ensure himself that the results will be valid, and it is a responsibility of the Jet Propulsion Laboratory to see [that] the experimenters do have [that] opportunity..."[3]

Newell's views were communicated to others among the original Ranger television experimenters.[4] The astronomer Gerard Kuiper wrote to Cunningham on June 18, expressing pleasure in the plans, and announcing that he hoped shortly to begin a separate assessment of the RCA television design. Such an assessment could be performed by a "competent engineer [and] a small staff of one, two, or three men" hired by and reporting to the experimenter team. "I would assume," Kuiper declared, "that we would work closely with the companies building the equipment so that our evaluation and studies are truly independent of the work at JPL... A nearly passive role as was that of the TV experimenters in the past," he added, "is unproductive." Kuiper promised to submit a formal cost proposal for this work to Headquarters just as soon as he returned from a meeting in Europe.[5]

The promised activity deeply troubled Schurmeier in Pasadena. He had his hands full requalifying the Ranger flight components and dealing with those problems already surfacing in the redesigned RCA television system.[6] With the design frozen, no time remained to consider any "independent" assessments by the experimenters. NASA could rename the experimenter team and expand their duties, but, Schurmeier advised Cunningham, to avoid complications Headquarters should carefully define the experimenters' role in relation to the project manager— and do so without delay.

During the next few weeks in Washington Cunningham devoted much of his attention to this sensitive and potentially schedule-busting issue. On July 10, with Oran Nicks and JPL's Campen and Schurmeier, Cunningham issued a policy statement on the organization and role for Ranger's visual experimenters. The new team, composed of "coexperimenters," would be chaired by a "principal investigator" responsible for "organizing the efforts of, assigning tasks to, and guiding the other members of his team." Led by the principal investigator, the team would engage in "planning the functional testing and calibration" of the television system, "assist" in preparing the space flight operations plans, and determine the "scientific merit" of flight options in the event difficulties were encountered during a mission. Finally, the principal investigator would supervise the scientific analysis and dissemination of the lunar photographs, and preparation of the reports to NASA describing the results obtained. The JPL project scientist would assist him in these duties—but beyond those specified, the principal investigator held no other functions in Project Ranger.[7]

This response to prior pledges of an expanded role for scientists hardly pleased Gerard Kuiper. Not only did he judge the role of the principal investigator to be too restrictive, but it had been formulated by others while he was in Europe. Schurmeier, on his part, sure that Kuiper would be named principal investigator for Ranger Block III, was also reasonably sure that, once named to the post, Kuiper would reject the July 10 policy statement. Worried, Schurmeier analyzed the difficulty for Pickering. If Kuiper pressed for a separate assessment of the television system, the conduct of project affairs was certain to be disrupted. Any "situation where there is direction of the TV system that is independent of JPL and the project," Schurmeier asserted, "is completely unworkable because the TV is such an integral part of the spacecraft and the mission." The problem had to "be resolved clearly and in a timely manner," and to this end Charles Campen had prepared another definition of the role of the principal investigator for review and possible adoption.[8]

While Pickering, Schurmeier, and Burke went over Campen's proposed redefinition of roles in Pasadena, Newell announced the names of the experimenters selected for the visual imaging flights. Newell had discussed the choice of a principal investigator with Eugene Shoemaker and Gerard Kuiper. All had agreed on the assignment: Kuiper, the Director of the Lunar and Planetary Laboratory at the University of Arizona,[9] would serve as principal investigator on Project Ranger; Shoemaker was promised the same assignment in the Surveyor soft-lander project. With that accomplished, Gerard Kuiper endorsed a list of Ranger coexperimenters (Figures 71 through 75):[10]

Eugene M. Shoemaker, U.S. Geological Survey

Harold C. Urey, University of California at San Diego

Raymond L. Heacock, JPL

Ewen A. Whitaker, University of Arizona (a colleague of Kuiper's in the Lunar and Planetary Laboratory)

Though an important step, naming the team of experimenters hardly solved the problem of knitting a solid relationship between the team and the project. The pertinent NASA General Management Instruction did not contain the terms "principal investigator" or "coexperimenter" and, hence, offered little guidance. Under the terms of another Management Instruction, Schurmeier himself had the responsibility and authority to direct Project Ranger.[11] The precise roles and prerogatives of a principal investigator and his coexperimenters within the wide umbrella of Schurmeier's authority had to be settled—before the flight of Ranger 6.

The experimenters as well as project engineers agreed that revision of the management instructions guiding flight projects and their scientific activities was urgently needed. But insufficient time remained to accomplish that difficult feat. Kuiper, who had been unwilling to accept the statement of role for the principal investigator formulated in his absence, now agreed with Schurmeier that everyone should meet and negotiate a "Memorandum of Agreement" to establish his respective responsibilities. Such a document would define responsibilities of the project office, the NASA Headquarters program office, and the experimenters. It would be ratified by each group, signed by a representative of each, and serve to

Fig. 71. Block III Principal Investigator Gerard Kuiper

order their relationships for the duration of Project Ranger. On July 31 and August 1 Kuiper, Cunningham, and Schurmeier met at JPL and hammered out a draft agreement.[12] With minor alterations, it was signed by all three men on September 11, 1963.

This agreement between scientists and engineers acknowledged that the project manager possessed "direct responsibility for complete project execution." Along with the authority granted by Headquarters in July, the principal investigator was to "establish the requirements" in space flight operations for the kind and format of data processing provided to experimenters, and act as "the sole arbiter" of the unanalyzed visual data for everyone within NASA and JPL who had an official function to perform "for a period not to exceed three months after

Fig. 72. Block III Coexperimenter Eugene Shoemaker

completion of the mission." The principal investigator and the coexperimenters were also granted "direct access to information at the TV Subsystem Contractor (RCA)," but, in return, "all contact must be coordinated with the [JPL] Space Sciences Division Project Representative;...all direction of the Contractor will be done solely by JPL." Finally, in the event of a dispute, the Ranger Program Manager at Headquarters would investigate any "appeals from the Principal Investigator" and refer them to the NASA Director of Lunar and Planetary Programs for a "final decision."[13]

        Though much still remained to be done before the launch of Ranger 6, Headquarters and JPL had at least reached a modus vivendi for visual science in the Block III missions.[14]

Fig. 73. Block III Coexperimenter Harold Urey

READY TO GO

By early December, project engineers had put the finishing touches on the Ranger 6 spacecraft and its six television cameras (Figure 76). All of the critical diodes had been replaced and systems tests completed in good time to meet the lunar launch period extending from January 30 through February 6, 1964. On December 6, Pickering wrote to Newell, informing him of the progress made and suggesting that NASA Headquarters "appoint a small review group, independent of JPL, to examine the entire Ranger 6 system...," including the launch vehicle and the Deep Space Network, and to certify its flightworthiness. A group composed of members of the Kelley Board, Pickering said, would meet the requirement of independence and also "minimize the amount of time required of the Ranger Project staff [to acquaint them] with the details of the project."[15] A cachet from such a board, he recognized, would not only serve to verify JPL's own

Fig. 74. Block III Coexperimenter Raymond Heacock

findings but provide insurance against political attacks should Ranger 6 also meet with disaster.

The letter arrived in Washington on Monday, December 9, at the beginning of an extremely busy and crucial week for the Office of Space Science and Applications. Given available and anticipated resources, Newell, Cortright, and Nicks had to decide on those flight projects to be continued, started, or terminated—decisions that would lead to the cancellation of Ranger Block V by Friday. An evaluation of the flightworthiness of Ranger 6 was important, but at that moment it could not compare with charting the future course of NASA's lunar and planetary programs then underway. The leaders of Space Science and Applications may also have sensed the political ramifications of Pickering's request. Whatever the case, Newell decided against a special NASA review board, and instructed Cunningham and his assistant Walter Jakobowski to represent Headquarters at the Ranger 6 flight acceptance review meeting.

Fig. 75. Block III Coexperimenter Ewen Whitaker

On December 17 Cunningham and Jakobowski met in Pasadena with representatives of the project office and members of the JPL Ranger Design Review Board. All evaluated the status of and test records for all Ranger system components, including the new RCA television system to be flown for the first time, to make certain that each was flightworthy and that a high confidence could be placed on a successful mission at the end of January. Everyone agreed: Ranger 6 was ready to go.[16] Two days later Ranger 6, carefully packed inside its air-conditioned van, left JPL. The spacecraft arrived at the Florida launch site on December 23 and joined the Atlas-Agena B launch vehicle to begin the final round of preflight tests. But however extensive the preflight care, the ultimate scientific return and the true and only test of the Ranger Block III vehicles, Cunningham explained to Webb, would come "in the last ten minutes of the sixty-six-hour trip to the moon," when the television cameras would be turned on.[17]

Fig. 76. At JPL Technicians Make Final Adjustments on Ranger 6

While the spacecraft and launch vehicle moved toward Cape Kennedy,[18] Ranger's scientists selected its photographic target on the moon. They consulted Apollo representatives at the Manned Spacecraft Center in Houston on the choice, Vrebalovich recalled, but "learned that the design of Apollo's landing gear was already frozen;" anything space science might find out about lunar surface slopes in 1964 "could only confirm or deny that design." Apollo's engineers in Houston "frankly didn't care where we put Ranger 6...so the first choice of a target was our own."[19] In October 1962 Apollo officials had succeeded in their claim to support from NASA's unmanned lunar missions. Their assistants in Houston now seemed entirely indifferent to the potential scientific and engineering product of Project Ranger, making of that claim a mockery.

In large measure, the Ranger scientists based their choice of a target on the lunar lighting conditions to be encountered during the January–February launch period. All of the Block III television missions had to be directed to areas on the moon near the terminator, or shadow-line, where the best contrast for picture-taking was to be found; however, there was a threshold. If a target was too near the terminator, the light would become so poor that closeup pictures were useless. Somewhere, in a north–south band parallel to the terminator, there existed an "ideal" impact distance that guaranteed the best contrast for television photography.

Some months before, Donald E. Willingham had devised a means of finding that ideal distance. An electrical engineer recently graduated from Purdue, Willingham was the experiment representative of Project Ranger. Together with Thomas Rindfleisch, a young mathematician in the JPL Space Sciences Division, Willingham developed a formula known as the lunar "figure of merit," which told the experimenters just how far away from the terminator a spacecraft had to land on any given day to achieve optimum contrast in photography.[20] Using the figure of merit scheme, Gerard Kuiper and his Ranger coexperimenters selected the Julius Caesar region in Mare Tranquillitatis as the lunar target during the first few days of the launch period. At the coordinates 8.5 degrees north latitude and 21 degrees east longitude this region also fell within 10 degrees of the lunar equator, the area of prime interest to Project Apollo. Although little actual interest in the target was apparent in Houston, in keeping with the mission objective, Ranger's scientists had nonetheless factored the manned flight requirement into their first target choice.[21]

Like the experimenters and project engineers, the Deep Space Network was also ready to go. The man at JPL in charge of this function, Eberhardt Rechtin, had been appointed Assistant Laboratory Director for Tracking and Data Acquisition a few months before. In his new position Rechtin directed all of the deep space tracking stations, the interstation communications net, as well as the processing of data sent to the Space Flight Operations Facility at JPL. But the special teams that conducted flight operations and the facility itself remained a responsibility of the JPL Systems Division.[22] The friction and the occasional confusion that sparked along the periphery between these two powerful jurisdictions, as well as NASA plans for two and three flights to be handled simultaneously in the years ahead, caused Pickering to make further changes in the organization. On December 24, 1963, he announced the formal creation of the Deep Space Network "combining the Deep Space Instrumentation Facility tracking stations, Interstation Communications, and all 'mission-independent' portions of the Space Flight Operations Facility." For Ranger 6 and thereafter, except for the ephemeral flight operations teams and special, mission-dependent project equipment in the control center, Rechtin would have charge of all three ground components in space flight operations.[23]

The testing of these components concluded on January 24, 1964, with a final operational readiness test. Both normal and abnormal flight conditions had been simulated, engineers subjected to unexpected equipment failures and communications breakdowns, and emergency procedures perfected. Personnel at the overseas tracking stations and the JPL Space Flight Operations Facility performed well;[24] the deep space net was judged ready to track and command Ranger 6 in every foreseeable situation.

As the first launch opportunity on January 30 neared, Cunningham and Schurmeier spent more and more time at Cape Kennedy. The test results pleased both men. Ranger 6, atop its Atlas-Agena B rocket at Launch Complex 12, completed the joint flight acceptance composite test at midmonth. The only significant problem encountered was a slow response of the spacecraft's roll gyroscope—a discrepancy quickly rectified. While the Atlas-Agena B continued launch readiness tests under the guidance of Himmel's engineers from the Lewis Research Center, JPL engineers disconnected Ranger 6 and returned it to its hangar for one more system test.[25]

Elsewhere around the country special preparations had been made for the flight of Ranger 6. NASA officials had established a "mission status room" at Headquarters, where up-to-the-minute news of the flight could be furnished to government leaders, press briefings held, and lunar photographs released to the news media.[26] On the West Coast, Pickering notified JPL employees that word of the mission's progress would be disseminated by flash bulletins, spot announcements over the public address system, and, at lunar encounter, by an announcer in JPL's von Karman auditorium, where representatives of the world's press and television would be gathered to view the first closeup pictures of the moon's surface.[27]

Unquestionably, at the beginning of 1964, Ranger 6 was specially important to Gerard Kuiper and the Ranger experimenters eager for a closeup look at the moon. It was also specially significant to Congressmen anxious for an American lunar "first" after twelve previous failures; to a Democratic President, who insisted that the nation's defense and space capabilities were second to none; and to Homer Newell and his associates in the Office of Space Science and Applications, who sought redemption in a flight success before the fiscal 1965 budget went to the Hill early in February.* And it was a special mission for Pickering and his colleagues in Pasadena, who were determined to erase the five-flight "Ranger black mark" on the otherwise untarnished escutcheon of Caltech's Jet Propulsion Laboratory.

---

* Indeed, in January 1964 NASA officials hastily directed that Ranger 6 "be called Ranger A prior to launch. Similarly, what have been Rangers 7 through 9 . . . will be called Rangers B through D. If this Ranger is successfully launched, it will be named Ranger 6" (NASA document, *Information Plan Ranger A;* 2-1800). This nomenclature was generally ignored by the news media and, to avoid confusion, it is not used here. Commenting on the confusion anticipated, *Life* magazine observed tartly: "Instead of numbering all the Rangers consecutively, thus keeping accurate count of the failures, [NASA officials] designated the next one Ranger A instead of Ranger 6. If it failed, the next try would be Ranger B. Only when the launch came off successfully would it be called Ranger 6. By this trick they hoped to soak people's memories in alphabet soup to convey the impression that there had been only five failures before the final success" (James Hicks, "Many A Slip Twixt Earth and Moon—And Measles Too," *Life*, August 14, 1964, p. 36a).

The Flight of Ranger 6

Launch operations began in the early morning darkness on Thursday, January 30, 1964. The countdown proceeded smoothly as Schurmeier and Cunningham observed the completion on time of each benchmark from their command post in Hangar AE. Communications lines across the country permitted launch operations and immediate postlaunch events to be monitored also at the JPL Space Flight Operations Facility in Building 125, and there, William Pickering, Robert Parks, and Gordon Kautz listened attentively to the proceedings. A new room upstairs in the JPL control center, carefully fitted out to meet the needs for viewing and analyzing photographs of the moon, awaited the momentary arrival of Ranger's experimenters.

The Florida dawn revealed an overcast sky and the stirring of a gentle onshore breeze. The prognosis for the weather from sea level through the stratosphere was, nevertheless, favorable. The count reached zero at 10:49 am EST, and the Atlas mainstage engines ignited. Those nearby heard the familiar thunder and felt the reverberations as Atlas 199D and its precious cargo disappeared into the overcast. Telemetry data flowing into the Florida command post and the control center in Pasadena were entirely favorable.

Right on course in the first leg of the lunar mission, the Atlas performed flawlessly, certifying the effort previously invested by Himmel and his Agena team. But two minutes after launch, upon shutdown and separation of the Atlas mainstage engines, project officials were startled to observe the telemetry monitoring the RCA television system in Ranger 6 unaccountably turn on and, sixty-seven seconds later, just as inexplicably turn off again. While the Atlas continued its programmed chores through burnout, Schurmeier and Cunningham scrutinized the Ranger telemetry returns. All of the spacecraft's subsystems appeared to be operating satisfactorily, and the tension eased.

Downrange, 160 kilometers (100 miles) above earth, the rocket engine of Agena B 6008 ignited for the first time, inserting the vehicle into an elliptical earth orbit. As the Ranger-carrying Agena swung eastward over the South Atlantic twenty-five minutes after launch, the second burn was recorded. The ensemble, now moving at 11,820 meters per second (24,270 miles per hour), rose above the horizon of the South African Tracking Station at 11:19 am EST, and Ranger 6 separated from the second-stage Agena launch vehicle. The television telemetry, South Africa reported, had turned on properly—no other malfunctions could be detected. Within minutes Ranger 6 extended its solar panels and high-gain antenna, acquired the sun and earth, and stabilized in attitude. Wary project engineers in Florida and California began shaking hands—once again NASA and JPL had an operating Ranger on its way to the moon.

Schurmeier left the command post and caught a plane for the West Coast, arriving at JPL late in the afternoon. There he met with Robert Parks, Ranger Flight Operations Director Patrick Rygh, and other members of the project staff to consider plans for the midcourse maneuver and, among other matters at hand, discuss the curious telemetry events at launch. Initial calculations showed that the Atlas-Agena guidance had functioned in near-perfect fashion. Ranger 6 was traveling a path only a few thousand miles wide of the intended point of impact, an error well within the correction capability of Ranger's midcourse motor. After engineers confirmed its precise course, midcourse maneuver commands would be prepared and transmitted.

But the question of the telemetry malfunction remained unanswered. Had the event adversely affected the television system? Telemetry now arriving at the Space Flight Operations Facility plainly stated that normal conditions prevailed. To perform a checkout now, Schurmeier would have to order the entire television system turned on. And, if for some reason the system failed to turn off, the battery-driven television cameras would be left without power when Ranger 6 reached the moon. Schurmeier and his lieutenants weighed the evidence and agreed: such a test with its attendant risks was unwarranted; the mission would proceed according to the flight plan.

Early Friday morning flight controllers verified and transmitted the commands for a midcourse maneuver, instructing Ranger 6 to adjust course toward its lunar target in the Sea of Tranquility. The spacecraft responded at the designated time, pitching and rolling into the specified attitude, and the midcourse motor burned properly. A few minutes later onboard sensors reacquired the sun and earth, and, still speeding moonward, Ranger 6 resumed normal cruise operations. Another milestone had been successfully passed. Pleased members of the orbit determination group went to work at JPL analyzing the new trajectory. Except for the single anomaly during launch, it was so far a textbook flight.

Later that day Schurmeier and Cunningham met briefly with newsmen to discuss the midcourse maneuver and prospects for the mission. The JPL project manager drew a circle about the Julius Caesar region on a large map of the moon, and explained that the circumference would shrink in the hours to follow as engineers refined the trajectory data. But a first look at the flight path, Schurmeier said, indicated that Ranger 6 would land "near" its intended target. Now midway through the mission, he seemed willing to endorse the terse hope already expressed to newsmen by JPL Director Pickering: "I am cautiously optimistic."[28]

As Ranger 6 approached the moon on February 2, Cunningham, Schurmeier, and Rygh reviewed the actions to be taken at lunar encounter. In an adjacent room, Pickering and Parks were joined by Newell and Cortright from NASA Headquarters; all watched the activity on the floor of the control center and listened intently to Richard Heyser at the Goldstone Tracking Station announce the time remaining and describe the events to be completed. Representatives of the news media and other project personnel in the JPL auditorium listened to Walter Downhower. Chief of the Systems Design Section, Downhower had volunteered to perform Heyser's function publicly as the "voice of mission control" for Ranger 6. In a few minutes he was to be the focus of national attention as all awaited receipt of the first closeup pictures of the moon.

The velocity of Ranger 6 increased as it plunged towards the Sea of Tranquility. Trajectory data had revealed that the machine would impact within a few miles of the planned target. Analysis of the spacecraft attitude showed that the alignment of the camera axis relative to the moon was within acceptable picture-taking tolerances. Schurmeier and Rygh discussed the option of a terminal maneuver to improve the resolution of the final pictures with Gerard Kuiper and the Ranger coexperimenters. So close to success, the scientists elected not to attempt it. They would not chance a possible last-minute malfunction. Accordingly, Schurmeier ordered the two full-scan television cameras and their associated transmitter turned on automatically by the TV backup clock timer, and the four partial-scan telephoto cameras and their transmitter by a real-time command from the ground.

At impact minus eighteen minutes the spacecraft dutifully signaled that the two full-scan cameras had begun a five-minute warmup period. A few minutes later it flashed similar word for the four partial-scan cameras. The two sets of cameras were to switch automatically to full power and begin transmitting pictures at impact minus thirteen and ten minutes, respectively. Personnel at the Goldstone Tracking Station switched on the video receiving equipment. "Thirteen minutes to impact," Downhower announced from his broadcast booth to a waiting world, "there is no indication of full power video." Then, after a pause, amid growing murmurs in the audience below him: "ten minutes to impact, we are still awaiting transmission from the spacecraft of full power video."

From the JPL control center Schurmeier hurriedly directed that backup commands be transmitted from Goldstone to the spacecraft in an attempt to turn on the television system. Minutes later, however, Heyser reported no confirmation of the action from the otherwise chattering Ranger 6. A final command also proved futile. In the JPL auditorium, Downhower continued to report the time and, with growing anxiety, repeat the phrase: "We still have no indication of full power video." The sing-song tones of the Ranger 6 analog telemetry which filled the auditorium at JPL abruptly ceased. "We have our first report of impact," Downhower exclaimed at 1:24 am PST on February 2, and, for the last time, "still no indication of full power video" (Figure 77).

Stunned officials in Washington recoiled at the next words piped into the auditorium at NASA Headquarters when a technician inadvertently switched lines: "Spray on Avon Cologne Mist and walk in fragrant beauty..."[29] In Pasadena, Cunningham pushed his chair away from the console in the control center and stood up. "I don't believe what's happened," he said, repeating the words again and again as he left the room looking straight ahead. Pickering, who had walked onto the floor shortly before impact, turned to those nearby, declaring "I never want to go through an experience like this again—never!" Schurmeier wondered silently to himself how many more project managers Ranger might yet claim.

Across the Laboratory a subdued group of newsmen and employees began to stir inside von Karman Auditorium. Walter Downhower relaxed his grip on the microphone, arose from his chair, and went home. At least he could avoid the

Fig. 77. The Audience in Von Karman Auditorium Hears Downhower Describe the Final
Moments in the Flight of Ranger 6

postflight press conference (Figure 78). But during the drive he turned on the car
radio and was treated to a replay of his own voice broadcasting the final moments
in the life of Ranger 6. He listened in disbelief—once more through to the end.
Downhower would never again agree to serve as the voice of any NASA space
flight project.[30]

Logged as the twelfth successive American lunar flight failure, Ranger 6
blemished an impressive record of space successes cited by President Johnson a few
days before.[31] It also opened a week in which NASA officials appeared before
Congress to testify in favor of a $5.3 billion budget—much of it earmarked for
manned and unmanned lunar exploration—for fiscal year 1965. Under these
circumstances, in the days immediately following, the news media were most
temperate in their public accounting of the disaster. The Ranger spacecraft was
acknowledged to have performed flawlessly even if the television cameras had
failed.

Fig. 78. Homer Newell, William Pickering, and Harris Schurmeier Answer Newsmen's Questions
at Ranger 6 Postflight Press Conference

Chapter Sixteen

THE WORST OF TIMES

AMID the more temperate media reaction, William J. Coughlin, the pro-business editor of *Missiles and Rockets*[1] pronounced Ranger 6 a "one-hundred percent failure." JPL's record on Project Ranger, he insisted, was a disgrace; more offensive, NASA had paid the California Institute of Technology a $1.2 million fee in fiscal year 1964 to staff and operate the nonprofit laboratory responsible for the lunar debacle:

> This may be exactly the heart of the problem. An academic environment is neither comparable nor conducive to the kind of hard-driving industrial atmosphere required to make complex space hardware function in a highly reliable manner. A price of $1.2 million for a leisurely university atmosphere and little else is exorbitant...We think Congress should reopen the whole question of the JPL–Caltech relationship.[2]

Of course Ranger 6 had not been a complete failure, but to other close observers, the television difficulty had also raised serious doubts about the supervision of Ranger by Homer Newell's Office of Space Science and Applications. And like Coughlin, many insiders wondered about NASA's methods of contracting for and managing research and development efforts through an institution of higher learning. Coughlin had branded the project itself a loser. Those in the space agency who could, carefully avoided any direct association with Ranger. A few publicly declared Ranger's lunar pictures dispensable; Apollo, they asserted, could proceed without the assistance of NASA's unmanned missions.[3]

With the very justification of the remaining Ranger flights now called into question, the future of the project hung by the most tenuous of threads. Should it be canceled or completed in 1964? The answer to that question depended upon another investigation of the project by JPL and NASA officials, and upon the response from Congress (Figure 79).

"YOU'RE SHY—I'M SHY 28 MILLION BUCKS!"

Fig. 79. "You're Shy–I'm Shy 28 Million Bucks" (Courtesy Fort Wayne [Indiana] *News-Sentinel*)

THE RANGER 6 INQUIRY AT JPL

On Sunday morning, a few hours after the impact of Ranger 6, William Pickering appointed a JPL–RCA board of inquiry to investigate the malfunction of the television system. Television project engineer Donald Kindt, placed in charge, was directed to find the cause of the failure and recommend action to prevent its recurrence on Ranger 7. The next day, on February 3, 1964, Pickering appointed a second investigative body composed of JPL section chiefs and chaired by the "voice of Ranger 6," Walter Downhower. This group was to check on the "completeness" of the Kindt-led project investigation and act as "a review board to evaluate [its] conclusions and recommendations."[4] Pickering expected to have the findings and recommendations in time to modify Ranger 7 before its scheduled launch at the end of February.

The JPL investigation began auspiciously, with words of encouragement from Representative Joseph Karth, the subcommittee chairman of the House Committee on Science and Astronautics, who commented to reporters that it seemed reasonable for the project to continue. If only two of the three remaining Rangers were successful, he said, the effort would be "worth the money and time spent on it." Committee Chairman and California Democrat George P. Miller sent a personal note to Pickering reaffirming faith in NASA and JPL, and offering his opinion that the performance of Ranger 6 and the accuracy of its flight path truly had made that mission "an accomplishment of the first order."[5] To be sure, both men had become closely associated with and approved funds for Ranger over the years. With Laboratory morale at its nadir, the Congressional encouragement was most welcome.

Kindt and the JPL investigators quickly isolated the time of the failure: the sixty-seven seconds when the television telemetry had unaccountably turned on during ascent through the earth's upper atmosphere. Electrical arcing in that critical pressure region,* they determined, had destroyed the high-voltage power supplies to the television cameras and their transmitters. But what caused it, and the enigmatic cessation and subsequent normal operation of the television telemetry on the way to the moon, defied simple explanation. Opinion at JPL was divided over several possible causes and, consequently, over what remedial action to take.

On February 11 Pickering notified NASA Headquarters of the impasse and its potential negative effects. JPL's investigation had not uncovered a single "definitive cause" of the television failure, he informed Newell, and the next Ranger launch should be postponed. "A firm decision on the rescheduling of Ranger 7 must await results of a study effort of several possible preventive steps and the effort and time required to mechanize them." Newell and Ranger's managers in Washington had no choice but to agree.[6]

A few days later, on February 14, the JPL investigating board issued its final report.[7] Although providing more details, it came no closer to a consensus on the precise chain of events or their cause. Low battery voltages proved that power had been used by more than simply the cruise mode telemetry during part or all of the sixty-seven seconds. The television system, it was concluded, had indeed come on prematurely, perhaps because the internal command switch had advanced several steps, or because of shorting across exposed pins contained in the external umbilical connector on the Agena nose fairing. In the latter instance, arcing between a battery pin and the nearby pin of the sensitive television command circuit used to monitor the television system on the launch stand could have activated the system. But neither electrostatic discharge nor any other mechanism that might trigger arcing across these pins had been uncovered. The most probable explanation, the JPL board therefore concluded, was that the system "turned on by

---

* Gaseous breakdown to a state of electrical conduction occurs most readily at the densities and pressures prevailing in the upper atmosphere.

a movement of the command switch into its warmup mode... Following the turn-on, three additional movements of the command switch returned it to position zero, disconnecting the arcing components from the battery and conditioning the command system for a normal sequence of events ..."[8]

On the basis of these findings, the board recommended to Schurmeier's project office six measures to prevent a recurrence of the failure on subsequent flights:

1. Modify the Ranger television turn-on circuitry to lock out any possible turn-on signal. Mechanical separation of the spacecraft from the Agena vehicle in earth orbit should activate the circuitry, permitting turn-on to be accomplished. (A second option included eliminating the four-step command switch and replacing it with a different one.)

2. Reduce to a minimum the number of electrical leads from the Agena nose fairing umbilical connector through the spacecraft to the television system. (All leads that could trigger a turn-on were to be eliminated or have resistors added.)

3. Modify the Ranger television telemetry to transmit data on system performance during the warmup period near the moon.

4. Conformally coat all exposed terminals in the television system, and pot or sleeve any exposed connector pins to minimize the possibility of an accidental short.

5. Perform a pin retention test on all connectors.

6. If found necessary, provide venting of the television subsystem assemblies.[9]

The investigative report and its technical recommendations, approved by Downhower's watchdog group, were then turned over to the NASA board of inquiry that would make the final recommendations for Ranger's future.

THE NASA INQUIRY: DISPARATE FINDINGS

NASA's Ranger 6 Board of Review was formed on February 3, 1964, by Associate Administrator Robert Seamans rather than by Homer Newell, whose Office of Space Science and Applications was responsible for the faltering Ranger. A brief news release stated that the NASA board would review, independently, "the failure analyses and corrective procedures being developed by JPL," and make "appropriate recommendations on the conduct of the remainder of the Ranger flight program."[10] The four men whom Seamans named to the Ranger 6 board had little or no direct association with the project or the Office of Space Science and Applications. Francis Smith was the respected chief of the instrument research division at NASA's Langley Research Center in Hampton, Virginia. Herman LaGow, a former member of the Kelley Board and physicist with twenty years at the Naval Research Laboratory and NASA, headed the systems review group at the Goddard Space Flight Center in Greenbelt, Maryland. Eugene Dangle, the board's secretary, brought experience as a Technical Programs Officer in the Headquarters Office of Program Plans and Analysis. Walter Jakobowski, Bill Cunningham's Ranger Program Engineer, served ex-officio to help familiarize board members with the project but, as Seamans made clear, not as a voting member. As chairman, Seamans selected his own Deputy Associate Administrator for Industry Affairs, Earl D. Hilburn.[11]

A self-made 43-year-old executive, Hilburn had acquired a reputation as an effective manager in industry, most recently as vice-president and general manager of the Electronics Division of the Curtiss-Wright Corporation. Seamans had picked Hilburn to consolidate and oversee NASA procurement policies and practices, particularly to encourage the use of incentive contracts instead of the cost-plus-fixed-fee type. Almost immediately after Hilburn had joined NASA six months earlier,[12] Webb assigned him the task of reviewing the contractual relationship between Caltech, JPL, and NASA. The prevailing contract reflected JPL tradition, which was to operate rather independently of its governmental sponsor under the academically managerial protection of Caltech. Hilburn recommended including a number of terms in a new contract to tighten NASA's control over the Laboratory and, with Webb's backing, forced most of them through against vigorous objections from Caltech and JPL. Now Hilburn, something of a bête noire at the Laboratory, was to command NASA's evaluation of Ranger 6, much to the dissatisfaction of the scientists and engineers in Pasadena (Figure 80).

The NASA Headquarters review board—shortly to become known as the "Hilburn Board"—convened on Thursday, February 6, in Washington. The next day the board members met in Pasadena, and began one week of evaluating and discussing JPL's Ranger 6 findings and recommendations. Hilburn, a manager before he was a technician, soon found himself deeply troubled by the various explanations under debate at JPL. The absence of a single "definitive cause" of the television failure suggested to him that more than a simple technical problem was involved. If JPL project and division personnel had overlooked a marginal design—indeed a design flaw—in the umbilical connector on the Agena nose fairing, might not other parts of the RCA television system and the JPL spacecraft likewise be impaired? Project Manager Schurmeier and his JPL associates remained convinced that the electrical arcing, triggered by a single unknown event, had "cascaded" in the low pressure and density of the upper atmosphere so as to burn up the television high-voltage power leads. But was it not just as reasonable to suppose that *two or more* discrete failures resulting from design and testing deficiencies had occurred nearly simultaneously?

On February 14, the day that JPL investigators issued their final report, the members of the NASA Ranger 6 Review Board presented their preliminary findings and recommendations to officials of the space agency in Washington. Deputy Administrator Hugh Dryden and Deputy Associate Administrator Walter Lingle, along with Newell, Nicks, Cortright, Cunningham, and Jakobowski,

Fig. 80. NASA Deputy Associate Administrator for Industry Affairs Earl Hilburn

listened intently to the presentation. The board, Hilburn reported, had uncovered numerous deficiencies in the design and testing of the "split" television system used in the Ranger Block III vehicles. This system was not entirely redundant, he observed, having been hastily redesigned after the flight of Ranger 5. These television systems also had recorded instances of inadvertent turn-ons at RCA during tests. The Ranger 6 television system, furthermore, had not been checked out at full power on the pad just prior to launch.

The test history of the Ranger spacecraft appeared hardly better. Torsional vibration of the spacecraft and television system at JPL had been conducted with the proof test models but not with the flight articles. JPL also preferred to conduct mission tests using numerous hard wires to monitor the performance of the spacecraft subsystems, another example, he continued, of procedures that in the board's opinion jeopardized a thorough understanding of how the vehicle would perform in space.[13] These conditions, along with the lack of any telemetry data that pinpointed the failure, led the board to conclude that more than one cause was responsible for the loss of the Ranger 6 television system.

A quick technical fix, such as JPL's proposal for electrically "locking out" the television system during ascent through the atmosphere, would therefore be insufficient to guarantee success. Ranger, the board said, would have a very low chance of achieving its objectives if NASA were to follow that course; much more stringent measures were necessary. The RCA television system, the Hilburn group concluded, had to be redesigned and retested, with a delay of twelve months or more before the next flight. Moreover, the board wished to broaden its investigation at JPL to include an evaluation of the JPL spacecraft, with further recommendations to follow.

Dryden, who had remained silent throughout most of the meeting, advised the board members that another prolonged delay for Ranger might well result in NASA's canceling the project altogether. Under the proposed approach, Ranger's lunar pictures would arrive too late to benefit Project Apollo measurably, and the added costs of redesign and retesting might better be applied to the upcoming Surveyor Project or Lunar Orbiter. He asked the board members to discuss the preliminary findings and recommendations further with Newell's staff and the JPL project team on February 19 and 20, in a meeting to chart recovery plans for Project Ranger.[14]

A PUBLIC ACCOUNTING

Unprepared for and startled by the asperity of Hilburn's oral report, the members of Newell's office were truly distressed. If the board did not alter its position, the cancellation of Ranger appeared a distinct possibility. In separate memos, Cunningham and Jakobowski condemned the board's findings and recommendations in the strongest terms. The usually reticent Jakobowski refuted the findings point by point and openly urged positive lockout of the televison system as the best approach for NASA to take. He judged the "ultra-conservative approach" adopted by the board to be ill-founded and untimely, and he found the manner in which the board presented its recommendations "to top management" to be most objectionable. Newell had not been given the opportunity to rebut Hilburn at the meeting. Now, he worried, management might "accept a pessimistic report without even asking for the views of the Program team."[15] Nicks endorsed Jakobowski's memo, as did Cunningham. "Ranger," Cunningham added, "is in the unfortunate position of being not only in the public spotlight because of its mission and its failures, but also in the Agency and Congressional spotlight because of the continual publicity which started with the Kelley Board of Inquiry. I honestly believe that it is possible for a program to be reviewed to death, and this is very likely in the case of Ranger if the investigation is allowed to continue."[16]

Newell was in no position to stop an investigation authorized by Webb, Dryden, and Seamans, but he shared many of the same reservations expressed by his subordinates. The project, he decided, must continue. In the two-day meeting on February 19 and 20, Hilburn, Nicks, Cunningham, Schurmeier, and their associates argued and discussed the various alternatives for proceeding with

Ranger. After compromise on all sides, Cunningham reported, "a general approach was agreed upon":

1. The Ranger 7 television system would be modified, reworked, and retested at the RCA plant in Hightstown, New Jersey, while the spacecraft remained at Cape Kennedy.

2. A proof test model of the modified television system would be requalified, then returned to JPL for further tests.

3. Changes to the spacecraft would be limited to the wiring necessary to accommodate the modified television system. Both the flight spacecraft and television system would be retested at the Cape.

Although the exact changes and detailed plans remained to be completed, this course would permit the launch of Ranger 7 at the end of May.[17] Newell, meantime, urged Pickering to consider the recommendations of the Hilburn Board "objectively" in arriving at a detailed plan. Until Headquarters approved all of the retest and requalification procedures, he said, "the official flight schedule will remain under study."[18]

Still, for all of Newell's diplomatic intercession and the tentative compromise on Ranger 7, Hilburn pulled no punches when on March 17, 1964, he submitted the board's final report to Webb. "In view of the evidence that *two or more* failures must have occurred within the spacecraft's TV subsystem," the report averred, "the Board broadened its investigation to include an evaluation of any general weakness in the Ranger design, testing philosophy, and procedures which might have contributed to, or enhanced the possibilities of, in-flight failure." The Board, the report continued, had found a good deal wrong:[19]

1. The two video chains and their associated power and control systems were unnecessarily complex... Furthermore, the two TV systems were not entirely redundant; there were many boxes, plugs, junctions, cables, and control circuits which were common to the two chains and in which a single failure would lead to the disablement of both television systems...

2. Hazardous conditions resulted from certain practices employed in the design and construction of the spacecraft. These included the use of a male umbilical connector, a multiplicity of circuits through the umbilical, exposed terminals where a foreign particle could produce a short circuit, and the use of unvented and unsealed boxes.

3. Most of the Ranger testing was carried out in such a manner that potentially dangerous situations which could have enhanced accidental triggering of critical control circuits could not be vigorously assessed. Because of approximately 300 wire electrical connections between the operating circuitry of the Ranger

spacecraft and its associated external test equipment, the ground test conditions are quite apart from the true space flight conditions, where only radio communications tie the spacecraft to its ground control systems.

4. The Ranger 6 spacecraft contained a directional antenna...that was never system tested with the high-power TV transmitters.

5. Prelaunch systems verification was incomplete (e.g., the last operation of the complete TV subsystem on Ranger 6 was twelve days prior to launch). This resulted from the necessity for removing the TV payload from the bus to fuel the midcourse motor, and the Jet Propulsion Laboratory's reluctance to risk possible damage to the space vehicle after it was reassembled, with the [pyrotechnic] squibs in place under the shroud.

Project officials at JPL and Headquarters had conceded only portions of the second finding (one actually associated with the television subsystem and not, as claimed, with the spacecraft). The rest of the findings, which concerned JPL design and test philosophies, they had rejected. In substance and wording, those findings strongly implied an oversight at NASA, malfeasance at JPL, or both. The JPL staff had dissented with particular sharpness from the board's recommendations for corrective action to modify and retest Ranger 7.

Although JPL had agreed to change the television system and discuss other modifications, it had steadfastly refused for Ranger 7 to accept three important changes in testing practices demanded by the Hilburn Board, viz.,

1. Eliminate all ground wire connections during vibration testing, and use "absolute minimum wires" during mission testing.

2. Eliminate "microinspection" of certain components after environmental tests (JPL had agreed to do so only for Ranger 7).

3. Use minimum wires during prelaunch checkout.

Laboratory officials had insisted that, with the limited engineering telemetry available on Ranger, wire connections to the spacecraft to sample subsystem performance should be maintained in tests. In view of this disagreement, the board, which had no authority to impose its will on Schurmeier's JPL project office, urged NASA management to monitor closely all the Ranger 7 tests, target the launch of Ranger 7 for the end of June "but hold the official launch date 'under study' pending evaluation of the validity and results of the Ranger 7 tests in Pasadena," and detain Ranger 7 at JPL instead of the Cape until NASA "concurs that the spacecraft is flightworthy."[20]

James Webb received the final report but took no immediate action. The disquisition posed disrupting procedural problems. On the one hand the news media and the Congress expected word of the investigation's results and of NASA's plans for Ranger. On the other hand public release of the report was liable to stir a massive wave of contention within the space agency. On March 26 Robert Seamans sent copies of the Hilburn final report to Pickering, DuBridge, and the inventor, businessman, and chairman of the Caltech Board of Trustees, Arnold O. Beckman. The classified report, Seamans explained in his covering letter, was an "internal document not for public distribution or discussion with the press..." Access was being limited to the three men while "we...consider very carefully what our responsibility is with respect to reporting to the Congressional Committees...and to the press."[21] Under the terms of transmittal, the document could not be shown to other JPL officials.

In Washington on March 31, Webb, also disturbed by the continuing Caltech–JPL resistance to NASA direction, sent identical letters to Clinton P. Anderson, Chairman of the Senate Committee on Aeronautical and Space Sciences, and George P. Miller, Chairman of the House Commmittee on Science and Astronautics. Copies of the letter were telegraphed to Pickering and Beckman. The four-page, single-spaced missive explained the work of the Hilburn Board and reported on "the present assumptions regarding the Ranger 6 failure and our current plans for future flights." Although excluding the board's recommendations, all five of the disputed findings were reproduced virtually word for word. They indicated "a number of deficiencies in design, in construction, and in the testing of this [Ranger 6] spacecraft..." at JPL. The Hilburn report itself, Webb asserted, was classified Confidential. It had been prepared for NASA use and represented only a portion of the available material. "Consequently, this report will not be released, especially since it does not represent NASA's complete judgment and final implementing plans."[22]

But however the letters were intended, members of Newell's office worried about Congressional demands for an investigation of Ranger based on the debatable charges. In Pasadena, the letters sharply increased exasperation over the seeming byzantine maneuvers of NASA Headquarters in its dealing with Caltech–JPL.

ACTION AND REACTION

Copies of Webb's letters arrived in the Office of Space Science and Applications the same day that they were posted to Capitol Hill. No one had been prepared for them; indeed, it seemed inconceivable that letters containing the Hilburn Board charges could have been sent forward without prior consultation, or the chance to append documents of rebuttal.[23] It appeared as if Webb's office had made a grievous administrative error or, worse, the Administrator had lost all confidence in those at NASA responsible for Project Ranger. Morale plummeted.

On Monday, April 6, Newell consulted with Cortright, Nicks, and Cunningham. At Newell's request, Nicks prepared for his signature a memorandum to Seamans. It expressed strong reservations over sending the disputed charges to Congress; charges in these letters, the memo averred, often made no technical sense and thus would be "subject to misinterpretation." The action, moreover, had had a serious "impact on the morale in the Office of Space Science and Applications and the Jet Propulsion Laboratory."[24] A lengthy attachment set forth Office objections to the technical aspects of the board's findings conveyed to Congress:

1. The two RCA video chains might have been complex, but the Hilburn report implied that any system not entirely redundant was improperly designed. "Even if the two TV channels had been completely redundant, the spacecraft itself would have represented a non-redundant series of systems of far greater complexity... A single failure in the attitude control system, for example, could result in complete loss of television data, as could the failure of a booster engine."

2. The report's description of practices employed in the design and construction of the spacecraft conveyed "no other conclusion than that the spacecraft was assembled with complete disregard for normal quality control practices..." For example, "the method and degree of venting unpressurized [electronic] boxes will always be a matter of engineering judgment. Insofar as the Ranger flights are concerned, we have never encountered a problem attributed to venting."

3. JPL's use of wire electrical connections during spacecraft testing was also a matter of judgment. Ranger communicated over great distances on a wideband radio-frequency link which limited the telemetry bit rate. Therefore, "because of this lack of capability to adequately monitor...the spacecraft with the RF link, cables are required during environmental tests." Care nonetheless had been taken to check for sneak circuits in the ground equipment which might have caused or obscured failures during testing.

4. The Ranger 6 directional antenna was "not normally used during 'system tests;' " however, it was "a simple, fixed geometry, 1.2-meter (4-foot) dish which is always functionally tested after installation to ensure that it is transmitting properly, and video is transmitted by means of this antenna with the spacecraft in launch configuration on the launch vehicle." It was not radiated at full power within the shroud atop the Agena for fear of damage to the spacecraft and launch vehicle.

5. Although it was true that JPL never exercised the television transmitters at a full 60 watts power on the launch pad, "their reasons seem very sound. Should the midcourse motor squibs fire, igniting the midcourse motor, it would cause complete destruction of the spacecraft, and if this occurred on the launch pad, probably also the Atlas-Agena... On the other hand, the low-power test which was performed functionally checked out the complete TV system, except for the application of high

voltage ... which had been checked out a few days earlier" in the hangar at Cape Kennedy.[25]

The Hilburn report, to be sure, did contain a number of charges based upon inconclusive evidence and erroneous technical assumptions. No one could be sure, for example, that "two or more failures must have occurred" aboard Ranger 6. And, because of the limited engineering data available from Ranger's analog telemetry, the hard wire electrical connections that JPL used between the spacecraft and its external ground test equipment really made technical sense. The board members knew along with everyone else that something catastrophic had gone wrong, though whether it involved the spacecraft as well as the television subsystem and the umbilical connector on the Agena nose fairing remained a question mark. But it was also true that more spacecraft testing, including full-power television tests without wire connections to the test consoles, could have been accomplished as a doublecheck, and NASA and JPL officials had to consider it.

A draft of Nicks' rebuttal to the Hilburn charges was shown to Robert Seamans. Since Congress had already received the findings, he considered formal criticism sensitive at this point. "We were authorized to put it in the files," Cortright later recalled, "but not to send it."[26] Seamans, in turn, took up the issue with Webb. Leaders of the Office of Space Science and Applications, Webb learned, would likely take issue with various aspects of the Ranger 6 findings in any Congressional investigation.

And so, it seemed, they would. The Hilburn report, Nicks informed Newell bitterly, not only contained inaccuracies, but, since it contained no military secrets, it had also been classified in a highly irregular fashion. "Allowing only limited distribution, such as preventing the Director of JPL from showing the report to his staff, indicates clearly that NASA did not establish the review board for the purpose of feedback to project personnel." If the only purpose of the investigation had been to establish "a basis for a critical letter to Congress," Nicks concluded, "we in the Program Office were naively misled initially into supporting it as a constructive endeavor."[27] At JPL, where the report iself had been viewed as a final canard from Hilburn, the NASA investigation, the report *and* the letters to Congress now appeared connected in a devious attempt to bludgeon the Laboratory's management.[28] Exasperation gave way to outrage. Pickering instructed JPL officials to refuse interviews with newsmen and issue no public statements.

Publicly, Project Ranger was increasingly treated as a manifestation of the differences between NASA and Caltech–JPL. Confusing reports and rumors multiplied, especially in Southern California. "If Ranger 7 fails," the Los Angeles *Herald-Examiner* proclaimed, "management of the government owned laboratory may be taken out of the hands of the California Institute of Technology ... Director William H. Pickering may be removed from the post he has held 10 years."[29] But if NASA wished to separate JPL from Caltech, the *Los Angeles Times* added editorially, "it could do this without first resorting to a campaign of defamation, which not only damages JPL but reflects unfavorably on one of the country's very great schools of science and technology . . . "[30]

Responding to queries at a news conference a few days later, Webb said that his letter to Congress had at least been an administrative mistake. He had been traveling when he completed and sent the letter back to Headquarters. Presuming that Newell's office already had seen and approved the draft, he signed it upon his return. For that he was sorry. Nonetheless, Webb concluded for the newsmen, even though some at NASA and JPL "may...see problems created by such a letter, I take it pretty seriously that we have responsibility for giving [Congress] a report, and that is what I did."[31] But the Administrator's explanation and apology came too late to avoid serious problems on Capitol Hill, for on April 9, 1964, Congressman George P. Miller, chairman of the House Committee on Science and Astronautics, had announced that the Subcommittee on NASA Oversight would investigate charges that the Ranger failures were "due to faulty design and inadequate testing by the Jet Propulsion Laboratory."[32]

CONGRESS INVESTIGATES RANGER

A Congressional investigation of Project Ranger gave NASA leaders good cause for apprehension. Although they considered California's Democratic Congressman George Miller to be generally a friend of NASA and JPL, the Congressman he designated to head the inquiry was Joseph E. Karth. From Minnesota, Karth had few space interests to nourish or protect, and, despite his generous comments after the flight of Ranger 6, he had been more often a strong critic both of Caltech–JPL performance on Ranger and of the manner in which NASA managed its solitary contract field center.[33] Webb immediately warned Karth of the "unfortunate" timing of the Ranger hearings, and that the inquiry might dangerously affect morale and the program. Webb particularly cautioned the Congressman against taking the Hilburn report as NASA's final word on the difficulties with Ranger. It was, he emphasized, "only one working document."[34]

But the witnesses—for NASA: Webb, Seamans, Newell, Nicks, Cunningham, and Cortright; for JPL: Pickering, Parks, and Schurmeier—quickly learned that the Minnesota Democrat took the report at least as a significant preliminary word. After the critical Ranger letter, Karth declared when he opened the four days of hearings on April 27, "it was hardly reasonable to expect that...a Congressional investigation was not in the best interests of the country and Congress." The Subcommittee on NASA Oversight intended to review any possible Ranger "technical deficiencies," but its "primary interest" in this hearing would be devoted to "problems of management" at NASA and JPL.[35]

The technical issues focused on whether the Ranger equipment had been adequately tested before flight. Some critics, Karth contended, alluding to the NASA investigation of Ranger in 1962, charged that instead of thorough testing before launch, JPL relied on a general philosophy of " 'shoot and hope it works.' "

That had never been the NASA or JPL approach, the Ranger witnesses countered. If the Congressman had received the contrary impression, from whatever the source, it was because of confusion about what constituted a proper test. True, Ranger's high-gain antenna, for example, though tested separately, had not been system tested with the high-power TV transmitters after installation on the spacecraft. No such test was needed and, in any case, none could be reliably conducted before launch in the hangar at Cape Kennedy. Even though the space agency and JPL witnesses repeatedly insisted that the Rangers had been adequately tested, however, they also conceded that the remaining Ranger spacecraft would be subjected to additional tests, including a number of those recommended in the Hilburn report. Cortright forthrightly summarized the situation:

> I believe that the project people both at JPL and in NASA Headquarters felt that the testing program...was quite adequate for this job. The reason we are looking at it with a second view now, and instituting even additional tests, is that we don't know what happened to Ranger 6, and we are trying to do everything we can think of which might improve the system.[36]

The focus of the managerial issue was the responsiveness—or what Karth suspected as the lack of responsiveness—of JPL to the direction of NASA. Pickering flatly declared that, while JPL felt free to dispute NASA technical and organizational judgments, the Laboratory always accepted ultimate decisions and directions from Headquarters. The problem was to remain responsive to Washington while maintaining the spirit of free inquiry characteristic of a university laboratory. In fact, the new NASA–Caltech contract, Pickering explained, permitted NASA increased control over JPL.[37] "It would be quite unwise," Webb added in support of Pickering's observation, for subcommittee members "to start out on an effort to convert this into a civil service laboratory..." NASA and Congress must never "forget that a very, very large proportion of [JPL's] work is entirely satisfactorily handled—in fact, is outstandingly handled—and that there are men there who have done and are doing work more technical, more complex, more difficult than is being done in almost any other place by the human race."[38] The latest contract with Caltech, he assured the legislators, afforded the space agency all the controls and authority it needed.

Despite Webb's praise of JPL and announced confidence for the future, Karth was dissatisfied. "NASA is the contracting agency of the Government; [it] should be, in fact, the boss of the program. NASA provides the money, and therefore should have more to say about how this work is to be done, and by whom it should be done..." And considering Ranger's disappointing flight results, Karth supposed that at least NASA should have installed a strong technical team in Pasadena, "to oversee or supervise, not just management practices at JPL, but technical approaches as well." Webb had to say that closer control from NASA might have produced better results, but he doubted it. Indeed, under such conditions JPL might have lost some of its best people. As to Karth's proposal for a strong technical team at JPL, it was NASA policy, Webb asserted, to avoid that kind of exceptional "intervention and monitoring" of its contractors.[39]

All the same, the Karth subcommittee's report came down hard on the management of Project Ranger by JPL and the Office of Space Science and Applications. Relying on the portions of the Hilburn report made public, the previous NASA investigation in 1962, and on the testimony, the subcommittee charged that JPL had failed to establish rigid and uniform testing and fabrication standards for the Ranger spacecraft. The Laboratory had also poorly organized the project at the outset, then demonstrated an "embarrassing unwillingness" to accept NASA direction and make necessary changes. NASA's supervision of the work was judged inadequate because of a tendency at Headquarters to regard the Laboratory as a field center instead of a contractor. But in specific practical terms the Karth report scarcely threatened the Caltech–JPL relationship. It advised the space agency to take four steps: provide "appropriate" supervision of JPL either as a contractor or field center; install a general manager at JPL as a deputy director responsible for flight projects such as Mariner and Ranger; reform the institutional relationship so as to achieve responsiveness to NASA direction; and consider executing a one-year contract with Caltech instead of a longer-term agreement.[40]

Save for settling the precise period of the contract extension,[41] so far as NASA and JPL were concerned all the other arrangements were going into effect, including the recommendation for a general manager, one of NASA's longstanding—and long resisted—demands of JPL. On June 29 Pickering announced that, effective August 1, retired Air Force Major General Alvin R. Luedecke, General Manager of the Atomic Energy Commission, would replace Brian Sparks as Deputy Director of the Jet Propulsion Laboratory. The appointment, approved by NASA and Caltech, gave Luedecke general manager responsibility for the day-to-day technical and administrative activities of the Laboratory.[42]

The Ranger inquiry, William Coughlin railed in *Missiles and Rockets,* approached a whitewash. "One cannot lay the blame for this at the door of the able Chairman... It seemed that his hands were tied by what appeared to be almost a conspiracy to soothe the wounded feelings of the Jet Propulsion Laboratory."[43] *Science* heaved a public sigh of relief. In the face of NASA's own original charges, the journal mused, Webb's unexpected support of JPL in the hearings had led many "to conclude that,...like the grand old Duke of York in the nursery rhyme, [he] had marched his soldiers up the hill, then marched them down again."[44] Of course, the original NASA charges had only been the Hilburn report, which the agency, including its director, had not endorsed. Considering that, Webb's support of JPL should have surprised no one. Webb himself was certainly pleased with the outcome. He told Representative George Miller that Karth had recognized "we are dealing with an extremely delicate situation, much like walking down Fifth Avenue in your BVD's." He for one was also gratified that Karth had resisted the pressure to "look for scapegoats."[45]

To Webb's mind, there were no scapegoats. There was a lunar flight project at JPL still to be completed successfully. At this stage in the nation's evolving space program, furthermore, he had little choice but to back the Caltech-operated laboratory. He also had no reason not to, considering the JPL concessions: greater control from NASA, a new general manager, acceptance of more spacecraft testing, and greater attention to the RCA television system. Now it remained to be seen how the Laboratory and NASA would fare with Ranger 7.

Chapter Seventeen

RANGER 7: A CRASHING SUCCESS

A LL of Ranger's participants "very clearly understood" that personnel changes were likely in Pasadena *and* Washington should Ranger 7 also fail, and the Hilburn Board's contentions be proved accurate. The tension was correspondingly magnified, and the pressure to succeed now "was unbelievable."[1]

REWORKING RANGER 7

In New Jersey, efforts at reworking Ranger's television system encountered an unexpected snag. For Ranger Block III, JPL had virtually relied on RCA as an extension of its Systems Division. For space-borne television applications, RCA was the acknowledged expert.[2] RCA's Astro-Electronics Division was designated the "subsystem supplier," and, subject to JPL approval and acceptance, authorized by contract to design, fabricate, test, and deliver all of the television subsystems and associated ground support equipment for Ranger. In Hightstown, New Jersey, JPL television project engineer Donald Kindt and two assistants had monitored the work, concentrating on the subsystem's functional performance. The impact of Ranger 6 and the investigations that followed, however, radically changed attitudes on all sides. At JPL, Schurmeier established a special "task team" to conduct a detailed review of the RCA design changes planned for the television subsystem, and "of the TV subsystem [proper] to assure ourselves that there are no other areas of major concern."[3]

A special deputy was assigned to assist Kindt as the JPL engineers in residence at Hightstown swelled from three to approximately twenty-five. RCA likewise multiplied the engineers at work on a subsystem suddenly in the public limelight, and its Astro-Electronics Division Ranger Chief Bernard Miller acquired a deputy as well.[4] RCA leaders brought in more "specialists, even got people from their Princeton Research Laboratory, which was separate from Hightstown, and put them on, eliminated the car pools which took people home at 4:30, and really turned to." The project also created a special interlocking RCA–JPL problem/failure reporting procedure, requiring the signature of Kindt as well as the RCA reliability representative on every television problem or failure analysis report.[5] Finally, NASA Headquarters dispatched a special team of engineers to Pasadena to oversee the testing of the Ranger 7 spacecraft at JPL. "I will expect these NASA representatives," Oran Nicks instructed Cunningham, "to write their own reports and not reassign to JPL personnel any tasks relating to their monitoring functions."[6] Careers hinged on the next flight, and all the participants would settle for nothing less than double-checking each other.

As recovery plans jelled, Schurmeier rescheduled the flight of Ranger 7 for the lunar launch window of June 27 through July 2, 1964, which, along with modification of the television subsystem and the basis for a spacecraft test program, was approved by Headquarters.[7] But the launch of Rangers 8 and 9 had to be postponed six months or more. NASA had committed two Mariner C spacecraft for flight during the Mars launch opportunity in the fall of 1964. They held priority, and they were to occupy the Atlas-Agena launch facilities at Cape Kennedy from August through November. Then, on March 27, 1964, quality control personnel at RCA opened one of the sealed electronic modules in the modified Ranger 7 television subsystem. Inside, the inspectors found a small polyethylene bag containing 14 screws and a lock washer.

RCA began an investigation immediately, and submitted the findings to Schurmeier in Pasadena four days later. This incident, the RCA security office reported, was an accidental one caused by the pressures of working around the clock in three shifts seven days a week on the camera electronics.[8] But others in the RCA project office, including supervisors, argued that this kind of "accident" was highly improbable. From their point of view, a bag of screws simply could not have slipped through the inspection process undetected. They believed that even though there was no direct evidence, the incident had to be an act of sabotage—with all that that implied for the preceding television failure. To make matters worse, the quality control records for this module in the twelve-hour period when it was sealed were missing.

If sabotage of the project could be established, all the Ranger 7 television hardware would have to be reopened and thoroughly reinspected, which would delay the flight at least a month and increase project costs some $500,000.[9] And only four weeks remained before Mariner C would claim the launch facilities at the Cape. Any delays, furthermore, would work to the benefit of the Soviet lunar program. No Soviet lunar missions had been recorded since Luna 4 in April 1963. NASA and JPL expected another launch momentarily, and credit for the first closeup pictures of the moon's surface would belong to the first nation to take them. A decision had to be made about the television subsystem—soon.

At mid-month, Schurmeier, who had expanded the investigation of the bag of screws incident, sent his deputy, Gordon Kautz, to Hightstown for a final evaluation. Kautz reviewed all the records and personally interrogated the RCA project participants. "It looked like sabotage...," Kautz later mused, "[but] I had the inherent and intuitive feeling that we didn't have any saboteurs there...I made an assessment that this was indeed an accident—the people were so fatigued back there that this had occurred." As Congress prepared to investigate Project Ranger on April 22, Kautz told Schurmeier his findings and opinion. After a long telephone conversation, Schurmeier agreed that the Ranger 7 television hardware should not be reopened and reinspected. Work was to proceed on schedule.[10]

On the Trail of Ranger 6

All the same, the fact remained that no one at NASA, JPL, or RCA really understood why the television cameras on Ranger 6 had failed to work. The mystery remained a topic of concern inside the project and out. Cartoonist Chester Gould dispatched detective Dick Tracy and the mechanical genius Diet Smith to the moon's Sea of Tranquility in Smith's "space coupe" to find the answer for themselves (Figure 81). Alighting beside debris in the crater formed by Ranger 6, Diet Smith exclaimed: "It's easy to see *why* the cameras failed. They were never turned on—look." Holding aloft a tangle of wires, the bemused detective replied: "People won't believe us." Erroneous as the account was, *Life* magazine remarked, "millions of Americans didn't find it unbelievable at all."[11] Another popular rumor held that NASA and JPL officials had forgotten to take the dust covers off the television camera lenses.

The turn-on of the television subsystem during ascent through the earth's upper atmosphere haunted the JPL project office. Far more painful than Gould's mocking cartoon, however, were the allegations contained in the Hilburn Board report, and its presumption "that *two or more* failures must have occurred *within* the spacecraft's TV subsystem."[12] That assertion directly contradicted opinion at JPL and RCA. Even though the formal Ranger 6 investigations had failed to

Fig. 81. Dick Tracy Investigates, With Note From Burke to Schurmeier (Created by Chester Gould © Chicago Tribune-New York News Syndicate)

uncover any certain cause, something was responsible, and Schurmeier was determined to know what it was. On March 23, a few days after NASA issued the Hilburn report, he directed the Launch Vehicle Systems Section to continue with the investigation at JPL.[13]

Maurice Piroumian, an electrical engineer who had migrated to America from France after World War II, was placed in charge. Suspicions by this time had shifted from the command switch and electronics within the television subsystem to the male pins contained in the external umbilical connector on the Agena nose fairing. This cluster of pins permitted different elements of the television subsystem to be exercised and monitored on external power by means of a plug and extension cord linking the spacecraft to the launch complex. The spacecraft switched to internal power when the plug was pulled away at liftoff, and a small door then swung closed over the pins and latched mechanically. But the pins had been placed very close together, and a "hot" pin connected to the television's battery was located 6.4 millimeters (0.25 inch) away from the pin of the sensitive television command circuit. Twenty volts were present on the battery pin, and only 3 volts were needed to actuate the television command circuit and step the switch. The inadvertent turn-on, the engineers now reasoned, had been caused by something bridging the gap between these two pins (Figure 82).

During the next few months, Piroumian and his team of JPL engineers, working closely with others at RCA, concentrated on finding that mechanism. Tests were conducted to see if the umbilical door could have warped during the ascent so as to cross the pins, or, in the overcast above the Cape, if electrostatic

Fig. 82. Exposed Pins in the Agena Umbilical Connector

charge and discharge could have occurred on the surface of the nose fairing, or whether a shock wave of sufficient force to ionize gas had been produced during the staging of the Atlas launch vehicle.

Of the three hypotheses, the shock wave theory was perhaps the most intriguing. It was propounded by Alexander Bratenahl, a physicist in JPL's Space Sciences Division, who had tied it to the peculiar manner of staging of the Atlas launch vehicle.[14] Struck by the simultaneity of the staging sequence and the television turn-on, Bratenahl reasoned that the two events had to be closely linked. Telescopic movie films of other Atlas flights showed a large white cloud momentarily envelop the rockets during main-stage engine shutdown and separation.

Personnel at General Dynamics Astronautics, the Atlas manufacturer, informed the physicist that 112 kilograms (250 pounds) of liquid oxygen were released into the atmosphere when the oxygen lines were severed upon jettisoning the main stage engines. Bratenahl's preliminary calculations indicated that this much liquid oxygen dumped into the upper atmosphere could possibly have generated a blast wave whose mechanical force could "buckle the umbilical door and momentarily short the pins activating the command switch," or whose "ionization effects could have produced a high conductivity region around the pins permitting the shorting." Bratenahl transmitted this supposition, endorsed by Hans Liepmann of Caltech's Aeronautical Department and Robert Mackin of JPL's Space Sciences Division, to Schurmeier in late May.[15]

On June 29, Piroumian's investigators presented their findings to the JPL project office. Schurmeier learned that, regrettably, all of the hypotheses had been disproved. Though most RCA and JPL investigators favored the electrostatic charge and discharge thesis, the experimental evidence was inconclusive. The decrease in the dielectric strength of air with altitude made it highly unlikely that appreciable spacecraft voltage could build up at the very high altitude where the turn-on had occurred. Theoretical calculations prepared by James Kendall, Jr., Bratenahl's colleague in the Fluid Physics Section, had proved the shock wave theory false. Lacking a positive cause after weeks of fruitless study, Schurmeier terminated the investigation.[16] Apparently the failure of Ranger 6 was to be relegated to that category of glitches that appear once but can never be duplicated again in the laboratory.

Most scientists and engineers have experienced the tantalizing suspicion that the solution to a problem must lie hidden in the data at hand, and that disquieting state now afflicted Bratenahl. As Ranger 7 proceeded toward its launch, he set aside other duties and pushed on with the investigation alone. He obtained more movie film and had the frames enlarged. Careful restudy of the Atlas separation sequence revealed brilliant luminous flashes sparking within the white cloud of liquid oxygen surrounding the rocket. Another call to General Dynamics in mid-July, and the physicist had his answer.

What he had at first believed to be a shock wave was in fact a powerful detonation flash wave. The severing of the main-stage Atlas engines spilled not only 112 kilograms (250 pounds) of oxidizer into the atmosphere but 67 kilograms (150 pounds) of kerosene as well. The sustainer rocket engine, burning all the while, ignited the vaporized mixture, which rapidly moved forward over the missile at a rate of 180 meters (600 feet) per second. "The spherical shape of the flash wave advancing against a Mach 3.5 flow," Bratenahl observed, "suggests considerable energy in whatever is going on in the flash, otherwise the wave would be badly distorted by this high-velocity flow field." And since the umbilical door was mechanically latched rather than hermetically sealed, and normally bowed outward, there was ample room for the plasma to enter and short the pins. The sequence of Ranger 6 events strongly implied that this was exactly what had happened: The Atlas engines had separated at 140.008 seconds after liftoff; the television subsystem had turned on at 140.498 seconds, coinciding with the advance of the flash wave. A subtle design error in an umbilical connector had cost the project Ranger 6.[17] Were the project still interested, Bratenahl added almost as an afterthought, a few relatively inexpensive experiments could be conducted to confirm this hypothesis.[18]

Having satisfied his intellectual curiosity, Bratenahl took his time with the report. The importance of his findings—discrediting as they did the allegations of the Hilburn report—escaped him completely. So did the anxiety that beset other scientists and engineers on the project. Thus, instead of bothering Schurmeier, he sent the memorandum to the Space Sciences Project Representative, Charles Campen, who was busy with preparations to receive and process Ranger 7's lunar photographs. Delivered shortly after the flight of Ranger 7, Bratenahl's memorandum got buried on Campen's desk, disappeared in the turmoil which followed, and was all but forgotten.[19]

PREPARING TO GO AGAIN

When the Congressional hearings drew to a close on May 4, 1964, Schurmeier took stock of the project's expanded test program and of plans to launch Ranger 7. The spacecraft, returned to JPL from Cape Kennedy, had had its electrical system altered for the RCA payload. The television subsystem, modified, tested, and accepted by the project office, had arrived a few days earlier.[20] Plans for additional spacecraft testing had been outlined during the hearings, but the details remained to be worked out. With the launch set at the end of June, barely seven weeks remained to establish acceptable procedures and conduct the tests. That schedule was too tight. On May 11, Schurmeier requested a delay of four weeks—a move calculated to consume all of the time left before the Mariner Project claimed Launch Complex 12 at the Cape. Homer Newell approved the delay immediately.[21] Now no margin for error and subsequent recovery was available. Should another serious problem be encountered, all of Ranger's flights would be postponed indefinitely.

JPL engineers and NASA personnel ironed out details of the test program for Ranger 7 in mid-May. Besides a "no wire" test, they approved a minimum wire test of Ranger 7 at JPL inside a thermal vacuum chamber. Also, procedures to qualify and conduct a full-power test of the television subsystem with Ranger 7 fully fueled at Cape Kennedy were established.[22] All of this activity would be monitored by the team of NASA engineers appointed earlier by Oran Nicks.

Making doubly certain that the test standards expected were in fact met, Newell created a special ad hoc group to determine Ranger 7's flightworthiness, authorize its shipment to the Cape, and, finally, approve the launch. Termed the Ranger 7 "Buy-Off" Committee, which meant the acceptance of the vehicle for flight, its membership bespoke the importance Headquarters accorded this mission. Chaired by Newell, the committee consisted of Edgar Cortright, Oran Nicks, Herman LaGow—representing the Hilburn Board—Robert Garbarini, a Newell troubleshooter and director of engineering for the Office of Space Science and Applications, and Francis Smith, Assistant Director of the Langley Research Center.[23] On June 15 and 16, NASA's Buy-Off Committee convened in Pasadena; after a review of the records, the committee members determined that Ranger 7 and its modified television subsystem had met or exceeded established acceptance criteria. The committee authorized JPL to ship Ranger 7 to Cape Kennedy to begin the final round of testing.[24]

Ranger's experimenters, meanwhile, chose a lunar target for Ranger 7. Maxime Faget, Assistant Director for Engineering and Development at the Manned Spacecraft Center, had recommended flying Ranger 7 to the identical Ranger 6 impact site so that the freshly-made crater might be observed and the moon's surface hardness extrapolated from the crater's depth.[25] But lighting conditions at that location were no longer acceptable. Using the Willingham–Rindfleisch "figure of merit" scheme, Gerard Kuiper and his experimenter associates selected a target 11 degrees south of the lunar equator, along the northern rim of the Sea of Clouds. Shadow contrast here would be optimum for picture taking; what was more, this little-known mare region appeared suitable for manned landings, *and* it was traversed by fascinating rays from the crater Copernicus. Apprised of the target, Apollo officials in the Office of Manned Space Flight offered no further comment.[26] Newell approved the selection, and in early July the Ranger experimenters reconvened at JPL to participate in a final series of television tests employing the Ranger proof test model.

The flight hardware had already begun final tests at Cape Kennedy. Ranger 7 completed a second "no cables" test on July 6, whereupon it was moved to the launch complex and mounted atop the Atlas-Agena launch vehicle for countdown tests. Three days later NASA's Buy-Off Committee reconvened at the Cape. After an all-day meeting, committee members agreed that Ranger 7 and its launch vehicle were "flight ready."[27] They authorized the project leaders to proceed toward a launch on July 27, the first opportunity in the five-day launch period. By July 26, all of the flight machinery at the Cape was deemed ready, and so was the new nerve center of the Deep Space Network.

For the first time, command and control would be exercised from the Space Flight Operations Facility at JPL[28]—a three-story, bunker-like, rock and concrete edifice that concealed spacious and functional quarters. Ranger's Space Flight Operations Director Rygh took pride in the new control center and its large operations hub, known as the net control area. Blackboards, pinboards, and makeshift telephone links had disappeared. Instead, teletype printers displayed Ranger's encoded telemetry. Keyboard tape perforators and tape readers interpreted, verified, and dispatched commands. An operations voice control system provided a "hands off" talk and listen capability on the communications nets. Rear projection screens positioned before the controllers displayed pertinent information. Closed-circuit television was available for viewing other data, and for face-to-face discussions with Ranger's experimenters, and guidance and flight analysis teams in nearby rooms (Figure 83).[29]

Fig. 83. Net Control Area in the New Space Flight Operations Facility

By July 26, the Deep Space Network completed the last net integration and operational preparedness tests. Everything was again ready. But the confidence of early Ranger days was missing, displaced by an unspoken, gnawing anxiety. Had every conceivable test been made? Had contingencies for every mischance been devised? Ranger participants had thought all possibilities of failure eliminated before. Failure this time would mean certain disaster for Ranger, with profound repercussions for JPL and NASA. Virtually every project engineer and scientist harbored his own particular concerns for this mission, including especially Gordon Kautz, beset by second thoughts over the recommendation to proceed without reinspecting all of the electronic modules in the RCA television subsystem. "That recommendation," Kautz recalled looking back over a legion of flight project choices, "was the most significant one I ever made."[30]

THE FLIGHT OF RANGER 7

Newsmen from around the world gathered at the JPL auditorium for the launch of Ranger 7—three days ahead of the scheduled lunar impact. There, closed-circuit television permitted them to view the activity in the nearby control center. George Nichols, replacing Downhower as the voice of Ranger 7, kept the newsmen posted on the status of launch operations at Cape Kennedy. Across the country, in the marble and glass NASA Headquarters building in Washington, the small sixth-floor auditorium was again equipped to receive Nichols' account for assembled officials and visiting dignitaries.

At 6:47 am EDT on July 27, 1964, countdown for the launch of Ranger 7 began. Newsmen were apprised of steady progress to meet the day's launch window, which extended from 12:32 to 2:42 pm. Schurmeier monitored events from the new control center at JPL, while project engineer Wolfe replaced him at the Cape. The project manager, it had been decided, could ill afford to spend any time on a plane returning to the West Coast after the launch when he might well be needed for important decisions. Fifty-one minutes before liftoff, Atlas personnel called a hold to replace a faulty telemetry battery. After an hour's delay the count resumed, but halted again when engineers detected a malfunction in the Atlas ground guidance equipment. This problem was not so easily resolved, and at 2:20 pm George Nichols told his audiences: "We are sorry to announce the launch for July 27 has been scrubbed." Though the weather at Cape Kennedy remained favorable for another launch attempt the next day, the unexpected postponement seemed a bad omen.

On Tuesday morning, July 28, the countdown recommenced at Launch Complex 12 and proceeded smoothly. Exactly at 12:50 pm EDT, just a few seconds into the launch window, Altas 250D roared to life and, slowly gathering momentum, bore Ranger 7 upward into clear Florida skies (Figure 84). When the Atlas mainstage engines shut down and were jettisoned a few minutes later, the booster was enveloped by the resultant detonation wave. Ranger's telemetry continued without perturbation; the television subsystem appeared to be safe and

Fig. 84. Launch of Ranger 7

operating properly. Above Antigua, Agena B 6009 separated from the Atlas and ignited its engine, inserting the craft into the prescribed elliptical earth orbit. Shortly thereafter the tracking station on Ascension Island reported all was well as the vehicle arced eastward over the South Atlantic. At 1:20 pm, over the West Coast of South Africa, the Agena engine fired again, injecting the spacecraft onto its lunar trajectory. The deep space station at Johannesberg reported separation of the spacecraft from its carrier rocket: Ranger 7 was on its way to the moon.

As flight controllers set to work determining the trajectory and comparing it to the one desired, Ranger 7 extended its solar panels and high-gain antenna, and acquired the sun and earth. Later in the afternoon, Pickering, Parks, and Schurmeier met with newsmen to answer questions and inform them personally of the mission's progress. The launch vehicles, Pickering said, had placed Ranger 7 on a very accurate trajectory which, if no adjustments were made, would cause the spacecraft to graze the leading edge of the moon and impact on the backside. Commands for the midcourse maneuver, to correct the flight path toward the target in the moon's Sea of Clouds, were to be radioed to Ranger at 3:00 am PDT the following morning. Though many hours still remained before this mission was over, the JPL officials were now willing to concede a 50–50 chance of success. "But the memory of the letdown in the final minutes with Ranger 6," Richard Witkin reported in the New York Times, "was too fresh for anyone here to feel very confident."[31]

Early Wednesday morning, on July 29, the coded maneuver commands were transmitted to Goldstone in the Mojave desert and relayed to Ranger 7. At the appointed hour the spacecraft responded, rolling and pitching into the desired attitude. Ranger's midcourse engine ignited at 3:27 am PDT, slowing the spacecraft and correcting its course to bring it down on the moon at a planned 21 degrees West Longitude, 11 degrees South Latitude, in the Sea of Clouds. At JPL pleased flight controllers awaited the tracking data that would be used to predict the precise point of impact. A few hours later Pickering, Parks, and Schurmeier met again with newsmen in the JPL auditorium. Ranger 7's midcourse maneuver had gone flawlessly, the elated project manager announced; "there have been no anomalies whatever from launching up to now." The craft, he said, would crash almost exactly on target at 6:25 am PDT Friday morning. Would the television cameras work this time, a newsman asked. "Very definitely," Schurmeier responded. "It is still a long time until Friday," Pickering hastened to interject, but he agreed that the worst hurdles were over, and the odds for success had improved measurably, perhaps to 80–20 instead of 50–50.[32]

Cruise operations continued without a hitch on Thursday, with telemetry confirming Ranger 7's health. Project experimenters, led by Gerard Kuiper, had by now convened in the experimenters' room at the JPL control center, where they watched flight controllers nearby, and kept their fingers crossed. Newell, Cortright, and Nicks boarded a plane for Pasadena, there to join Pickering, Schurmeier, and Cunningham for the final phases of the mission. In Washington, others would "listen-in" at JPL inside the auditorium at NASA Headquarters. Lunar impact,

Washington time, would occur at a respectable 9:25 am, and a number of Congressmen had indicated their intention to join NASA officials and reporters on this occasion. With James Webb on a long-delayed vacation overseas, George E. Simpson, NASA Assistant Deputy Administrator, agreed to host this gathering on the East Coast.

Ranger's engineers and scientists, on duty or off, got little sleep Thursday night. By 6:00 am PDT on Friday morning, July 31, Newell, Cortright, and Pickering had taken their places in the visitors' gallery in the Space Flight Operations Facility. Below and in front of them, flight controllers at their consoles monitored Ranger 7's incoming telemetry. Directly beneath them in a small project office, Schurmeier and Rygh supervised the final events in the life of Ranger 7. Unlike the preceding flight, the television cameras were to begin warmup twenty minutes before impact instead of eighteen, and the period for warmup had been halved, from three minutes to ninety seconds (Figure 85).

As Ranger 7 neared the moon, the attitude of the spacecraft with respect to the surface was determined to be acceptable for picture taking. As before, Kuiper advised Schurmeier that Ranger's experimenters wished nothing done to jeopardize its chances. Schurmeier agreed; no terminal maneuver to improve picture resolution would be attempted. At 6:07 am, George Nichols informed tense newsmen and JPL employees packed into the auditorium that the command to

Fig. 85. Newell, Pickering, and Cortright Confer at the Flight Control Center
early on July 31, 1964

warm up Ranger's two full-scan television cameras had been radioed to the spacecraft. Ninety seconds later he reported excitedly: full video power "strong and clear" on the full-scan cameras. Applause erupted from the audience. In the control center, standing behind the flight controllers with Kautz and NASA's director of physics and astronomy programs John Naugle, Cunningham was ecstatic. "It's too soon," Kautz warned. These pictures would be no better than the ones obtained from earth-based telescopes. Naugle nodded, yes, wait.

They did not have long to wait. The command to warm up the four partial-scan cameras equipped with telephoto lenses quickly followed at 6:10 am. Another ninety seconds, and Nichols relayed the news from Goldstone: full video power was reported on the partial-scan cameras. Inside the auditorium, applause turned to cheers (Figure 86). Thereafter, Nichols raced to keep pace with what looked to be burgeoning success: "Twelve minutes before impact, excellent video signals continue... Ten minutes... no interruption of excellent video signals. All cameras appear to be functioning... all recorders at Goldstone are 'go'... Seven minutes... all cameras continue to send excellent signals... Five minutes from

Fig. 86. Cheering

impact . . . video signals still continue excellent. Everything is 'go,' as it has been since launch . . . Three minutes . . . no interruption, no trouble . . . Two minutes . . . all systems operating. Preliminary analysis shows pictures being received at Goldstone . . . One minute to impact . . . Excellent . . . Excellent . . . Signals to the end . . . IMPACT!'' The hum of Ranger's chromatic telemetry inside the auditorium abruptly ceased at 6:25 am PDT. The soft hiss of space static that replaced it was drowned by the roar of approval from newsmen and employees alike. Papers in the air; amid the shouts of many, a few openly wept. (Figure 87).[33]

At the Space Flight Operations Facility, controllers abandoned their consoles for handshakes, backslapping, and more papers in the air. It was bedlam. Rygh left the project booth and made his way through the jubilant throng to a colleague, Bert Dickinson, who, probably more than anyone else, knew the meaning of every Ranger telemetry measurement. They shook hands silently. "I was," Rygh recalled, "emotionally choked up, and I don't get that way often." "For Bud and Jim Burke, Gordon Kautz, and those of us who had lived Ranger for so long, it was kind of a spiritual happening."[34] For all the Ranger leadership, no other space flight mission at JPL would quite duplicate the sensation.

NASA and JPL project leaders soon left the net control area and adjourned to the experimenters' room, where the elated scientists eagerly awaited delivery

Fig. 87. And Weeping

from Goldstone of the first closeup pictures of the moon. Ranger's past trials, the frequent differences between scientists and engineers, even institutional relationships and the Congressional investigation, all were forgotten. Several cases of champagne that Schurmeier had purchased for the flight of Ranger 6 were broken open now, and together engineers and scientists toasted Project Ranger. Whatever their individual feelings at that moment, they could be sure of one thing: Ranger 7 was a resounding, a crashing success.[35].

Chapter Eighteen

KUDOS AND QUESTIONS

MIDSUMMER 1964—the World's Fair in New York, lush prosperity, black riots in Harlem and Rochester, the emerging "white backlash," an intensifying Presidential campaign—the Republican candidate Barry Goldwater promising to place direction of the space program largely in the hands of the military—President Lyndon Johnson announcing the commitment of 5000 more American advisory troops to Vietnam—and, in the middle of it all, Ranger 7: "We had made a historical achievement," *The New York Times* asserted, "and everything else for a while seemed small by comparison."[1]

NASA, the nation, and the world possessed the first closeup pictures of the moon's surface on video tape and 35-millimeter film. This success, even more than the first American manned orbital flights, it was widely believed, had redeemed the country from the humiliation of Sputnik 1, and signaled its ascendence in space exploration.

JUBILANT DAYS

Within an hour of Ranger 7's lunar impact, Newell and Pickering made their way down the hill from the Space Flight Operations Facility to a press conference in the JPL auditorium. At the entrance they were informed of an important phone call—the White House was holding open a line, the President wished to speak with them. Taking the call in a nearby office, the two men received President Johnson's felicitations. Ranger 7, the President said, was a "magnificent achievement." "On behalf of the whole country, I want to congratulate you and those associated with you in NASA and the Jet Propulsion Laboratory and in the industrial laboratories. All of you have contributed the skills to make this Ranger 7 flight the great success that it is... This is a basic step forward in our orderly program to assemble the scientific knowledge necessary for man's trip to the moon."[2] Would they, Johnson inquired, be willing to brief him on Ranger 7's lunar findings the next day? They would indeed, he was assured.

Now joined by Project Manager Schurmeier, Pickering and Newell entered the JPL auditorium to a standing ovation (Figure 88). "This is JPL's day, and truly an historic occasion," Newell observed, adding, "the entire mission was a textbook operation." "We have had our troubles," Pickering reflected, "...[but] this is an exciting day." How did he view the Laboratory's future after the success of Ranger 7, a newsman asked? "I think it's improved," Pickering replied with a wide smile, evoking laughter and applause. JPL's director passed the laurels for the success to Schurmeier and those directly involved with Ranger at JPL, NASA, and in industry. Schurmeier, in turn, thanked Ranger's team of engineers and scientists.

The focus of everyone's attention, however, was on Ranger's pictures, and when they might be made available. Schurmeier tantalized the audience with hints that the pictures were very good. All of the camera elements and ground recording equipment had functioned perfectly. NASA and JPL officials had been in touch with those at Goldstone who had seen polaroid contact prints, but beyond stating

Fig. 88. Newell and Pickering Shake Hands Before the Ranger 7 Postflight Press Conference

that the final pictures contained details never seen before, Schurmeier declined further comment. Just as soon as the pictures were processed, the eager newsmen learned, a number would be made public, it was hoped before the day was out.[3]

The pleasure and jubilation were no less exuberant in Washington, where other space agency officials, a large Congressional delegation, and reporters listened to the proceedings from the auditorium at NASA Headquarters. "Everyone here is happy as hell," an agency spokesman declared. NASA Assistant Deputy Administrator George Simpson called Newell and Pickering after the press conference to offer his own congratulations. Also on hand for the occasion, House space committee chairman George Miller assured newsmen that Ranger 7 "puts us well ahead of the Soviets in the exploration of space." Then, referring to the recent Congressional investigation, he added, "I want to make it crystal clear that the Jet Propulsion Laboratory is doing a splendid job." Elsewhere the news was the same—and all of it was good.[4]

Even as the early morning news conference proceeded at JPL, the Ranger 7 film footage was flown from Goldstone to the Hollywood-Burbank Airport. Under guard, it was taken to nearby Consolidated Film Industries for processing. During the next six hours positives were made from the master negatives, then duplicate negatives from the positives. By late afternoon on July 31, the prints and slides were rushed to the Laboratory in Pasadena. While the experimenters pored over one set of pictures, Pickering, Newell, and members of the JPL Executive Council examined duplicates. The crispness of detail in the later frames—exposing the lunar surface in terms of a few meters instead of a few hundred kilometers—moved everyone first privileged to see them. They were truly superb. Newsmen were notified that selected pictures would be shown publicly by the experimenters in the JPL auditorium that same evening. With the long sought lunar data in their hands, it was the scientists' turn at center stage (Figure 89).

At 9 pm Pickering introduced the Ranger experimenters before a nationally televised news conference. Principal investigator Gerard Kuiper opened the presentation by explaining the magnitude of Ranger's success: "This is a great day for science, and this is a great day for the United States," he declared. "We have made progress in resolution of lunar detail not by a factor of 10, as [was] hoped would be possible with this flight, nor by a factor of 100, which would have been already very remarkable, but by a factor of 1000."[5] Kuiper proceeded to show and discuss a representative assortment of slides from among the 4316 pictures obtained. These closeup pictures, he stressed, speaking as an astronomer, covered but a limited area of the moon, making generalization difficult. However, the 1132 scan lines per frame, in contrast to the 500 or so lines found in conventional home television sets, made the quality of Ranger's pictures "extraordinary" (Figure 90). In the discussion period that followed, reporters peppered the experimenters with questions. Having already had the pictures for two or three hours, the scientists were pressed to furnish "the most significant findings" to come from them, and asked whether one could be assured that Apollo astronauts could walk safely on the moon's surface. Many more days of analysis remained before any formal

Fig. 89. Experimenters Heacock, Kuiper, and Whitaker Examine Ranger 7 Pictures at the Flight Control Center

conclusions would be advanced, Kuiper stated. All of the experimenters present were impressed by the continuity of features from the large craters observable with telescopes down to smaller and smaller ones. "I think the rounded features of the large number of secondary craters are new," he added, "also their large number is new."

As for the lunar surface, geologist Eugene Shoemaker ventured to guess, it was "composed of debris— it is littered with debris." This material, Kuiper further suggested in keeping with his own theory of the moon's evolution, was most likely finely pulverized lava, of low density, although affected by a "rain of small particles so that the upper few centimeters and maybe as much as 1/2 of a meter would be modified by these impacts."[6] The information gained from these pictures, Shoemaker responded to another question, encouraged "the belief that the moon's surface, in at least certain parts of the maria, will not represent a particular problem [for manned landings] as far as roughness is concerned." But the actual bearing strength, he cautioned, could not be measured without first putting

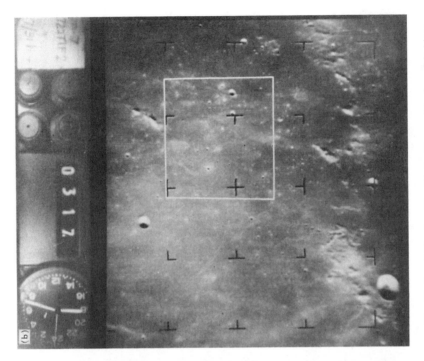

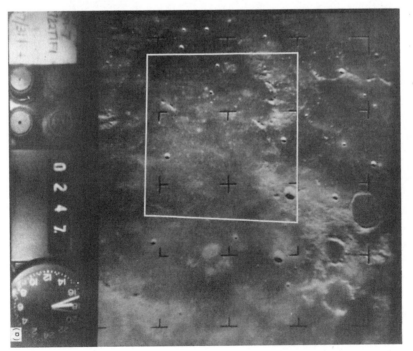

Fig. 90. Ranger 7 Closeup Pictures of the Sea of Clouds. North Is at the Top. The Clock and Number Identify the Frame. The Lens Markings Are Used for Scale Measurements. The White Rectangle Outlines the Next Full Picture To Be Viewed: (a) an Area 370.6 km on a Side; (f) an Area 2960 m on a Side, With the Large Crater in the Upper Left About 100 m in Diameter.

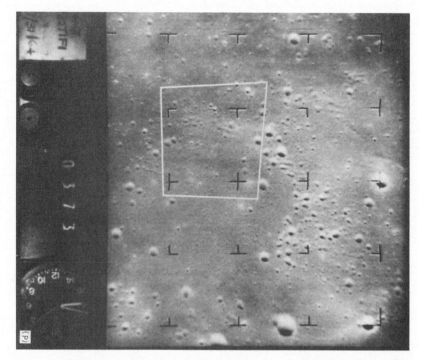

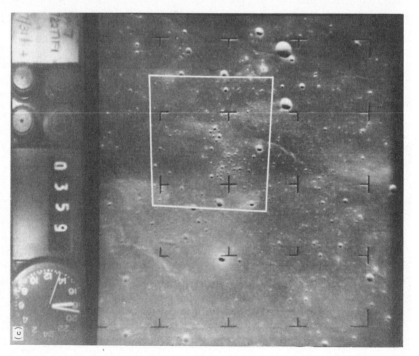

Fig. 90. (contd)

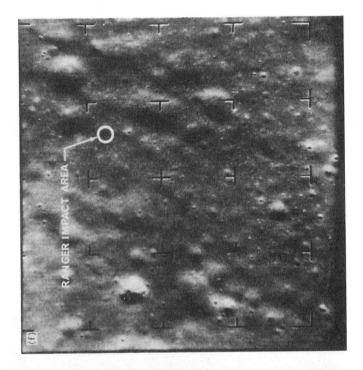

Fig. 90. (contd)

something down on the surface. Nevertheless, the surface appeared at first look to be compact enough to walk upon.

None of the theories concerning the moon's origin, Kuiper advised the reporters, however, seemed to be affected by the pictorial returns from Ranger 7. These pictures, he concluded, shed light on the evolution of the surface and its history after the moon's formation, not on the origin of the moon itself.[7]

Early the next morning, August 1, the "Redeye Special," a late evening jet flight from the West to the East Coast, touched down in Maryland with Newell and Pickering, who were shortly to see the President. On newsstands at the airport the papers acclaimed Ranger 7 and its lunar pictures in superlatives—on the front page and in editorials. Indeed, across the land, from the *Seattle Post Intelligencer* to the *Miami Herald,* or the *Boston Globe* to the *San Diego Union,* the praise was unanimous (Figure 91).[8] Overseas the foreign press had responded in a similar vein, seeming to agree that the United States had at last forged ahead of the Soviet

"Howdy"

Fig. 91. "Howdy" (Courtesy Tom Little in *The* [Nashville] *Tennessean*)

Union in space exploration. Even the Soviet press accorded the flight modest plaudits, though pointing out that the U.S.S.R. had photographed the moon from a spacecraft five years before.[9] The glowing accounts frequently heralded Ranger 7 as the greatest advance in space research since Galileo had trained his telescope on the heavens; it was heady stuff.[10]

Later that morning the two space officials were ushered into the White House Cabinet Room, where they explained for the President the Ranger 7 mission and the importance of its pictures to Project Apollo. President Johnson learned that portions of the moon's Sea of Clouds were suitable for Apollo landings, and that the design of the Apollo lander satisfactorily met observed surface conditions (Figure 92). With newsmen in attendance, the President soon queried Newell: He was satisfied with the return on the investment in Project Ranger, was he not? Newell said he was delighted. "Elated?" Johnson prompted. "Elated," Newell responded. Did the Ranger 7 "adventure" leave any doubts about the wisdom of going to the moon? "Not in my mind, not at all," Newell replied. "I would feel that we were backing down from a real challenge." He was

Fig. 92. Newell and Pickering Brief President Johnson

urged to continue. "We would lose leadership," Newell suggested. "In the world?" the Chief Executive asked. "In the world," Newell averred, sustaining the catechism. "This is a battle for real existence in the world, isn't it—for survival?" Newell agreed that it was, and said he was now "hopeful" that Project Apollo could proceed to a lunar landing on schedule. Responding to further questions, Newell assured the President that Ranger 7 confirmed the planning and soundness of America's civilian space program. "You don't expect a Congressional investigation then?" Johnson said with a smile.[11]

Afterwards, an obviously pleased President addressed both men, offering them his assessment of the progress achieved in space exploration and the thanks of a grateful nation. "We started behind in space...," he asserted, "[however] we know this morning that the United States has achieved fully the leadership we have sought for free men." Ranger, he went on, represented a weapon of peace rather than one of military might, and "a victory for peaceful civilian international cooperation..." Neither Newell nor Pickering was to be decorated with medals, Johnson concluded, "but they do have, and all of their associates from Mr. Webb down to the fellow who sweeps out the dust in the remote test laboratory, the gratitude and the admiration of all Americans of all faiths, of all parties, of all regions."[12]

NEW INTEREST IN A NEW ERA

No longer a liability, Project Ranger had vindicated American space policies and presaged accomplishments yet to come. A satisfied Representative Joseph Karth, chairman of the Ranger investigating subcommittee, inserted foreign editorial acclaim in the *Congressional Record*.[13] House space committee Chairman George Miller, at a special Ranger briefing for members of the House of Representatives on August 5, termed the mission "one of the greatest accomplishments that NASA has ever made..." It was, he said, "really the beginning." Addressing the legislators after Miller, Kuiper agreed that "a new era has begun in the exploration of the moon." Ranger's pictures, he declared, confirmed the design of the Apollo lander, indicated a smoother surface than had been expected, and tended to discredit the "deep dust" theories of the lunar surface.[14] That same day, NASA's long-delayed $5.3 billion appropriation for Fiscal Year 1965 sailed through the Senate by a voice vote, while efforts to reduce it were rebuffed 69-20 in a roll call. Senator J. William Fulbright decided against submitting a prepared amendment to cut 10 percent of the funds earmarked for Project Apollo, telling a reporter crisply: "The climate has changed."[15]

Enthusiasm over the lunar pictures was just as pronounced at the Manned Spacecraft Center in Houston. During the Ranger 7 mission a technical assistant at the center had been detailed to JPL to obtain prints of any lunar pictures for analysis by Apollo officials. But after the considerable improvement in resolution, the quantity of pictures, and preliminary findings were announced at the experimenters' press conference on July 31, the Houston plans changed overnight. Joseph F. Shea, the Apollo Spacecraft Program Director, and the Director of the Manned Spacecraft Center Robert R. Gilruth announced that they would personally visit JPL for a first-hand briefing by Ranger's experimenters.[16] Shea and Gilruth subsequently informed newsmen that an Apollo landing appeared "a lot easier than we thought." Much of the mare region photographed by Ranger 7, Shea observed, looked "relatively benign," adding that the 0.93-meter (3-foot) diameter pads on the legs of the Apollo lander would be able "to take hold" in this type of terrain.[17] If previously Homer Newell's office and the project experimenters had been left virtually alone to select Ranger's targets on the moon, now Apollo officials claimed priority in the choice of future lunar targets for Ranger.[18]

The Apollo group in the Office of Manned Space Flight at Headquarters wanted Ranger 8, scheduled for launch in February 1965, to answer a new question: Did lunar plains areas other than the Sea of Clouds exhibit similar, gently undulating features where crater "rays"* were less prominent? Such areas would be even more hospitable landing sites. In October, Air Force Major General Samuel C. Phillips, NASA Deputy Director of the Apollo Program, queried Oran Nicks on the matter. The target selected for Ranger 8 by the Office of Space Science and Applications, he advised, "should be within the permissible Surveyor and Apollo landing zones," and "in a ray-free marial region...."[19]

In a letter to JPL a month later, Shea in Houston had another request: "Our first suggestion" for targeting Ranger 8, he informed Schurmeier, "would be to return to the impact site of Ranger 7." "Much could be deduced about the mechanical properties of the surface from a crater produced by a known mass traveling at a known velocity." If that were not feasible, he preferred a highland site because "we believe the greatest gain in confidence would be achieved by predicting and finding a poor [landing] site."[20] The desires of the various Apollo directorates, not to mention Ranger's own experimenters, had to be reconciled, and for that purpose a meeting was convened on November 19 at JPL.

On the appointed date, Apollo representatives from the Manned Spacecraft Center and the Headquarters Office of Manned Space Flight joined JPL project personnel, all five Ranger experimenters, and a contingent from Newell's office. Schurmeier opened the proceedings, explaining that everyone's target preferences were to be discussed; but the final list of candidate sites, he made clear, was to be submitted by the Ranger experimenters and approved by Newell. When the question of sites arose for discussion, the Ranger experimenters endorsed the recommendations advanced by the Apollo group at Headquarters. Targeting Ranger 8 in a "blue" or dark mare along the lunar equator, Gerard Kuiper observed, should confirm the Ranger 7 findings supportive of Apollo, permit

---

*The ray systems photographed by Ranger 7 appeared to be largely composed of "secondary craters"—created by rocks and debris hurled outward across the lunar surface during the impacts that made the primary craters.

computation of the incidence of cratering in a ray-free area, and improve predictions of other similar regions on the moon. Shoemaker and Whitaker concurred, as did Harold Urey, who suggested Mare Tranquillitatis as the most likely candidate. If Kuiper's lava flows were in fact present on the moon, he added, they should be observable here.

The representatives of the Manned Spacecraft Center conveyed word of Shea's continued preference for a return to the point of impact of Ranger 7 or, barring this possibility, aiming for a "rough" highlands site. Surveyor Project personnel recommended sites of interest for Surveyor landings. The target preferences of the Apollo group at Headquarters and those of Ranger's experimenters, Schurmeier said in closing the meeting, appeared nearly identical. Since the launch period extended from February 17 through February 24, and in the absence of a consensus, it seemed advisable to consider a dark, ray-free area in the early part of the launch period, and a Surveyor landing site if the launch were delayed to the fifth or sixth day. Targets in these regions would be submitted before Christmas to NASA Headquarters.[21]

On January 19, 1965, Newell notified George E. Mueller, Associate Administrator for the Office of Manned Space Flight, of the Ranger 8 targets approved by his office and NASA's Space Science Steering Committee.[22] Conforming to the recommendations discussed at JPL in November, they were:

Launch Date	Lunar Target	Latitude	Longitude
February 17	Mare Tranquillitatis	13.5°N	24.0°E
February 18	Mare Vaporum	14.5°N	12.0°E
February 19	Sinus Medii	0.5°N	1.0°W
February 20	Secondary craters near Copernicus	4.0°N	15.0°W
February 21	Surveyor landing site No. 5	15.0°S	30.5°W
February 22–23	Crater Gassendi	18.0°S	40.0°W
February 24	Domes near Crater Marius	12.0°N	56.0°W

Mueller responded immediately. Though in agreement with the stated scientific objectives, he advised Newell that some of the aim points chosen for Ranger 8 were not entirely satisfactory for the needs of Project Apollo. Latitudes much closer to the moon's equator were preferred by the Office of Manned Space Flight. "In light of this, I would like you to consider the possibility of changing to the following targets..."[23]

Launch Date	Lunar Target	Latitude	Longitude
February 17	Mare Tranquillitatis	0.5°N	24.0°E
February 18	Highland region	0°N	13.0°E
February 19	Sinus Medii	0.5°N	1.0°W
February 20	Mare region	1.5°N	14.5°W
February 21	Mare region	3.0°N	28.5°W
February 22–23	Ray-free mare	1.0°S	42.0°W
February 24	Oceanus Procellarum	3.0°S	57.0°W

To Newell, Apollo's needs and wishes simply could not be denied. A few days later he apprised Ranger's scientists and engineers of Mueller's request, and of his decision to reconsider "targets within the proposed landing zone for the Apollo LEM spacecraft" on February 6 at NASA Headquarters.[24] This time, however, the meeting included Newell and Mueller, as well as Newell's science chief John Clark, and Donald Wise of the National Academy of Sciences' Space Science Board. The compromise list of Ranger 8 targets finally agreed upon emphasized flat lunar areas that "are not only of interest to a better knowledge of the moon, but provide the opportunity for observing sites in the current Apollo landing zone:"[25]

Launch Date	Lunar Target	Latitude	Longitude
February 17	Mare Tranquillitatis	3.0°N	24.0°E
February 18	Mare Vaporum	14.5°N	12.0°E
February 19	Sinus Medii	0°	1.0°W
February 20	Near Crater Gambart	4.0°N	15.0°W
February 21	Near Crater Reinhold	3.0°N	28.25°W
February 22–23	Oceanus Procellarum	3.0°S	44.0°W
February 24	Oceanus Procellarum	12.0°N	56.0°W

One week before launch, Ranger 8 had a destination. At the National Academy of Sciences, however, Space Science Board Chairman Harry Hess and his colleague Donald Wise viewed this targeting exercise and its outcome with alarm. Hess wrote directly to Mueller, urging restraint in any more demands by Apollo upon Project Ranger.[26] Ranger, Wise in turn counseled, was an unmanned instrument that could profitably be directed to lunar regions other than flatlands:

"It would be a great mistake to simply keep the Ranger 8 target list alive and use [it] for Ranger 9." In the future, "for reasons both of pure science and landing engineering," attention should be given to highland areas and the problems of lunar "surface processes."[27] Ranger's experimenters could no doubt agree with Hess and Wise, even though they had begun to advance disparate conclusions about the nature of the moon's surface.

HARD QUESTIONS FOR SPACE SCIENCE

In August 1964, Ranger's experimenters had begun assembling an atlas of Ranger photographs and constructing their individual scientific findings. Perusing the data as the work progressed, Kuiper marveled at the accomplishment: "To have looked at the moon for so many years, and then to see this...it's a tremendous experience. The whole thing was like a perfect symphony."[28] No less pleased for reasons of state, President Johnson announced the dispatch of Ranger photographs to leaders of 110 countries on August 16, asserting that "men of all nations recognize this as one of the greatest extensions of human knowledge about the lunar surface to occur in many centuries."[29] But how much the photographs had actually contributed to Apollo or man's knowledge of the moon was now opened to question.

On August 28, a month after Ranger 7's impact, the Interim Scientific Results Conference convened at NASA Headquarters.[30] Nicks chaired the meeting. Ranger's experimenters, he informed newsmen, were to leave immediately for Hamburg, there to present these scientific findings before the International Astronomical Union. Gerard Kuiper, Eugene Shoemaker, and Ray Heacock were on hand to discuss their results and answer questions. Harold Urey was unable to attend because of an injury; Ray Heacock would discuss Urey's analysis. Ewen Whitaker, already in transit to Hamburg, was also absent.

Kuiper displayed and discussed a number of slides of the moon taken both from earth and from Ranger's full- and partial-scan cameras. He also showed a five-minute motion picture. Composed of successive picture frames taken as the moon rushed ever closer, the film gave the viewer the novel sensation of being aboard the spacecraft as it plummeted to impact. Many of these pictures, Kuiper asserted, revealed ridges and flows typical of lava fields. Geologist Shoemaker agreed with Kuiper, and reaffirmed a prior observation benefiting Apollo. Even though ray systems—secondary and tertiary craters—from the craters Tycho and Copernicus criss-crossed this mare region, the lunar slopes appeared more gentle and the surface "smoother than I dared hope. I think it is good news for manned exploration."[31]

But Harold Urey, speaking through Ray Heacock, was not so sure that the news was good for manned exploration or for the presence of lava flows on the moon. To be sure, smaller and smaller ridges on the moon could be observed as the resolution improved. What caused them, he asserted, could not be ascertained from Ranger's pictures. Likewise, "the question of whether the lunar maria consist

of lava or finely divided material is, in my opinion, unanswered by these pictures."
The bearing strength and depth of the surface material, certainly, could not be
determined. Earth-based measurements of the thermal conductivity of the outer
parts of the moon, he reminded the audience, contradicted predictions of a firm
and compact surface. These measurements were "very low," and could, as
postulated by Cornell astronomer Thomas Gold, be accounted for "by a soft,
spongy, fairy-castle structure for some depth."[32] The Apollo lander, the chemist
implied, could be in for trouble.

Answering questions afterwards, Kuiper and Shoemaker were pressed hard
by reporters to justify their opinions in light of Urey's skepticism. The layer of
fragmented material on the moon's surface, Kuiper maintained, would most likely
be "crunchy," as one might expect of pulverized lava, and probably no deeper
than 1.5 meters (5 feet). Shoemaker concurred, but extended the depth to a
possible 15 meters (50 feet) of compacted debris, depending upon the amount of
erosion that had occurred over time. Ranger's television experiment, he explained
again, was designed to provide information on the surface features, particularly the
roughness of these features. That much it had done, confirming the design of the
Apollo lander. The precise bearing strength of the surface remained to be
investigated by the Surveyor machines to follow Ranger (Figure 93).[33]

Science, Shoemaker concluded in response to further queries, advanced one
step at a time, and with each step new questions arose naturally in a continuous
exploratory process. "We are not going to solve the origin of the moon [even] with
Project Apollo ... And," he added for emphasis, "we will not solve the origin of
the moon with a hundred years of post-Apollo exploration." Oran Nicks
interrupted to bring an end to the conference. Other scientists around the world, he
said, would in due course have an opportunity to study these pictures and to
advance their own theories. "It is to be expected that there will be different
theories evolving from these data ... That is the nature of science."[34]

The International Astronomical Union meeting in Hamburg, Nicks
reported to Newell a few days later, "was truly an American show." The Ranger
photographs were the talk of the conference, and the showing of the approach
sequence film brought an ovation. Commission 16 of the Union also named the sea
photographed by Ranger 7 "Mare Cognitum," the sea that has become known, a
move greeted with approval on all sides. The Ranger papers had sparked lively
and altogether natural disagreements in the scientific discussions that followed.[35]

But upon returning from Europe, Ranger's experimenters were jolted by an
article in the prestigious *Saturday Review*. Science editor John Lear charged that,
instead of assuring the safety of Apollo astronauts, Ranger's data confirmed the
dangers to man of ray-filled lunar maria. Scientific assertions to the contrary,
furthermore, raised the spectre of government-supported "instant science," in
which "a competent and if possible eminent scientist is captured by granting him
tax monies to finance his favorite type of experiments ... [and] he is called upon to
report findings he is not yet certain of." If he resisted this pressure, the scientist
was reminded of duty to country and science, and of the possible loss of federal

Fig. 93. Shoemaker and Kuiper Answer Newsmen's Questions at the
Interim Scientific Results Conference

support. If the scientist complied, "his reports become the basis for more generous
appropriations by the Congress, the research he loves is supported more richly
than ever, and he may even be invited to the White House to shake the hand of the
President and answer politically loaded questions on which the future of the entire
planet is professed to hang."[36]

Whatever the political response of the President and Congress, Lear had
clearly misread announced findings favorable to Apollo landings as politically
motivated, and was unaware of past struggles simply to ensure a place for science
in Project Ranger. Though Homer Newell had encountered the Presidential
questioning, Gerard Kuiper, aroused and very irate, saw himself as the supposed
eleemosynary captive in this instance. Worse, Kuiper had granted Lear the
interview on which the article was based. In a bitter letter to Norman Cousins, the

magazine's editor, the astronomer excoriated the interviewer, his thesis, and the attack on federally supported science:

> In the exhilarating hours after the Ranger 7 impact, I coined the term "instant science" in a jocular fashion to imply the need for scientists who had just received a mass of exciting new data to come up with some reasonable comment in a short while for a first news broadcast ... I casually quoted it to Mr. Lear who, without my knowledge, adopted it and wove it into a sinister pattern of government-supported science, thereby completely missing the meaning of the term ... [Mr. Lear implies] that I have given him information to the effect that scientists are under pressure to make statements favorable to the continuance of large government projects. This is complete fantasy. I have never myself been under pressure from anyone in the formulation of my own recommendations and would obviously resist such pressure had it been applied.[37]

In the weeks preceding the launch of Ranger 8, doubts about the project's value appeared more frequently in influential journals. Some commentators misconstrued the objectives of Ranger, others found its support of Apollo to be inconsequential: "Despite wide publicity to the contrary," *The Wall Street Journal* advised readers, "it is becoming increasingly apparent that Ranger 7 and its unprecedented closeup pictures of the moon have not yet solved the major mystery in the way of manned landing: will the moon's soil support the space ship or engulf it in a sea of dust?"[38] In the absence of incontrovertible findings, Ranger 7's contribution to planetary science was likewise suspect. Ranger might be "a magnificent engineering achievement," *The Christian Science Monitor* asserted, but what good were the visual data if no firm conclusions could be drawn? "The plain fact is, three months after impact the photos themselves have added very little to our knowledge ... Scientifically [they] rank well below the [non-visual] data gathered by the early, less costly, space probes which revealed the Van Allen radiation belt."[39]

Although disturbing to each of the scientific experimenters, this criticism particularly perplexed and troubled the principal investigator, Kuiper. He hardened the line on his own conclusions, and, seeming to make John Lear's pontification a self-fulfilling prophesy, began a spirited public defense of the project and of visual imaging for planetary science. With Ranger pictures in hand at the International Conference on Earth Sciences in Boston on September 30, he argued "with new forcefulness" that much of the lunar surface was eroded lava, "sand-blasted" by high-speed particles from space.[40] The adverse assessment of the project in *The Christian Science Monitor,* for example, Kuiper now warned Ranger coexperimenters privately, "points out the necessity for the scientific team to come up with a strong and coherent Final Report, which will enable NASA to counter or prevent articles of this type ... Clearly, the responsibility of the scientist has increased when projects of the magnitude of Ranger are reviewed."[41]

Having pushed so long and hard for a Ranger success, neither NASA, JPL, nor the scientists, it appeared, could adequately explain exactly what that success meant. With all other experimental instrumentation having been stripped from the Block III machines in favor of the television cameras a few years before, Ranger could be attacked as being of little value to science. Conversely, it could also be faulted for failing to answer questions about the moon's surface of great interest to Apollo. To the interested public, in fact, Project Ranger was fast falling between the two stools: space science on the one side and Apollo support on the other.

Ranger's scientists would perforce share any fall in esteem—a prospect Kuiper clearly found unsettling. And, like the engineer-managers before him, he also found that tension and frayed nerves accompanied the notoriety and excitement in projects of the magnitude of Ranger. Time remained, neverthless. Additional closeup photographs of other lunar regions from Rangers 8 and 9 could perhaps lend credence to one of the theories of the moon's surface and help dispel the doubts of Ranger's worth to science and Apollo. Then again, they might muddy the issue further.

Chapter Nineteen

THE RANGER LEGACY

A S the new year began, in an open letter to *The Christian Science Monitor,* Gerard Kuiper declared: "I definitely know of no better or cheaper way to get high-resolution lunar photographs--if I did, I would propose it." Project Ranger, he insisted, was unquestionably worth a final price tag of $270 million.[1] To be sure, the solitary experiment remaining aboard the Ranger spacecraft, the visual imaging television system, had fulfilled the lunar mission objective specified by NASA. But the first closeup pictures had seemed to generate as much heat as they did light.[2]

ONE MORE FOR APOLLO

On January 4, 1965, NASA's Ranger 8 "buy-off" committee met in Pasadena. The committee members found Ranger 8's test record to be "quite clean," and they approved shipment of the spacecraft to Cape Kennedy.[3] At the Cape during January, Ranger 8 and its Atlas-Agena launch vehicle methodically moved through the prescribed series of prelaunch tests. The often frustrating organizational and technical problems that had beset this process in earlier years were gone; ground testing space flight hardware at the Cape, as at JPL, had become routine. On February 2, NASA officials found the launch vehicle and the spacecraft flight-ready.[4] Final tests began immediately to meet the first day of the February lunar launch period.

As the hour of launch neared on February 17, the familiar drama was repeated: project officials at the JPL Space Flight Operations Facility supervising last-minute details; newsmen gathered for the occasion at the JPL auditorium; early rising agency officials and dignitaries watching and listening from the NASA auditorium in Washington. At 6:05 am EST, just a few minutes into the launch window, the engines of Atlas 196D came to life. Before the cameras in a nationally televised launch, the rocket majestically bore the 364-kilogram (808-pound) spacecraft moonward. Ranger 8's target was on the lunar equator in Mare Tranquillitatis, the flatland region of prime interest to Project Apollo, and the machine was expected to reach that destination in 65 hours.

Except for a curious drop in telemetry power during the midcourse maneuver, Ranger 8 performed flawlessly, and its managers decided to exercise the spacecraft completely. By a terminal maneuver at lunar encounter, the television cameras could be pointed directly along the flight path. That would reduce the smearing of the visual images while improving the resolution of the final closeup pictures (Figure 94). Schurmeier and Cunningham advised the experimenters accordingly. But the midcourse telemetry anomaly, JPL project scientist Vrebalovich later recalled, caused "the experimenters to become engineers again, and they decided against the terminal maneuver because it was too risky." The project officials reluctantly agreed to the experimenters' decision.[5] Early Saturday morning, on February 20, without any change in the attitude of Ranger 8, project officials commanded Ranger's television cameras to warm up. Once again, as the cameras came on and video signals began arriving at the Goldstone tracking station, George Nichols described the rush of impending success. And again, at 1:57 am PST, lunar impact brought the Ranger audience in the JPL auditorium, the flight control center, and in Washington to its feet with enthusiastic cheers and applause.[6]

Pickering, Schurmeier, Cunningham, and Cortright met with newsmen at a 2:30 am press conference in the JPL auditorium. It had been "a great satisfaction to see the project go so smoothly," Pickering declared. Schurmeier noted that Ranger 8 had landed within 24 kilometers (15 miles) of the preselected aim point in the moon's Sea of Tranquility. "It was another textbook flight," he said. Then, alluding to the loss of telemetry during the midcourse maneuver, he added, "Every textbook I've read has had a few small errors." Reporters learned that Ranger's pictures would be processed quickly, and some of them released at the experimenters' news conference that same afternoon (Figure 95).[7]

At the experimenters' news conference Kuiper showed and discussed for reporters a few representative pictures from among the 7137 provided by Ranger 8. The closeup pictures of Mare Tranquillitatis revealed surface features virtually identical to those in Mare Cognitum. In fact, Kuiper observed with the last slide, "If you did not know that this was taken with Ranger 8, you would think it was one of the Ranger 7 pictures." The remarkable similarity between two different lunar maria most impressed Ranger's scientists, who now supposed that "probably all lunar maria are pretty much this way."[8] However, there was one new interpretation. Drawing on a supposition made by Harold Urey, the experimenters had termed a feature appearing in many pictures from Rangers 7 and 8 "dimple craters." This kind of crater resembled the sort of subsidence one found in the sand of an hour-glass draining into a cavity below. Consequently, Kuiper asserted, it could be speculated that a similar phenomenon applied in certain regions of the lunar maria, much like collapse depressions on earth, with material dropping beneath the surface (Figure 96).[9]

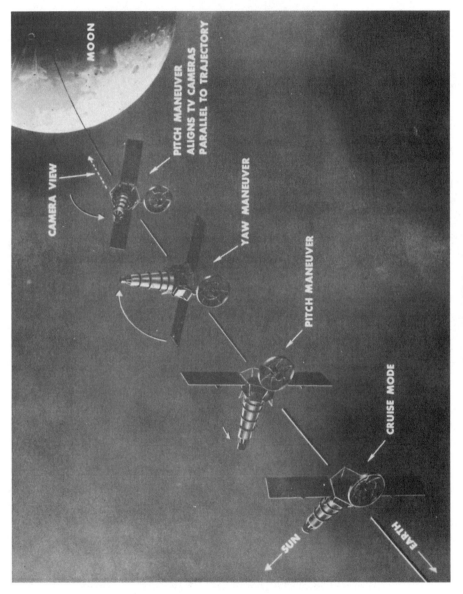

Fig. 94. Ranger Block III Terminal Maneuver

Fig. 95. Kuiper, Heacock, and Whitaker Examine Ranger 8 Pictures at the Flight Control Center

In the brief question and answer period, newsmen asked Kuiper to explain further the ramifications that dimple craters might be expected to have for the safety of manned landings. These depressed areas, he responded, might be regions of localized collapse, as often happened in lava fields. The surface itself, he continued to maintain, probably had a bearing strength similar to rock froth, melted and solidified in a vacuum. It should prove adequate to support manned landings, although, he now added, it also could be "treacherous" in spots. Eugene Shoemaker believed that dimple craters were nothing more than degraded secondary impact craters, and that the model of a lunar surface composed of fragmented debris had proved consistent; the similarity in fine detail especially was "one of the most striking results" to come from Ranger 8. That could be considered "very encouraging from the point of view of Project Apollo." But Ranger's pictures alone, Harold Urey reiterated, could not answer these questions. Until a Surveyor machine had made a soft landing, the actual bearing strength of the moon's surface simply could not be determined with certainty.[10]

Had Ranger, therefore, "speeded the day when we know man can land on the moon?" a reported inquired pointedly. Of course, Pickering responded; Ranger had added "a great deal of information" about the moon. "What specific areas of information had been added?" another quickly asked. Cortright answered for Headquarters: "We actually didn't know whether there was any spot on the

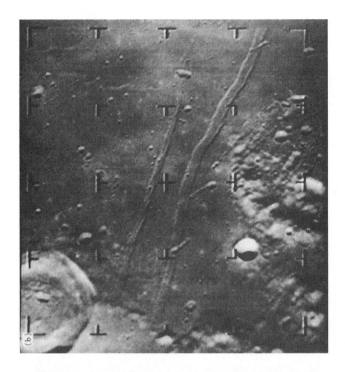

Fig. 96. Ranger 8 Pictures of the Sea of Tranquility: (a) an Area 107.8 km (N–S) by 123.9 km (E–W); (f) an Area 1360.3 m (N–S) by 1296.3 m (E–W). The Band of Radio Noise at the Right of (f) Occurred Upon Cessation of Transmission After Impact. North Is at Top. The Lens Markings Are Used for Scale Measurements.

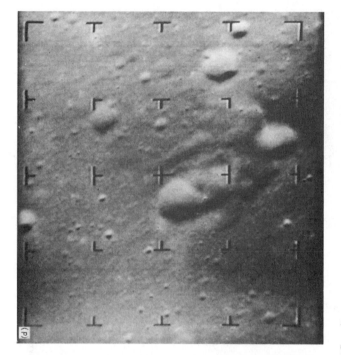

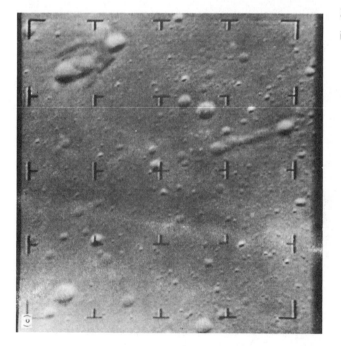

Fig. 96. (Contd)

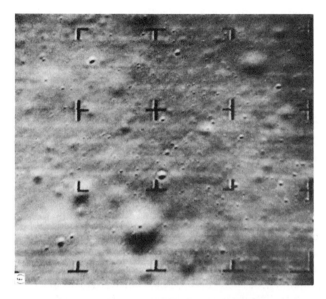

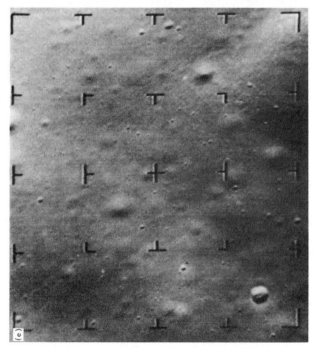

Fig. 96. (Contd)

moon that was level enough and smooth enough for the Apollo spacecraft to land," he stated. "On July 31 [1964] we found there was at least one [site] ... Today we know that there are at least two, and we have reason to believe that this view ... is probably typical [of lunar maria]. The big question now is: 'is it hard enough?' "[11]

The New York Times headlined this successful Ranger mission as one of support for Apollo.[12] But what the Ranger pictures meant for man on the moon, the *Times* concluded in an editorial the next day, was confirmation of uncertainties that "strengthen still more the case for making progress slowly, without any deadline on Project Apollo."[13] As for science, the experimenters' news conference had only added to public skepticism; the Los Angeles *Herald-Examiner* summed up the situation for its readers: "The Moon: Rock? ... Dust? ... Porous? ... Spongy?"[14] Virtually all the proponents of different lunar theories could and did find support in the pictures. Indeed, the very close similarity of the surface detail in the final pictures from both flights meant that no consensus was possible. Not now at any rate.

Scientific consensus in Project Ranger was far from the mind of President Lyndon Johnson on February 25, as he took his seat in a large conference room at NASA Headquarters. Victorious in the November elections, his inauguration already a warm memory, he was to be briefed on Ranger 8 and the current NASA program (Figure 97). The former Congressional leader and prime mover of the National Aeronautics and Space Act, Johnson was the first American President ever to visit NASA Headquarters. Instead of summoning NASA officials to the White House, the President explained afterwards, he had come over personally to convey to *them* his sincere thanks. Their productive efforts, he said, were "an example for all the rest of our Government and for public servants everywhere."[15] Confronted with increasingly difficult demands at home and in Vietnam, the Administration contemplated the moon and took obvious pride in the national achievements represented by such space projects as Gemini, Mariner, and Ranger (Figure 98). Next up was Ranger 9.

THE LAST ONE FOR SCIENCE

On February 27, 1965, one week after Ranger 8 landed on the moon, project experimenters and officials met at JPL to select the lunar targets for Ranger 9. Ranger's capabilities and restraints for the lunar launch period of March 19 to March 25 were first reviewed with Apollo and Surveyor representatives in attendance, then the experimenters recommended that Ranger 9 be directed to a lunar highlands region or other area of specific scientific interest. The Apollo officials concurred, and suggested a landing at the edge of material ejected from a crater in the lunar highlands. The proposals of Surveyor personnel, who hoped that Ranger 9 might return to another mare for purposes of Surveyor landing site selection, were overruled.[16]

Fig. 97. At Headquarters, NASA Administrator James Webb Explains Lunar Surface Model for President Johnson and Vice-President Humphrey. Constructed From a Ranger 7 Picture. Scale Models of an Apollo Lander and a Surveyor Machine Appear on the Surface (Courtesy World Wide Photos).

Ranger coexperimenters Heacock and Whitaker, designated to handle the preparatory work for their colleagues, met again on March 2 to consider a set of preliminary lunar targets. Their set emphasized large craters in the highlands. Separate meetings with the other experimenters during the next few days produced a firm list: [17]

Launch Date	Lunar Target	Latitude	Longitude
March 19	No agreement on a target reached among the experimenters		
March 20	No agreement on a target reached among the experimenters		
March 21	Crater Alphonsus	13.3°S	3.0°W
March 22	Crater Copernicus	10.0°N	19.5°W
March 23	Crater Kepler	8.2°N	37.8°W
March 24–25	Schroter's Valley	24.5°N	49.0°W

Fig. 98. "Successful Launch" (Courtesy Gene Basset, Scripps-Howard Newspapers)

When the list was submitted to Headquarters on March 10, Heacock had already strongly urged Cunningham to accept it. The two preceding flights, he said in a letter to the Ranger Program Chief, had made a "very valuable" contribution to Apollo, "but this is not generally recognized or appreciated." Moreover, "two impacts into mare areas have created a bad impression in terms of Ranger's ability to provide useful [scientific] data." Therefore, he concluded, "it seems only reasonable for NASA to allow the Ranger experimenters a free choice on this last mission. From all the inputs I have received, the Crater Alphonsus would be an almost unanimous first choice."[18] On March 10 Nicks wholeheartedly endorsed the targets recommended by Ranger's scientists, and Newell approved the selection a few days later.[19] Ranger 9 would fly to the lunar highlands for science—specifically, into the Crater Alphonsus if launched on the first acceptable day of the March period.

But the March 21 date conflicted directly with the planned launch of the first manned Gemini spacecraft. A precursor of Apollo, designed to test the space rendezvous docking procedures needed for manned flight to the moon, Gemini had priority. Gemini also required that the ground computers needed by Ranger at Cape Kennedy be available for its use one full day before launch. Still, the Gemini flight had already been postponed a number of times, and it might be delayed again. In that event canceling the March launch of Ranger 9 could be a costly mistake. On March 15 Associate Administrator Robert Seamans stayed the Gemini launch for one day, to March 23, permitting Project Ranger a chance at a single launch attempt on Sunday, March 21.[20]

At Cape Kennedy on March 18, Ranger 9 and its Atlas-Agena launch vehicle moved smoothly through the final prelaunch tests. Half a world away, at Baikonur in the Kazakhstan, U.S.S.R., two Soviet cosmonauts rocketed into space aboard the Voskhod 2. Before descending from earth orbit the next day, one of them, clad in a spacesuit and tethered to the spacecraft, floated in the void of space and "successfully carried out prescribed studies and observations." In Moscow, *The New York Times* reported, "a Soviet space official said 'the target now before us is the moon, and we hope to reach it in the not too distant future.' " In Washington, confidence in America's growing space capabilities was noticeably shaken.[21]

Inclement weather swirled over Southern Florida, complicated launch preparations for Ranger 9, and added further to the gloom. When the countdown began early Sunday morning, angry clouds obscured the sky, and winds gusting to 13 meters per second (30 miles per hour) buffeted Launch Complex 12. Project officials monitoring events from the flight control center at JPL, therefore, were pleased to learn that the winds above the overcast were acceptable for the flight—if only those at the surface would subside a little. The count was delayed. The winds slackened. At 4:37 pm EST, as the launch window neared its close, the countdown for the last Ranger mission concluded. The low-scudding overcast absorbed the clouds of steam and smoke as Atlas 204D, Agena B 6007, and Ranger 9 rose above the Cape and were quickly lost to view.

In the bright sunlight above the clouds, the Atlas-Agena performed perfectly. Injected on its lunar trajectory, Ranger 9 moved irresistibly toward the moon and the Crater Alphonsus. Named after a 13th-century king of Castile and patron of astronomers, the 112-kilometer (70-mile) wide crater was of particular scientific interest because its large central peak appeared to contain evidence of vulcanism. In 1957 Dinsmore Alter, the Director of the Griffith Park Observatory, had reported sighting fluorescent gas inside the crater, and a year later the Soviet astronomer Nikolay Kozyrev had succeeded in obtaining spectrographs of these emissions. Shortly after Project Ranger began, Urey had urged Newell to consider Alphonsus a lunar target of moment for this very reason.[22] Any overt plutonic activity inside the crater was likely to be revealed in closeup pictures.

On Monday, March 22, Ranger 9 successfully completed the midcourse maneuver and was correctly aimed into the crater Alphonsus. Other engineers and scientists at JPL, meantime, finished modifying a Surveyor "electronic scan converter" for use in Project Ranger. A last-minute idea agreed to by the experimenters, this device could accommodate Ranger's pictures for broadcast on commercial television across North America. It consisted essentially of two sets of television vidicon tubes facing each other. One set read the images appearing on its opposite number, effectively converting the lines per frame received from Ranger's cameras into conventional numbers of lines per frame. When Ranger 9 approached the moon early Wednesday morning, March 24, the scan converter was installed and checked out at the JPL Space Flight Operations Facility.[23]

For this mission, Ranger's experimenters had also agreed to a terminal maneuver to improve the resolution of the pictures. One half hour before impact, at 5:31 am PST, Ranger 9 executed the coded maneuver commands transmitted earlier from Goldstone. The spacecraft completed the necessary pitch and yaw movements, repositioned its high-gain antenna — once again pointed towards earth—and assumed the terminal orientation desired for picture taking. In von Karman auditorium at JPL, Ranger coexperimenter Heacock took his place in the broadcast booth to describe the pictorial events to the watching nation. At 5:48 am PST, twenty minutes before impact, Ranger's television cameras were warmed up. Moments later Goldstone announced the receipt of video signals. At the same time pictures selected from among those being taken on Ranger's two full-scan cameras filled television screens in the JPL auditorium and around the country. During the next eighteen minutes Heacock described the prominent lunar features that sprang into view as the spacecraft plummeted into the Crater Alphonsus. The transmissions ceased at 6:08 am, when Ranger 9 crashed beside the 1050-meter (3500-foot) high central peak within the crater (Figure 99).[24]

Afterwards, Heacock made his way as best he could through excited and inquiring newsmen to join his colleagues in the Space Flight Operations Facility. In the net control area, Ranger flight operations personnel and engineers had already left their consoles and were shaking hands and offering each other congratulations. Although their pleasure in the moment was undeniable, the mood was subdued. For them, Project Ranger had ended.[25]

At the following news conference, Nicks read from Silverstein's guideline letter that had established Project Ranger on December 21, 1959. The letter prominently listed closeup photography of the lunar surface among the experiments then planned for this unmanned scientific flight project, and that objective had been met. But, Nicks told the newsmen: "What you couldn't see and what we couldn't see in the beginning were some of the other things that Ranger would do. It was not in the guideline letter, for example, that Ranger should provide the capability for doing Mariner 2 to Venus [1962] in a very short time. It

Fig. 99. Urey, Whitaker, and Shoemaker, in Foreground, Watch Ranger 9 Pictures "Live" at the Flight Control Center

wasn't in the letter that Mariner 4 [1964], which is now on its way to Mars, would be a direct descendant of Ranger and use much of its technology. It also wasn't in the letter that Ranger should develop the attitude control system, the tracking capability, the midcourse maneuver system, and the TV advancements which are so obvious today.'' Pickering added his own thoughts to those already expressed: "The project we reflect on today has been a long and difficult road since 1959. We had our problems in the early days...[but] the achievements of the last three flights have shown that Ranger [could] carry out these deep space missions under remote command, that Ranger has indeed demonstrated the soundness of the basic system design, and that the closeup photographs...have opened a new field of the exploration of the moon.''[26] Most could agree with these observations—including Earl Hilburn, whose congratulatory telegram was to be found among others fast arriving at the Laboratory, hailing the successful completion of the project.[27]

The Ranger 9 pictures, Kuiper informed newsmen at the experimenters' press conference later that same day, held no surprises for science (Figure 100). To

Fig. 100. Urey, Kuiper, and Shoemaker Confer Before Ranger 9
Experimenters' Press Conference

be sure, dark "halo craters" on the floor of Alphonsus appeared of volcanic origin, but besides that, and the similarity of other surface features to those in the lunar maria, little else novel or unusual had appeared. Urey now agreed with his colleagues on one point. The dark halo craters probably were "due to some sort of plutonic activity beneath the surface," although he doubted their precise counterparts existed on earth. But, he added, evoking laughter, until more chemists were brought into the program, newsmen should not expect Ranger's experimenters to achieve "any reasonable interpretation [of the surface] that we can agree on (Figure 101)."[28]

Withal, the "live" television coverage of Alphonsus was popularly judged Ranger 9's most impressive accomplishment.[29] Everyone in North America with access to a television set had been able to watch the event and, as if holding a visual subscription to the *National Geographic*, experience firsthand the thrill of exploring the unknown. Relenting somewhat of its earlier opinion, *The Christian Science Monitor* offered "awed congratulations to all those involved in giving the public—'live,' in 'real time'—the sequence of pictures sent back by Ranger." "High and historic drama," *The New York Times* added, terming the mission "astronomy for the masses," and "the finest type of space research at this stage of history."[30] The people had participated, though the meaning and the medium of the experience remained scrambled. One local correspondent said it succinctly: "For most of us the pictures didn't look like much, but the mere fact of seeing them gave us a front row seat on science."[31]

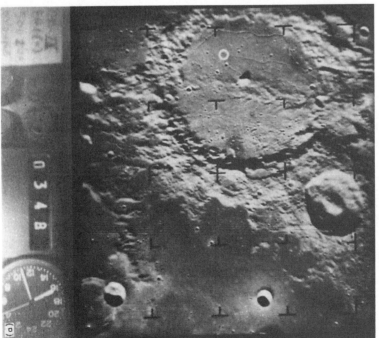

Fig. 101. Ranger 9 Pictures of the Crater Alphonsus: (a) an Area 214 km (N–S) by 202.7 km (E–W); (d) an Area 3.2 km on a side. North Is at Top. The Clock and Number Identify the Frame. The Lens Markings Are Used for Scale Measurements. The White Circle Denotes the Point of Impact.

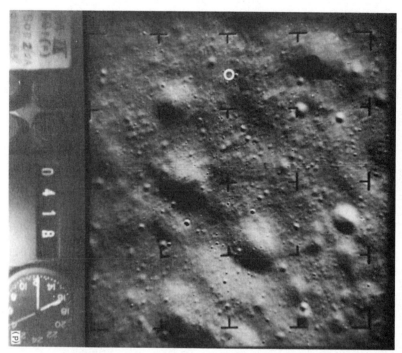

Fig. 101. (Contd)

President Johnson watched on a television set at the White House. Ranger 9, and the manned Gemini spacecraft launched successfully for a three-orbit mission around the world during Ranger's flight to the moon, assuaged much of the concern generated by the "spacewalking" Russian a week before. After the impact inside Alphonsus, the President issued a public statement exuding fresh confidence: "Ranger 9 showed the world further evidence of the dramatic accomplishments of the United States space team ... Steps toward manned flight to the moon have become rapid and coordinated strides, as manned space maneuvers of one day are followed by detailed pictures of the moon on the next (Figure 102)."[32] Called to the White House forthwith, James Webb briefed President Johnson and the members of his Cabinet on Ranger's photographs and their implications for Project Apollo.[33]

On Friday, March 26, Schurmeier joined Seamans and Gemini astronauts Grissom and Young at the White House. There, in special ceremonies presided

Ranger 9 Touch

Fig. 102. "Ranger 9 Touch" (Courtesy Tom Little in *The* [Nashville] *Tennessean*)

over by the President, the four men were honored for their contributions to the exploration of space. Webb unreservedly lauded Schurmeier's courage and competence (Figure 103). Memories of past trauma, of Ranger 6 and its aftermath, rose to mind. Hardly twelve months before Schurmeier had wondered if the project would survive. Now, surrounded by throngs of cameramen and reporters, decorated with NASA's Exceptional Scientific Achievement Medal, he was acclaimed a hero.

RANGER: AN ANALYSIS

However much credit he deserved, Schurmeier would have been the first to admit that he had benefited considerably from what the Ranger Project had forced NASA, JPL, and leading space scientists to learn since Pickering had first urged U.S. lunar flights in response to Sputnik. Prompted perhaps by the need for caution in dealing with the U.S. Air Force, the service that possessed large launch vehicles when Ranger began, and by a desire to maintain more control at Headquarters, NASA leaders had at first attempted to manage Ranger through committees, coordination boards, and other anemic administrative devices. Experience soon drove home the point that project management had to be delegated to a project manager at the pertinent field center, JPL. Experience also made clear the advantages of bringing together agency scientists and engineers both at Headquarters and in the field laboratory. And experience pushed NASA to

Fig. 103. White House Awards Ceremony. Left to Right: Vice-President Humphrey, Ranger Project Manager Schurmeier, President Johnson, NASA Administrator Webb

extract from the Air Force the authority for the Ranger Project manager to direct all of the work.[34]

The first project manager, James Burke, did a good deal to point up the meaning of the early experience for the NASA leadership. At first recognized only as the spacecraft manager, he fought the battles for a hierarchically structured project organization, sensible procedures for launch vehicle procurement and launch operations, the chance to serve a single master at NASA—manned or unmanned—and for the rights of the project manager and the experimenters. For a considerable time, Burke also had his battles within JPL itself. Following traditional JPL practice, the leaders of the Laboratory attempted to direct Ranger through a small project office supported by an array of longstanding technical divisions. The post-Ranger 5 investigations proved that this organizational form— where technicians report to two bosses—tended to produce hard administrative and technical problems for the project. Well-established JPL division chiefs, busy tidying up details on the Sergeant missile program, tended to respond cautiously rather than promptly to the first project manager.

Had Burke been more forceful and demanding, perhaps he could have made the early Rangers work. He did occupy himself considerably with trying to carve out an appropriate area of authority. The cost included having to delegate spacecraft details to the Systems Division, where many of them were attended to improperly. But Burke, youthful and informal, having never even headed a JPL technical division, could not easily make headway against the established chiefs at the Laboratory, not at least without decidedly more support from Pickering. It took the post-Ranger 5 pressures from NASA Headquarters to curtail the autonomy of the technical divisions and to centralize functions to assure quality and reliability in the project.[35]

Whatever Burke's lack of authority, he shared the tendency of JPL officials to apply their experience in the Army missile program to the challenges of space research. They did not embrace the cavalier "shoot-and-hope" approach of which they were accused, but they did count on a sound design rather than on extensive ground tests to guarantee spacecraft reliability. Actual "shakedown" flights were expected to provide data on aspects of the spacecraft performance that could otherwise be obtained only in elaborate ground test facilities. On the early Rangers this tendency was reinforced because special test facilities for such spacecraft had yet to be built when the space program began. But the erroneous weight limitation crippled the original lunar spacecraft design, and heat sterilization of major components compromised its performance. Flight failures resulted from coupling all these difficulties with a hell-bent-for-leather approach to beat the Russians.

The desire to beat the Russians while minimizing costs clearly influenced Burke and other JPL leaders to hold schedules inviolate, freeze the design of the Block II lunar machines at the erroneous weight limit, and drop redundant engineering features. Most NASA officials, for their part, countenanced, even encouraged the headlong rush to demonstrate American technical supremacy. Though the performance of the Block II seismometer capsule subsystem might be questionable and evidence of the undesirable effects of heat sterilization of components mounting, they too approved launching on schedule. Homer Newell summed up the urgency that moved scientists and engineers alike:

> During the early years of NASA there was great pressure to catch up with and excel over the Soviets. This was perhaps best typified by Keith Glennan's instruction...that we should conduct ourselves and our programs as if we were at war and intended to win it. While this was never written down, we heard it, believed it, and endeavored to carry it out.[36]

The haste engendered by the "space race" syndrome was accentuated by the ambitions of Burke and his engineering colleagues at JPL to build a planetary, not merely a lunar spacecraft. Launches to the moon could be made monthly; to the planets at much less frequent intervals. "We consciously set out to accustom our people to meet fixed launch schedules in NASA's deep space program," Burke recalled. "Celestial mechanics fixed the schedules. We knew we couldn't afford the delays that the earth satellite projects often claimed." Of course, attempting to make Ranger a planetary rather than a simpler lunar machine entailed high risks, but the early Ranger was conceived and authorized as a fast-paced, high-risk undertaking. The risks might not have been so great, it must be said, had Burke not had to accommodate a number of space science experiments at the same time that his engineers were attempting a radical advance in spacecraft technology. Space scientists tended at first not to appreciate the requirements of the engineering task, and their lack of appreciation seemed to have been mirrored consistently in the NASA office of Newell and Nicks, who failed to perceive the soundness of Burke's protests until Ranger was on the verge of disaster. The unrestrained competition for the scarce commodities of spacecraft weight, space, and power nearly brought down the project.

Burke knew the risks, accepted them—and paid the price. But he had the satisfaction of staying on to see Ranger become the success he had always believed it could be. The things for which he struggled—straightforward, unchanging project objectives, experiments that could not be altered at a scientist's whim, recognized authority and responsibility in and from all the agencies participating—Burke won all these in defeat. NASA and JPL leaders granted all of them to his successor, Schurmeier, who, with new test facilities and procedures, used them skillfully to the advantage of Ranger.

THE RANGER LEGACY

Ranger was accounted to have cost $267 million,[37] and observers wonder whether the project was worth the time, the money, and the careers it claimed. Perhaps not for sky scientists, whose experiments were left out of all but the ill-fated Block I flights. Perhaps not, it seemed at the time, even for planetary scientists. The lunar photographs supplied no decisive evidence about the formation, structure, or strength of the lunar surface. "Ranger's pictures are like mirrors," the planetary scientist Thomas Gold mused after the flight of Ranger 9, "and everyone sees his own theories reflected in them."[38] So indecisive a scientific outcome had virtually been ensured by the redirection of Project Ranger to serve Apollo, the eventual elimination of all experiments save the television cameras and, with the cancellation of Ranger Block V, of any nonvisual experiments to probe the moon's surface composition and structure.[39]

But, along with besting the Soviet Union in sending back to earth the first television pictures of the lunar surface, Ranger had eliminated any doubts about the adequacy of the design for the Apollo lander. It had also taught many space scientists that in space exploration, engineering would often have to come first. Homer Newell, Mr. Science at NASA, had learned that lesson and now patiently explained to scientists and Congressmen alike the knowledge to be won from the spectacular engineering task of Apollo. Who first stepped on the moon, he insisted, was not the issue; the individuals could stay but a short time. The scientific instruments they would use, those they would leave behind on the surface, and the soil samples they would retrieve would all yield rich scientific dividends. He was, as events were to prove, absolutely right.

No less important, the Ranger project itself had already quickened the pace of planetary science. It had helped the research preferences of planetary scientists assume a preeminent place in the councils of space science. It had made visual imaging a basic exploratory tool of planetary science, an accepted antecedent to the planning of further experiments.[40] Ranger's pictures themselves provided detailed lunar maps and the means to construct three-dimensional lunar surface models. Scientists observed for the first time craters from one meter to a few hundred meters in diameter, and, from the makeup of the moonscape, deduced evidence of vulcanism. The steady-state distribution of small craters was also discovered, and detailed evidence was first observed for the aging and evolution of individual craters and other surface features as a result of repetitive bombardment of the lunar surface by solid particles.[41] In all, the television pictures acquired by Rangers 7, 8, and 9 created the foundation of a provocative new discipline—the science of the lunar regolith—the study of the fragmented debris that makes up the moon's surface.

In Project Ranger, the Deep Space Network generated tracking data that improved knowledge of the mass of the moon by an order of magnitude. From these same data scientists found the radius of the moon to be 3 kilometers less than the previously accepted value, and they discovered an offset between the geometrical center of the moon and its center of mass—a discovery with profound implications for understanding the lunar interior.[42] Kuiper correctly remarked after the flight of Ranger 7 that lunar exploration had entered a new era; the project had transformed the centuries-old study of the moon from subtle conjecture to an experimental science.

Perhaps more than any other flight project, Ranger proved the technologies and the designs for the automatic machines NASA would use for deep space exploration: attitude stabilization on three axes, onboard computer and sequencer, directional scientific observations, midcourse trajectory and terminal maneuver capability, and steerable high-gain antenna. With Ranger, NASA's Deep Space Network perfected the two-way doppler tracking and communications system, including the means to measure the velocity between the spacecraft and tracking stations—the key to accurate trajectory computation. Television camera improvements, such as fast erasing and shuttering technology, became available to other projects. Ranger also broke new technical ground for NASA by using the Atlas-Agena B launch vehicle, by the parking orbit technique, and, less happily perhaps, by heat sterilizing spacecraft components.[43]

In 1965 NASA officials also liked to claim that Ranger might possibly benefit the commercial and scientific market places.[44] Knowledge gained in building the impact-limiter capsule, they suggested, could be transferred and adapted to deliver sensitive instruments for earth exploration, airdrop supplies to disaster victims, and improve collision-proofing of vehicles and the packaging of parcels. By 1976, only one case of such uses was known—the modification of Ranger's single-axis seismometer and its capsule to function in earth gravity and at various pressures, which has served the University of California's Institute of Geophysics and Planetary Physics at La Jolla well for 13 years of seismological observations on the ocean bottom, in midwater, and on land.[45] But however strained NASA's suggestions of possible spinoffs, image enhancement through digital computer processing was a decidedly impressive gift from Ranger to commercial applications. This process removed spurious noise received with Ranger's picture signal and enhanced contrasts in the photographs of the lunar surface by shifting the mean intensity level and expanding it to cover the full range of the gray scale from black to white.[46] The technique profoundly affected diverse disciplines including astronomy, lunar cartography, medicine, commercial communications, and microspectroscopy. When applied to enhance X-ray photographs, it was selected by Industrial Research Incorporated as the single most important technical innovation of 1967 (Figure 104).[47]

And Ranger repaid more than pictures and hardware. It taught the chief participants, scientists and engineers alike, what demands could and could not be compromised among those of schedule, performance, and costs. It taught them the importance of hardware testing, and how to integrate scientific experiments into flight projects. The development of the spacecraft, communications network, and managerial techniques combined to make Ranger an essential prerequisite to the nation's future instrumented exploration of the deep reaches of outer space.

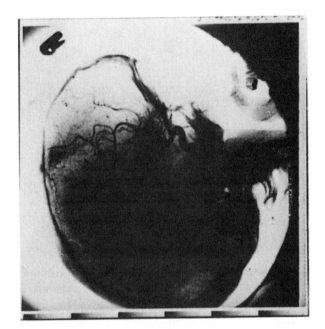

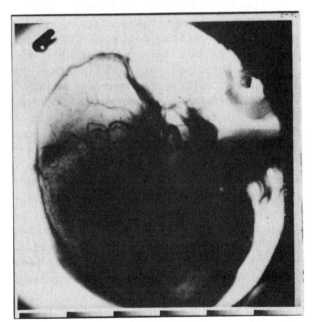

Fig. 104. X-Ray of Human Skull (Left) and X-Ray Enhanced by Computer Processing (Right)

SOURCES

Instrumented space exploration has just begun. Few bibliographic aids to historical research in this field have been published, and the bulk of the primary source material resides, unevenly indexed, in the archives and retired records of the responsible agencies.* In pursuing research for Ranger, I examined material at various locations. The NASA History Office Archives in Washington contained selected correspondence from a number of NASA offices, as well as special studies, minutes of meetings, surveys, and planning documents. Correspondence and other documents pertaining exclusively to Ranger, NASA relations with Caltech-JPL, and space science activities were found in the retired records of the NASA offices at the Federal Records Center, Suitland, Maryland. The minutes and correspondence of the President's Science Advisory Committee (PSAC) are in the custody of the National Science Foundation but housed for the most part in the National Archives in Washington. Regrettably, many of the PSAC documents are still classified, making them exceedingly difficult to view, much less to copy.

The retired records of the Project Office, Lunar Program Office, and the Director's Office at JPL, and in certain instances the minutes of the Caltech Board of Trustees at Caltech, yielded the information on the contractor's activity in Project Ranger. Financial data were obtained in contemporary progress reports and other material from the JPL project office and the NASA Financial Management Division. A few of the NASA and JPL Ranger records, particularly the investigative reports, remained classified when the research began; all of them, however, had been declassified by the time the manuscript was completed.

---

* These documents at NASA and JPL are open to the accredited scholar. In April 1976 the NASA History Office published a useful directory, *Research in NASA History: A Guide,* that is available on request.

Besides archival collections of paper and microfilm, personal interviews and movie films added insight and helped me comprehend the Ranger era. I taped and transcribed virtually all interviews. The person interviewed received a copy of the transcript, and was encouraged to correct and amplify the text. NASA officials answering questions prepared mostly beforehand were James Webb, Robert Seamans, Homer Newell, Edgar Cortright, Oran Nicks, William Cunningham, and Walter Jakobowski; those interviewed at Caltech and JPL were Lee DuBridge, John Hunt, William Pickering, Alvin Luedecke, Clifford Cummings, James Burke, Harris Schurmeier, Gordon Kautz, and Allen Wolfe.

As each person interviewed saw Ranger from a different level and perspective, each contributed to my understanding of the project and its place in a much larger, ongoing NASA program. I sensed what might be termed "purposeful reticence" on only a few occasions; I was impressed generally with the candor of the answers and the willingness of those interviewed to discuss issues that had once stirred institutional sensitivities and anguish. Movie film, existing in the form of JPL lunar and planetary program progress reports and postimpact news conferences, also proved helpful. Though sometimes encumbered by a sanguine sound track, they transported me back in time to view fabrication and test procedures and meet some of the principals around the machinery and in the facilities of the day.

Published sources used for the history of Ranger included commercial books, trade and professional journals, government publications, and newspaper articles. For the space sciences, J. Tuzo Wilson, *IGY, The Year of the New Moons* (1961); Walter Sullivan, *Assault on the Unknown* (1961); Patrick Hughes, *A Century of Weather Service* (1970); Homer Newell, *High Altitude Rocket Research* (1953); Lloyd Berkner and Hugh Odishaw, eds., *Science in Space* (1961); Samuel Glasstone, *Sourcebook on the Space Sciences* (1965); Wilmot Hess, ed., *Introduction to Space Science* (1965); and Robert Jastrow, *Red Giants and White Dwarfs* (1967), proved most illuminating. On the politics of research and space exploration, I found J. J. Penick, Jr., et al., eds., *The Politics of American Science* (1965); Daniel Greenberg, *The Politics of Pure Science* (1967); Jerome Weisner, *Where Science and Politics Meet* (1965); Vernon van Dyke, *Pride and Power* (1964); and John Logsdon, *The Decision to Go to the Moon* (1970), especially useful.

Surveying the creation of NASA and the midcentury state of missile and space technology, I consulted Alison Griffith, *The National Aeronautics and Space Act: A Study of the Development of Public Policy* (1962); Robert Rosholt, *An Administrative History of NASA, 1958-1963* (1966); Herbert York, *Race to Oblivion* (1970); Constance Green and Milton Lomask, *Vanguard: A History* (1971); Eugene Emme, ed., *The History of Rocket Technology* (1964); Michael Armacost, *The Politics of Weapons Innovation* (1969); and John Medaris with Arthur Gordon, *Countdown for Decision* (1960). The trade publications, professional journals, and colloquia used most often for this study included *Aviation Week and Space Technology, Missiles and Rockets, Astronautics, Journal of the British Interplanetary Society, Science, Journal of Geophysical Research, Proceedings of the Lunar and Planetary Exploration Colloquium* (1958–1961), *Annals of the International Geophysical Year* (1959), and the science sections of *Time* and *Saturday Review.*

Government publications proved indispensable. Congressional committee and subcommittee reports for the annual NASA authorization and appropriation hearings, as well as special studies, and investigative proceedings and findings, recorded how the space agency and JPL, Ranger, and the space science program fared in Congress. In the Executive Branch, special reports like *A Statement by the President and Introduction to Outer Space* (March 6, 1958), *Report to the President-Elect of the Ad Hoc Committee on Space* (January 11, 1961), and PSAC, *Report of the Ad Hoc Panel on Man-In-Space* (November 14, 1960), suggested preferred directions for space research. Government-solicited opinion of private consultants, such as that of the National Academy of Sciences-National Research Council, in *A Review of Space Research*, Publication 1079 (1962), and *Space Research: Directions for the Future*, Publication 1403 (1966), improved on this knowledge. *The Report to the Congress from the President of the United States, U.S. Aeronautics and Space Activities* (1959–1965) reviewed annually the progress of and plans for the nation's space program. Various NASA publications, the semiannual reports to Congress (1959–1965) and *Significant Achievements in Space Science 1965*, SP-136 (1967), also proved helpful. The annual NASA chronologies, *Astronautics and Aeronautics*, provided rapid and valuable reference to the forces operating within and upon the space agency.

Contemporary newspaper accounts, though sometimes at odds with the primary records, often captured the drama and excitement of a moment, or shed light on the motives and perceptions of Ranger's participants. For this purpose I most often used *The New York Times, Christian Science Monitor, Washington Post,* and local papers in Los Angeles and Pasadena. I also examined personal narratives covering JPL activities prior to the Laboratory's association with NASA;[*] regrettably, however, none that deal with the Ranger era exist. It is to be hoped that Keith Glennan, James Webb, Homer Newell, Lee DuBridge, William Pickering, and other NASA, JPL, and Caltech officials will consider writing their memoirs in the near future.

Except for some of the published secondary sources, all documents used in this history, wherever procured, were catalogued and entered by number in the JPL History Office archives. This number appears in parentheses at the end of each citation.

---

[*] Viz., Theodore von Karman with Lee Edson, *The Wind and Beyond* (1967); Frank Malina, "The Rocket Pioneers: Memoirs of the Infant Days of Rocketry at Caltech," *Engineering and Science,* February 1968; "Origins and First Decade of the Jet Propulsion Laboratory," in Eugene Emme, ed., *The History of Rocket Technology* (1964); "America's First Long-Range Missile and Space Exploration Program: The ORDCIT Project of the Jet Propulsion Laboratory, 1943–1946," *Spaceflight,* December 1973; and William Pickering with James Wilson, "Countdown to Space Exploration: A Memoir of the Jet Propulsion Laboratory, 1944–1958," as reprinted in *Lab-Oratory,* 1975/4.

Appendix A

LUNAR THEORY BEFORE 1964

The moon has long been a natural subject of inquiry. It is our closest celestial neighbor, and, next to the sun, the most prominent of heavenly objects viewed from earth. But the urge to examine the moon firsthand is heightened by other conditions. Although the moon is one of thirty-two known satellites that circle about the planets in our solar system, and other satellites are more massive, the moon is unique among them: it obtains an orbital angular momentum about the earth that exceeds the earth's rotational angular momentum, and no other planet possesses a satellite whose mass is so great a fraction of the mass of the primary body. One-quarter the size of the earth, the moon has one-eightieth as much mass, and possesses one-sixth the gravity of the earth. It rotates and revolves around the earth in equal periods, in a nearly circular orbit at a distance of approximately a quarter million miles, and on a plane inclined 5 degrees to the plane of the elliptic. Three principal theories have attempted to connect these intriguing conditions and explain the moon's origin.

The oldest hypothesis postulated a binary earth and moon system in which the two planets condensed from a single mass of interstellar gas and dust. This process of formation implied a similar chemical and physical constitution, and early in this century most scientists abandoned the idea because the relatively large size of the moon did not correspond to the difference in density between the two bodies, or to other important dynamical considerations. In 1954 Gerard Kuiper modified and revived this theory.[1] He held that the earth and moon originated as a double planet within a common gaseous envelope, and, since the heat generated by the decay of radioactive elements had been much greater early in the life of the solar system, the moon had undergone some internal melting, likely had a small iron core, and had swept up dust and debris remaining near the earth afterwards.

Before Kuiper revived the double planet thesis, George Darwin, son of the famous British naturalist, had in 1878 advanced an attractive alternative. Drawing on the pronounced tidal effects exerted between the earth and moon, Darwin speculated that the moon's mass had been ejected from a fluid and rapidly spinning protoearth when centrifugal force and solar tides, acting on matter in the earth's equatorial plane, exceeded the force of gravity. In time, the moon moved out to its present orbit and attained its coincident period of rotation and revolution as a result of tidal interactions between the two bodies.[2] In 1892 the Rev. Osmond Fisher suggested that the Pacific Basin marked the point of this separation, and that this material, having been drawn from the earth's mantle, explained the lower density of the moon.[3] The hypothesis, however, could not satisfactorily explain the cause for the moon's orbit at an inclination to the earth's equator. In 1930, moreover, Harold Jeffreys mathematically demonstrated that, given the earth's internal friction, damping of the required excessive tidal bulge would take place, preventing any separation of material.[4]

In the 1950s Harold Urey and Horst Gerstenkorn developed a third theory that rapidly gained wide acceptance. They proposed that the moon had accreted from gas and dust elsewhere in the solar system, was later captured by the earth at a close distance, and then moved out to its present radius from the radius of first capture.[5] More plausible than the Darwinian postulate, this theory also held the moon to be a primary body in the solar system, rather than a chunk of the earth's mantle, and it attracted many planetologists who hoped to find undisturbed on the moon clues to the formation of the solar system. Nevertheless, proponents of the capture model faced unresolved time-scale questions relating to the moon's present location and its gradual recession from the earth when contrasted with the proposed time and manner of its acquisition. Grove Karl Gilbert and Alfred Wegener,[6] and more recently Gordon MacDonald and others, suggested that the capture process involved a number of smaller moons, and that the orbital elements of the system changed at a rate proportional to the mass of the protomoons. Ultimately the largest moon increased in mass and moved outward as it swept up the smaller moonlets around the earth.[7]

Each of the preceding theories postulated an overall structure of our satellite that was hot—concentric and similar in form (though probably not in proportion) to that of the earth, or warm—essentially undifferentiated, or cold—partially heterogeneous but not layered. Corresponding scientific interest in surface morphology, reflecting one or another of these viewpoints, likewise led to differing explanations of the moon's features.

The most pronounced features on the visible sunbaked face of the moon consist of the great circular ringed maria, or lunar plains, clustered about and to the north of the lunar equator, and the intensely cratered ancient highlands to the south. Craters of varying sizes are everywhere in evidence. In addition, crater rays or streak systems can be seen associated with such major craters as Tycho and Copernicus, radiating out in all directions across the surface. Secondary features that appear under increased optical resolution include meandering cracks or valleys known as rilles, domical structures, and various wrinkle-ridges in the plains regions (frontispiece).

Two hypotheses sought to explain these conditions. The first, generally accepted for many years, accorded plutonic processes a central role.[8] Internal vulcanism, it was argued, shaped the twisted surface, formed calderas, and filled the lunar plains with darkened ash or lava flows. The second theory, first considered by Robert Hooke in 1665, credited meteoric impact as the principal agent responsible for sculpturing the surface morphology. Hooke conducted experiments with a mixture of pipe clay and water into which he dropped musket balls to form impact craters. But meteoric craters remained virtually unknown in contemporary terrestrial experience, and the impact hypothesis did not prevail.

In the 19th century, several years after Darwin proposed his origin of the moon, Grove Karl Gilbert reevaluated the impact hypothesis. Gilbert, the Chief Geologist of the United States Geological Survey, observed in 1892 that lunar craters simply did not conform in physical characteristics or in sheer numbers to terrestrial volcanic craters, and he reasoned that the moon's primary features, including the extensive maria, must be accounted for by impact cratering.[9] Alfred Wegener in 1921[10] and L. J. Spencer in 1932 and 1933[11] performed more detailed comparative studies that lent increased support to the impact thesis. These efforts culminated in the work of the industrialist-astronomer Ralph Baldwin in the 1940s.[12] Baldwin demonstrated conclusively that not only do lunar craters bear little or no physical resemblance to volcanic craters, but most lunar crater diameters as a function of their depth correspond closely to those of known terrestrial meteoric craters and artificial craters formed by explosive charges. Many proponents of the Baldwin thesis also believed that the material of the mare plains, rather than volcanic ash or magma, would be found largely composed of eroded dust.[13] Whatever the composition of the lunar soil, by 1960 the impact thesis was generally accepted among planetary astronomers and geophysicists as describing the primary process that shaped the lunar surface.

Beyond scientific advocacy purporting to explain the history of the moon and its surface features, a secondary current of interest eddied about the possible existence of life forms on the moon. This question, first seriously discussed after the advent of the telescope in the 17th century, found scientific opinion favorably disposed to the proposition. Some, such as Giovanni Riccioli, maintained, however, that the inhospitable environment—the apparent absence of water and any noticeable atmosphere—precluded higher life forms; he contended forcefully that the moon must resemble an arid desert. Sentiment in favor of sentient life on the moon, nevertheless, continued among authorities well into the 19th century. In 1780, for example, Sir William Herschel directed a communication to the fourth Astronomer Royal, Nevil Maskelyne, asserting that "there is almost an absolute certainty of the moon's being inhabited..."[14] The Director of the Vienna Astronomical Observatory, J. von Littrow, proposed in 1830 to establish communications with Selenites on the moon by constructing large geometric symbols on the Siberian steppes.[15]

Opinion disposed to intelligent life on the moon began to ebb after the "Great Moon Hoax" of 1835,[16] and reversed completely in 1837 with the publication of the definitive selenographic study *Der Mond* by Beer and Madler. But speculation concerning possible life forms, if not on the order of "reasoning Selenites" at least at a lower organic level, continued intermittently. In 1876 the British astronomer Edmund Neilson attempted to establish that the moon did in fact possess an atmosphere of sufficient density to support life.[17] And in 1924 W. H. Pickering, a Harvard College Observatory astronomer, reported observing possible flora and fauna.[18]

When space exploration began in the 1950s, scientific thinking generally concurred that organic matter probably was not present on the moon's surface, though it might possibly be found as spores in certain protected locations. Perhaps it could exist below the surface—a residue of some prehistoric period when the moon possessed a reducing atmosphere or was contaminated by material from the earth. The moon, after all, had acted as a gravitational trap for meteoroidal material accumulated from space over many eons. Detection of terrestrial-like organisms, for instance, would furnish strong support for the hypothesis of panspermia—that reproductive bodies of living organisms exist throughout the universe and develop wherever an environment is favorable. Opinion held, in any case, that care should be exercised not to introduce organisms from the earth that might accompany man-made instruments sent to the moon, and thus spoil a major scientific opportunity to discover and examine extraterrestrial life.[19] By the same token, assuming that lunar organisms might exist, Apollo astronauts returning from the moon would have to be quarantined to prevent "back contamination" of the earth.

The closeup pictures of the moon taken by Rangers 7, 8, and 9 could be and were appropriated to support each of the theories of the surface morphology, sharply escalating the scientific debate. It remained for other unmanned and manned lunar missions in the 1960s and 1970s to furnish tentative answers to many of these questions and settle the issue of life on the moon. Many more years and more missions will be necessary before a firm explanation of the moon's history is at hand.[20]

References

1.   Gerard P. Kuiper, "On the Origin of the Lunar Surface Features," National Academy of Sciences, *Proceedings*, Vol. 40, 1954, pp. 1096–1112.

2.   George H. Darwin, "On the Precession of a Viscous Spheroid," *Nature*, Vol. 18, 1878, pp. 580–582. Recent consideration in D.U. Wise, "Origin of the Moon by Fission," in B. G. Marsden and A. G. W. Cameron (eds.), *The Earth–Moon System* (New York: Plenum Press, 1966), p. 213.

3.   Rev. Osmond Fisher, "On the Physical Cause of the Ocean Basins," *Nature*, January 12, 1882; and, Communication, "Hypothesis of a Liquid Condition of the Earth's Interior Considered in Connexion with Professor Darwin's Theory of the Genesis of the Moon," *Cambridge Philosophical Society Proceedings*, Vol. 7, 1892, p. 335.

4.   Harold Jeffreys, "Resonance Theory of the Origin of the Moon, II," *Royal Astronomical Society Monthly Notices*, Vol. 91, November 1930, pp. 169–173.

5.   Horst Gerstenkorn, "Über Gezeitenreibung beim Zweikörperproblem" [Effect of Tide Friction on the Two Body Problem], *Zeitschrift für Astrophysik*, Vol. 36, 1955, pp. 245–274 (capture in a retrograde orbit); and Harold C. Urey, *The Planets: Their Origin and Development* (New Haven: Yale University Press, 1952), p. 25; also, "The Origin of the Moon's Surface Features," Parts I and II, *Sky and Telescope*, January and February 1956; and, "The Chemistry of the Moon," *Proceedings of the Lunar and Planetary Exploration Colloquium*, Vol. 1, No. 3, October 29, 1958, p. 1 (capture in a direct orbit).

6.   G. K. Gilbert, "The Moon's Face, A Study of the Origin of its Features,"
      Bulletin of the Philosophical Society, Washington, D. C., Vol. 12, 1893, p.
      262; Alfred Wegener, *Die Entstehung der Mondkrater* [The Origin of the
      Lunar Craters] (Braunschweig, Germany: Friedr. Vieweg & Son, 1921).
      About this time Wegener's theory of continental drift was published in
      English, and drew heavy fire from uniformitarian geologists. (Wegener, *The
      Origin of Continents and Oceans,* translation of the 3rd German Edition by
      J. A. G. Skerl, London: Methuen & Co., 1924).

7.   Gordon J. F. MacDonald, "Origin of the Moon: Dynamical Considera-
      tions," in *The Earth–Moon System,* op. cit., p. 198.

8.   The most recent detailed exposition of this theory is in J. E. Spurr, *Geology
      Applied to Selenology,* Vol. I (Lancaster, Pa.: Science Press, 1944).

9.   G. K. Gilbert, op. cit., p. 241; abstract in *American Naturalist,* Vol. 26,
      1892, p. 1056.

10.   Alfred Wegener, *Die Entstehung der Mondkrater,* op. cit. Wegener's
      important work, published in Germany, apparently remained little known
      in the United States. It was not cited by Baldwin, Urey, or other impact
      theorists in the 1940s and 1950s.

11.   L. J. Spencer, "Meteorite Craters," *Nature,* Vol. 129, 1932, pp. 781–784;
      and "Meteoric Craters as Topographical Features on the Earth Surface,"
      Geographical Journal, Vol. 56, March 1933, pp. 205–206.

12.   First consideration in Ralph B. Baldwin, "The Meteoritic Origin of Lunar
      Craters," *Popular Astronomy,* Vol. 50, August 1942, pp. 365–369;
      comprehensive treatment in *The Face of the Moon* (Chicago: University of
      Chicago Press, 1949); and *The Measure of the Moon* (Chicago: University
      of Chicago Press, 1963).

13.   See, for example, Thomas Gold, "The Lunar Surface," *Monthly Notices of
      the Royal Astronomical Society,* Vol. 115, No. 6, 1955, p. 585.

14.   Letter from Sir William Herschel to the Rev. Dr. Nevil Maskelyne, June 12,
      1780, as reprinted in Zdenek Kopal, *The Moon* (New York: Academic Press
      Inc., Publishers, 1964), pp. 119–120.

15.   Account as reprinted in David Lasser, *The Conquest of Space* (New York:
      Penguin Press, 1931), p. 34.

16.   This elaborate hoax, perpetrated by an enterprising American reporter,
      Richard Locke, ran as a serial in the *New York Sun* during August–
      September 1835 under the impressive title "Great Astronomical Discoveries
      Lately Made by Sir John Herschel, L.L.D., F.R.S., etc., at the Cape of Good
      Hope." The articles reported scientific observations supposedly made by
      Herschel (the son of William Herschel) with a new telescope in Africa that

permitted a lunar resolution of about 2 feet. Locke vividly described exotic lunar vegetation and simian life forms. Carefully laced with scientific "facts," the story was widely accepted and reprinted in Europe and America.

17.   Edmund Neilson, *The Moon, and the Condition and Configurations of its Surface* (London: Longmans, Green and Co., 1876), Chapter II.

18.   W. H. Pickering, "Eratosthenes No. 4," pp. 69–78; "Eratosthenes No. 5," pp. 302–312; and, "Eratosthenes No. 6, Migration of the Plats," pp. 393–404, in *Popular Astronomy,* Vol. 32, 1924.

19.   Cf., Joshua Lederberg, "Moondust," *Science,* Vol. 127, p. 1473, January 1958; and "Resolution of the Research Council, National Academy of Sciences," February 8, 1958, cited in *A Review of Space Research* (Report of the Summer Study conducted under the auspices of the Space Science Board of the National Academy of Sciences at the State University of Iowa, June 17–August 10, 1962), pp. 10–11. Opinion was by no means unanimous. Harold Urey, for one, argued that any contamination of the moon would be localized and not likely to affect other areas in the harsh environment: "This talk has caused a lot of trouble. The effort...to get the last bacterium off a missile or space ship is enormous, and probably futile. In the second place, it is very difficult to contaminate the moon, for the reason there is so much moon and the amount of chemical contamination that can occur there is so small." (Urey, *Proceedings of the Lunar and Planetary Exploration Colloquium,* op. cit., p. 31); see also CETEX report in *ICSU Review,* Vol. 1, 1959.

20.   A succinct review of the state of these theories immediately following the Apollo missions is contained in Richard S. Lewis, *The Voyages of Apollo: The Exploration of the Moon* (New York: Quadrangle/The New York Times Book Co., 1974); see also, Stuart Ross Taylor, *Lunar Science: A Post-Apollo View* (New York: Pergamon Press Inc., 1975).

Appendix B

LUNAR MISSIONS
1958 THROUGH 1965

Table B-1. Lunar Missions 1958–1965

U.S.A.			Results	U.S.S.R.		
Launch date	Name	Mission		Mission	Name	Launch date
Aug. 17, 1958	Pioneer 0	Flyby	Launch failed; booster explosion after liftoff			
Oct. 11, 1958	Pioneer 1	Flyby	Launch failed; uneven separation of second and third stages			
Nov. 8, 1958	Pioneer 2	Flyby	Launch failed; third-stage ignition failure			
Dec. 6, 1958	Pioneer 3	Flyby	Premature cessation of engine burn, apogee 107,500 km; discovered two radiation shells around earth			
			Passed within 5965 km of moon; measured lunar magnetic field and radioactivity	Undisclosed	Luna 1	Jan. 2, 1959
Mar. 3, 1959	Pioneer 4	Flyby	Missed moon by 59,000 km, tracked to 650,000 km; injection velocity low; measured cosmic radiation			
			First lunar impact; found moon to have virtually no magnetic field	Impact	Luna 2	Sep. 12, 1959
Sep. 24, 1959	Atlas-Able 4	Lunar orbit	Pad explosion during static tests, vehicle destroyed; planned for launch in October 1959			
			Circled moon; successfully photographed 60 percent of backside hemisphere of moon	Circumlunar	Luna 3	Oct. 4, 1959
Nov. 26, 2959	Atlas-Able (P-3)	Lunar orbit	Launch failed; shroud collapsed during launch destroying satellite			
Sep. 25, 1960	Atlas-Able (P-30)	Lunar orbit	Launch failed; second stage malfunctioned			
Dec. 15, 1960	Atlas-Able (P-31)	Lunar orbit	Launch failed; vehicle exploded at 70 seconds			
Jan. 26, 1962	Ranger 3	Seismic capsule rough landing	Missed moon by 36,589 km; launch vehicle guidance system failures. Ranger midcourse maneuver was mirror image of planned course correction			

Table B-1 (contd)

U.S.A. Launch date	U.S.A. Name	U.S.A. Mission	Results	U.S.S.R. Mission	U.S.S.R. Name	U.S.S.R. Launch date
Apr. 23, 1962	Ranger 4	Seismic capsule rough landing	Lunar impact; spacecraft computer-timer failure caused loss of mission			
Oct. 18, 1962	Ranger 5	Seismic capsule rough landing	Missed moon by 720 km; Ranger spacecraft power failure rendered midcourse correction and operation of most experiments impossible			
			Apparently intended as soft lander, missed moon by 8451 km; cause of failure not disclosed	Undisclosed	Luna 4	Apr. 2, 1963
Jan. 30, 1964	Ranger 6	Television impactor	Lunar impact (Sea of Tranquility); returned no pictures, television transmission failed			
Jul. 28, 1964	Ranger 7	Television impactor	Lunar impact (Sea of Clouds); returned first high-resolution pictures of lunar mare			
Feb. 17, 1965	Ranger 8	Television impactor	Lunar impact (Sea of Tranquility); returned additional high-resolution pictures of lunar mare			
Mar. 21, 1965	Ranger 9	Television impactor	Lunar impact (Crater Alphonsus); returned high-resolution pictures of highland crater			
			Lunar impact, intended soft landing failed; cause of failure not disclosed	Soft lander	Luna 5	May 9, 1965
			Missed moon by 160,000 km; midcourse correction failed	Soft lander	Luna 6	Jun. 8, 1965
			Solar orbit; returned first clear pictures of backside of moon	Photographic flyby	Zond 7	Jul. 18, 1965
			Lunar impact; intended soft landing failed; landing system failure	Soft lander	Luna 8	Oct. 4, 1965
			Lunar impact; intended soft landing failed; landing system failure	Soft lander	Luna 9	Dec. 3, 1965

Appendix C

SPACECRAFT TECHNICAL DETAILS

Table C-1. Spacecraft Technical Details

	Original Ranger		New Ranger
Block I	Block II		Block III

Objectives

Original Ranger — Block I	Original Ranger — Block II	New Ranger — Block III
Develop the engineering technology for spacecraft environment control, power conversion, attitude control, and communications to deliver scientific equipment in support of subsequent lunar and planetary missions. Produce scientific, engineering, and environmental data associated with deep space trajectories.	Collect gamma-ray data both in flight and at the vicinity of the moon; obtain lunar surface photographs; transmit, after landing, lunar seismic and temperature data; experiment with trajectory error correction and terminal attitude maneuver; continue the development of basic spacecraft technology; perfect sterilization techniques.	Obtain television pictures of the lunar surface for the benefit of the scientific program and the U.S. manned lunar program. Pictures to be at least an order of magnitude better in resolution than any available from earth-based photography.

Spacecraft description

Original Ranger — Block I	Original Ranger — Block II	New Ranger — Block III
Hexagonal magnesium frame base, 1.5-m diameter. Two trapezoidal solar panels, diametrically opposed, hinged to base; full span 5.2 m. Magnesium vertical supports connected together with A.1 angles and tubes supporting superstructure containing scientific experiments and omniantenna. Pointable high-gain antenna hinge-mounted to base. Attitude control jets mounted to base of frame. Overall height, 3.6 m.	Hexagonal magnesium frame base, 1.5-m diameter. Two trapezoidal solar panels, diametrically opposed, hinged to base; full span 5.2 m. Magnesium vertical supports connected together with magnesium angles and tubes. Block I superstructure replaced by lander capsule and equipment for gamma-ray and vidicon; omniantenna atop capsule. Pointable high-gain antenna. Radar altimeter antenna. Midcourse propulsion in center of space frame. Retrorocket in capsule assembly. Attitude jets on base of frame. Overall height, 3.6 m.	Hexagonal aluminum frame base, 1.5-m diameter. Two rectangular solar panels, diametrically opposed, hinged to base, full span 4.6 m. Television cameras mounted directly above base structure, with body-fixed optical axis. Pointable high-gain antenna hinge-mounted to base. Omnidirectional low-gain antenna mounted at apex of superstructure. Midcourse propulsion located in center of hexagonal structure. Attitude control jets mounted to base of frame. Overall height, 3.6 m.

Table C-1 (contd)

	Original Ranger		New Ranger
	Block I	Block II	Block III
Weight, kg (lb)			
Structures/mechanisms	60.8	Structures/mechanisms — 36.5	Structures/mechanisms — 43.0
Electrical (RF, TM, data)	27.0	Electrical (RF, TM, data) — 34.6	Electrical (RF, TM, data) — 26.2
Power	105.0	Power — 42.5	Power — 57.8
Attitude control (inert)	11.0	Central computer — 5.3	Central computer — 4.4
Controller-timer	4.4	Attitude control (inert) — 17.3	Attitude control (inert) — 25.9
Cabling	33.4	Cabling — 19.5	Cabling — 15.6
Friction experiment	9.5	Propulsion (inert) — 10.9	Propulsion (inert) — 10.4
Science (bus)	51.2	Lunar capsule system — 148.3	Television system — 173.0
Expendables	4.4	Science (bus) — 13.7	Expendables — 12.4
		Expendables — 7.6	
Allowable launch weight	306.7 (674.7)	Average allowable launch weight — 336.2 (739.6)	Allowable launch weight — 368.7 (811.1)

Control

Block I	Block II	Block III
10 N$_2$ jets	10 N$_2$ jets	12 N$_2$ jets
3 gyros	3 gyros	3 gyros
2 primary sun sensors	2 primary sun sensors	4 primary sun sensors
4 secondary sun sensors	4 secondary sun sensors	2 secondary sun sensors
1 earth sensor	1 earth sensor	1 earth sensor

Electrical power

Block I	Block II	Block III
4340 Si solar cells/panel (2)	4340 Si solar cells/panel (2)	4896 Si solar cells/panel (2)
Panels: 26.7 × 74.4 × 184.4 cm	Panels: 26.7 × 74.4 × 184.4 cm	Panels: 73.9 × 153.7 cm
Total area: 1.8 m^2	Total area: 1.8 m^2	Total area: 2.3 m^2
78.5 to 97 max. w/panel	78.5 to 97 max. w/panel	100 w/panel
Ag-Zn battery, 9000 Wh	Ag-Zn battery, 1000 Wh	Ag-Zn battery, 1000 Wh (2) S/C
Ag-Zn battery, 10 Wh	Note: battery rechargeable	Ag-Zn battery, 1200 Wh (2) TV
Note: no recharge capability for batteries		

Table C-1 (contd)

	Original Ranger		New Ranger
	Block I	Block II	Block III
Telecommunication			
	3-W L-band transponder	3-W L-band transponder	3-W L-band transponder
	250-mW transmitter	50-mW L-band capsule transmitter	60W L-band TV transmitter
	Quasiomnidirectional low-gain antenna, parabolic high-gain antenna	Quasiomnidirectional low-gain antenna	Quasiomnidirectional low-gain antenna, parabolic high-gain antenna
	Engineering data: 2 subcarriers, analog commutated, 1 sample per s	9 data subcarriers, analog commutated, sample rates of 0.1, 1.0 and 25 samples/s 2.5 and 25 bps. Video data at 2 kHz	Engineering data: 1 binary and 7 analog, commutated at 25., 1.0, 0.1 and 0.01 sample/s
	Science data: 4 binary, 2 analog and 1 special subcarrier		TV data: 1 analog, commutated at 1 sample/s
Propulsion			
	None	Monopropellant hydrazine	Monopropellant hydrazine
		222-N thrust	224-N thrust
		$\Delta V = 0.03$ m/s to 43.9 m/s	$\Delta V = 0.1$ m/s to 60 m/s
		$I_{sp} = 230$ s	$I_{sp} = 235$ s
		Total impulse = 14,000 N-s	Total impulse = 23,500 N-s
		4 jet-vane vector control	4 jet-vane vector control
Command			
	RTC-3 at 1 bps	RTC-7 at 1 bps	RTC-8 at 1 bps
	SC-10	SC-6	SC-6

Note: Real-time commands (RTC) for momentary switch closures, stored commands (SC) for maneuver durations.

Appendix D

RANGER EXPERIMENTS

Table D-1. Block I Scientific Experiments

Spacecraft	Instruments first approved	Agency and scientist	Instruments actually flown
Rangers 1 and 2	Electrostatic analyzer for solar plasma	Jet Propulsion Laboratory, M. Neugebauer, C. Snyder	Electrostatic analyzer for solar plasma

	Photoconductive particle detectors
	  CdS photoconductor
	  Thin-walled Geiger
	  Medium-walled Geiger
	  AU-SI counter
		State University of Iowa, J. A. Van Allen
		University of Chicago, J. A. Simpson
			Photoconductive particle detectors
			  CdS photoconductor
			  Thin-walled Geiger
			  Medium-walled Geiger
			  AU-SI counter

	Rubidium vapor magnetometer	Goddard Space Flight Center, J. P. Heppner	Rubidium vapor magnetometer
	Triple coincidence cosmic ray telescope	University of Chicago, J. A. Simpson	Triple coincidence cosmic ray telescope
	Cosmic ray integrating ionization chamber	Cal Tech, J. V. Neher Jet Propulsion Laboratory, H. R. Anderson	Cosmic ray integrating ionization chamber
	Lyman alpha scanning telescope (hydrogen geocorona)	Naval Research Laboratory, T. A. Chubb	Lyman alpha scanning telescope (hydrogen geocorona)
	Micrometeorite dust particle detectors	Goddard Space Flight Center, W. M. Alexander	Micrometeorite dust particle detectors
		LASL/Sandia Corp., J. A. Northrop	X-ray scintillation detectors (Vela Hotel)

335

Table D-2. Block II Scientific Experiments

Spacecraft	Instruments first approved	Agency and scientist	Instruments actually flown
Rangers 3, 4, and 5	Television camera FOV 1.0 deg 200 lines/frame	Jet Propulsion Laboratory, A. R. Hibbs, R. L. Heacock, E. F. Dobies U. of Arizona, G. P. Kuiper U.S. Geological Survey, E. M. Shoemaker UC San Diego, H. C. Urey	Television camera
	Gamma-ray spectrometer	UC San Diego, J. R. Arnold LASL, M. A. Van Dilla, E. C. Anderson Jet Propulsion Laboratory, A. Metzger	Gamma-ray spectrometer
	Single-axis seismometer Capsule temperature measurement Maximum deceleration at impact measurement	Cal Tech, F. Press Columbia University M. Ewing	Single-axis seismometer Capsule temperature measurement
		Jet Propulsion Laboratory, W. E. Brown, Jr.	Surface scanning pulse radar

Table D-3. Block III Scientific Experiments

Spacecraft	Instruments first approved	Agency and scientist	Instruments actually flown
Rangers 6, 7, 8, and 9	Television cameras (6) Full scan 　(1)　FOV 25 deg 　(2)　FOV 8.4 deg 　Each 1132 lines/frame Partial scan 　(2)　FOV 6.2 deg 　(2)　FOV 2.1 deg 　Each 290 lines/frame	U. of Arizona, G. P. Kuiper Principal Investigator U.S. Geological Survey, E. M. Shoemaker UC San Diego, H. C. Urey Jet Propulsion Laboratory, R. L. Heacock Coexperimenters	Television cameras
	Neher ionization chamber	Jet Propulsion Laboratory, H. R. Anderson, et al.	
	Geiger counter assembly	Jet Propulsion Laboratory, H. R. Anderson, et al.	
	Electron flux measurements	Applied Physics Laboratory, G. F. Pieper	
	Low-energy solar proton detector	Ames Research Center, M. Bader	
	Dust particle detector experiment	Goddard Space Flight Center, W. M. Alexander	
	Search coil magnetometer	Jet Propulsion Laboratory, E. Smith	
	Electron-proton spectrometer	U.C. Los Angeles, T. A. Farley, N. Sanders	
	Low-energy proton	Goddard Space Flight Center, G. P. Serbu, R. E. Bourdeau	

Table D-4. Block IV Scientific Experiments

Spacecraft	Instruments first approved	Agency and scientist	Instruments actually flown
Rangers 10, 11, and 12	Television cameras (6)	U. of Arizona, G. P. Kuiper Principal Investigator U.S. Geological Survey, E. M. Shoemaker Coexperimenter UC San Diego, H. C. Urey Jet Propulsion Laboratory, R. L. Heacock	Project cancelled
	Gamma-ray spectrometer	UC San Diego, J. Arnold	
	Surface scanning radar	Jet Propulsion Laboratory, W. E. Brown, Jr.	

Table D-5. Block V Scientific Experiments

Spacecraft	Instruments first approved	Agency and scientist	Instruments actually flown
Rangers 13, 14, 15, 16, and 17	Single-axis passive seismometer	Caltech, F. Press	Project cancelled
	Gamma-ray spectrometer	UC San Diego, J. Arnold	
	Surface scanning radar	Jet Propulsion Laboratory, W. E. Brown, Jr.	
	Television camera	U. of Arizona, G. P. Kuiper	

Appendix E

BLOCK III VISUAL SCIENCE: MEMORANDUM OF AGREEMENT

The impacting lunar television investigation with Dr. Gerard Kuiper as Principal Investigator has been selected by the Director of the Office of Space Sciences as the scientific experiment of the Ranger Block III space missions.

A clear and mutual understanding of the functions and authorities of the Principal Investigator, Project Manager, and Program Manager is required in order to achieve optimum scientific benefit and effective integration of the investigation with other portions of the project. This agreement stated herein follows the intent of NASA Management Instructions 4-1-1 and 37-1-1 and adapts or amplifies these instructions where necessary to meet the specific requirements of this investigation and project.

This agreement is a statement of the intentions of the signatories, is not binding upon their institutions or organizations, and involves no monetary considerations.

APPLICABILITY

This agreement is applicable to the proposal entitled, "Principal Investigator Program for Ranger Block III" numbered SC 7430 which was submitted to the Office of Space Sciences and subsequently forwarded to the Ranger Project Office, Jet Propulsion Laboratory, for execution and administration. The co-investigators are:

Mr. Raymond Heacock, Jet Propulsion Laboratory

Dr. Eugene M. Shoemaker, United States Geological Survey

Dr. Harold Urey, University of California, San Diego

Mr. Ewen Whitaker, University of Arizona, Tucson

RESPONSIBILITIES

Project Manager

The Project Manager has the direct responsibility for complete project execution and as such is the focal point for all activities relating to the project. The responsibility and authority of the Project Manager are described in NASA Management Instruction 4-1-1. In the event of discrepancies between this agreement and NASA Management Instructions 4-1-1 and 37-1-1, 37-1-1 governs. In general, it is the responsibility of the Project Manager to achieve the optimum balance between available resources and desired performance for each of the systems associated with the project so as to accomplish project objectives in the best possible manner. His major constraints are project objectives, total available manpower, facilities and funding, and schedule. Specifically with respect to the subject investigation and the associated interfaces, the Project Manager's responsibilities as implemented through the project organization include:

341

1. Optimizing scientific objectives in consonance with technical and resources constraints.

2. Provide the design, fabrication, and integration of the television camera system to execute the photographic mission as defined in the PDP.

3. Definition of project schedules, resources, risk, quality assurance, operations requirements, and change procedures.

4. Giving appropriate consideration to the responsibilities and authorities of the Principal Investigator in conducting his investigation.

5. Insuring that the Principal Investigator is provided with the necessary documentation covering the design and performance of the TV camera system and the pertinent spacecraft, trajectory, and flight operations information.

6. Establish mission operations plan, and control its implementation and execution.

7. Assuring that public information policy is established, and assuring timely preparation of mission scientific results for public release.

Principal Investigator

The Principal Investigator has the responsibility and authority described generally in NASA Management Instructions 37-1-1. Specifically with respect to the subject investigation, he is responsible to the Project Manager. He will consider decisions by the Project Manager as conclusive and will comply with them unless he promptly appeals to the Program Manager.

Consistent with the project constraints and organization, the Principal Investigator will conduct the following activities and, in the conduct of these, will report to the Space Sciences Division Project Representative as the designated representative of the Project Manager.

1. Define the optimum scientific objectives and functional requirements for the television system.

2. Approve the TV camera system functional specifications as a satisfactory interpretation of the scientific objectives and functional requirements.

3. Assure the adequacy of the qualification and acceptance test plans insofar as they apply to the instrumentation required for this investigation.

4. Participate in a) planning the functional testing and calibration of prototypes and flight instruments, b) calibration tests required prior to flight, and c) in analyses of the results of these tests.

5. Establish the requirements on space flight operations for the kind and degree of intermediate data processing required, and the format in which the data is to be provided the experimenters.

6. Assist in establishing the scientific merit of the various space flight options (standard or otherwise) that must be considered.

7. Participate as a consultant during flight operations as specified in the SFOP.

8. Organize the efforts of, assigning tasks to, and guiding the other members of his team of co-experimenters for the pre-flight, in-flight, and post-flight phases.

9. Conducting a scientific analysis of the data obtained through his investigation and disseminating the results. The Principal Investigator shall be the sole arbiter of the recipients of the unanalyzed data, other than his co-experimenters, those at JPL who perform the acquisition and transformation of the data, and those within NASA who have an official function to perform, for a period not to exceed three months after completion of a mission.

10. Provide, within three months of the flight date, a scientific report of the experimenter team describing the results obtained.

11. Meeting the established project requirements.

The Principal Investigator and his co-experimenters will have direct access to information at the TV Subsystem Contractor's (RCA), but all contact must be coordinated with the Space Sciences Division Project Representative; however, all direction of the Contractor will be done solely by JPL.

Program Manager

The Program Manager has the NASA Headquarters staff responsibility of insuring that all decisions on the project which will affect the scientific results of the project are coordinated with the appropriate Headquarters program scientists. Acknowledging this responsibility, it is agreed that the Program Manager will assume the following specific functions relative to this investigation and project:

1. Serving as the primary contact for necessary Headquarters co-ordination relating to the investigation.

2. Keeping the Director informed on project status.

3. Recommending to the Director actions to be taken on proposed changes or modifications in this investigation.

4. Investigating appeals from the Principal Investigator and referring these appeals to the Director for final decision.

CHANGES

Significant modifications to the scientific objectives are to be forwarded by the Project Manager to the Program Manager for referral to the Space Sciences Steering Committee for review and to the Director, Office of Space Sciences, for decision.

Changes to the complement of the investigator team desired by the Principal Investigator should be forwarded to the Program Manager, with the concurrence of the Project Manager, for referral to the Space Sciences Steering Committee for review and to the Director, Office of Space Sciences, for decision.

Dr. Gerard Kuiper (signed)

_____

Dr. Gerard Kuiper

University of Arizona,

Principal Investigator.

H. M. Schurmeier (signed)

_____

H. M. Schurmeier,

Jet Propulsion Laboratory,

Project Manager.

N. W. Cunningham (signed)

_____

N. W. Cunningham,

NASA Headquarters,

Program Manager.

Dated: 11 September 1963

Appendix F

RANGER SCHEDULE HISTORY

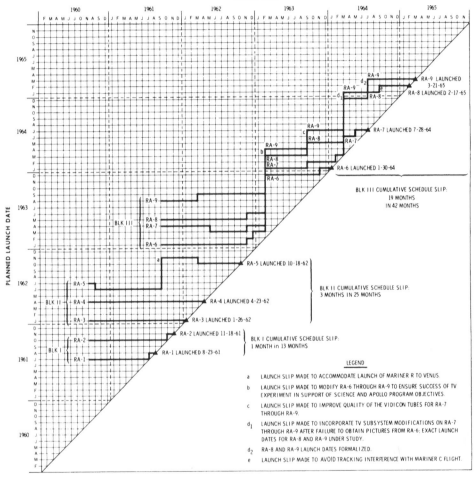

Appendix G

RANGER FINANCIAL HISTORY

RANGER FINANCIAL HISTORY

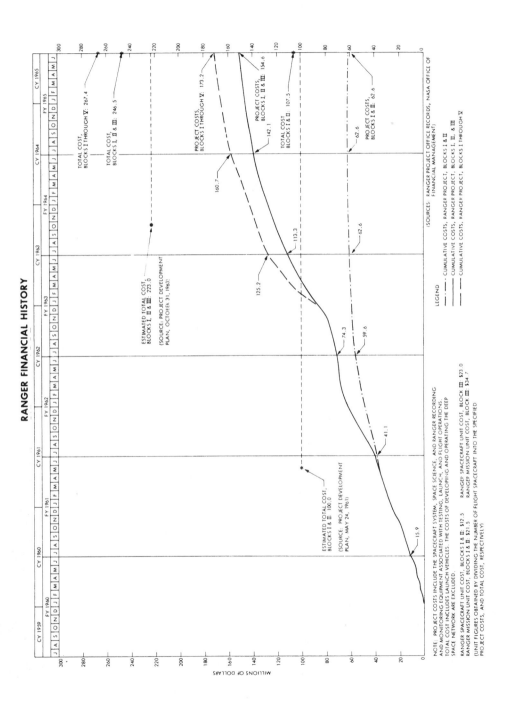

NOTE: PROJECT COSTS INCLUDE THE SPACECRAFT SYSTEM, SPACE SCIENCE, AND RANGER RECORDING
AND MONITORING EQUIPMENT ASSOCIATED WITH TESTING, LAUNCH, AND FLIGHT OPERATIONS.
TOTAL COST INCLUDES LAUNCH VEHICLES. THE COSTS OF DEVELOPING AND OPERATING THE DEEP
SPACE NETWORK ARE EXCLUDED.

RANGER SPACECRAFT UNIT COST, BLOCKS I & II: $12.5   RANGER SPACECRAFT UNIT COST, BLOCK III: $23.0
RANGER MISSION UNIT COST, BLOCKS I & II: $21.5   RANGER MISSION UNIT COST, BLOCK III: $34.7

(UNIT FIGURES OBTAINED BY DIVIDING THE NUMBER OF FLIGHT SPACECRAFT INTO THE SPECIFIED
PROJECT COSTS, AND TOTAL COST, RESPECTIVELY)

(SOURCES: RANGER PROJECT OFFICE RECORDS, NASA OFFICE OF
           FINANCIAL MANAGEMENT)

LEGEND

— · — · — CUMULATIVE COSTS, RANGER PROJECT, BLOCKS I & II
—————   CUMULATIVE COSTS, RANGER PROJECT, BLOCKS I, II, & III
—— —— ——   CUMULATIVE COSTS, RANGER PROJECT, BLOCKS I THROUGH V

351

Appendix H

RANGER PERFORMANCE HISTORY

RANGER PERFORMANCE HISTORY

FLIGHT	LAUNCH DATE	LAUNCH VEHICLE	SPACECRAFT	PRIMARY EXPERIMENT
BLOCK I				
1	8-23-61	O	◐	+
2	11-18-61	O	+	+
BLOCK II				
3	1-26-62	O	◐	+
4	4-23-62	●	O	+
5	10-18-62	●	O	+
BLOCK III				
6	1-30-64	●	●	O
7	7-28-64	●	●	●
8	2-17-65	●	●	●
9	3-21-65	●	●	●
BLOCK IV	CANCELLED			
BLOCK V	CANCELLED			

LEGEND

● SUCCESS

O FAILURE

◐ PARTIAL SUCCESS ACHIEVED THROUGH EXERCISE OF IMPORTANT SPACECRAFT SYSTEMS

+ NO TEST

Appendix I

A BIBLIOGRAPHY OF SCIENTIFIC FINDINGS

Heacock, R. L., "Ranger—Its Missions and Its Results." *TRW Space Log,* Vol. 5, Summer 1965, pp. 2–23.

Heacock, R. L., Kuiper, G. P., Shoemaker, E. M., Urey, H. C., and Whitaker, E. A., *Ranger VII, Part II, Experimenters' Analyses and Interpretations.* Technical Report No. 32-700, Pasadena, California: Jet Propulsion Laboratory, February 10, 1965.

Heacock, R. L., Kuiper, G. P., Shoemaker, E. M., Urey, H. C., and Whitaker, E. A., *Ranger VIII and IX, Part II, Experimenters' Analyses and Interpretations.* Technical Report No. 32-800, Pasadena, California: Jet Propulsion Laboratory, March 15, 1966.

Kuiper, G. P., *Interpretation of the Ranger Records.* Technical Memorandum No. 33-266, Pasadena, California: Jet Propulsion Laboratory, June 15, 1966. *Proceedings of the Caltech–JPL Lunar and Planetary Conference,* September 13–18, 1965.

Kuiper, G. P., *Interpretation of Ranger VII Records.* Technical Report No. 32-700, Pasadena, California: Jet Propulsion Laboratory, February 10, 1965, pp. 9–73. *Communications of the Lunar and Planetary Laboratory,* Vol. 4, pp. 58–59.

Kuiper, G. P., "Presentation of Ranger Moon Images." *Transactions of the International Astronomical Union; Vol. XIIB, Proceedings of the Twelfth General Assembly, Hamburg, 1964,,* London and New York: Academic Press, 1966, p. 658.

Kuiper, G. P., and Whitaker, E. A., "The Surface of the Moon." *Transactions of the International Astronomical Union; Vol. XIIB, Proceedings of the Twelfth General Assembly, Hamburg, 1964,* London and New York: Academic Press, 1966, pp. 658–662.

Kuiper, G. P., "Lunar Results from Rangers 7 to 9." *Sky and Telescope,* Vol. 29, May 1965, pp. 293–308.

Kuiper, G. P., "Advances in Earth Science." *Moon and the Planet Mars,* Hurley, P. M. Ed., Cambridge, Massachusetts: Massachusetts Institute of Technology, 1966, pp. 21–70.

Kuiper, G. P., "The Lunar Surface and the U.S. Ranger Program." *Moon and Planets: A Session of the Seventh International Space Science Symposium, Vienna, 10-18 May, 1966,* Amsterdam: North- Holland Publishing Company, 1967, pp. 23–44.

Kuiper, G. P., "A Lunar Evening with Dr. Kuiper." *Manned Spacecraft Center Technical Symposium,* Houston, Texas, April 26, 1965.

Kuiper, G. P., "The Surface Structure of the Moon." *The Nature of the Lunar Surface: Proceedings of the 1965 IAU-NASA Symposium,* Hess, W. N., Menzel, D. H., and O'Keefe, J. A., Eds., Baltimore: The Johns Hopkins Press, 1966, pp. 99–105.

Kuiper, G. P., "Volcanic Sublimates on Earth and Moon." *COMM. LPL,* No. 49, 1965, pp. 33–60.

Kuiper, G. P., "The Surface of the Moon." *New Scientist,* Vol. 30, No. 499, London, June 1966, pp. 633–634.

Kuiper, G. P., "The Lunar Surface and the U.S. Ranger Programme." *Proceedings of the Royal Society,* A, Vol. 296, 1967, pp. 399–417.

Kuiper, G. P., "Analysis of Ranger Photographs." *American Geophysical Union Spring Meeting,* Washington, D.C., April 21, 1966.

Kuiper, G. P., "Results from Ranger Series." *Symposium on the Lunar Surface,* Tucson, Arizona, January 27–28, 1967.

Kuiper, G. P., "Ranger Results and Panel Discussion on Moon." *NASA-University Program Review Conference,* Kansas City, Missouri, March 2, 1965.

Kuiper, G. P., "Ranger VII Results." *Ranger Symposium at University of Washington,* Seattle, Washington, December 29, 1964.

Kuiper, G. P., "Results of Ranger VII, VIII, and IX, Surveyor I, and Orbiter Data; Comparison of Moon and Mars." *American Institute of Aeronautics and Astronautics,* Denver, Colorado, November 18, 1966.

Shoemaker, E. M., "The Moon Close-Up." *National Geographic,* Vol. 126, No. 5, 1964, pp. 690–707.

Shoemaker, E. M., "The Geology of the Moon." *Scientific American,* Vol. 211, No. 6, 1964, pp. 38–47.

Shoemaker, E. M., "Interpretation of the Small Craters of the Moon's Surface Revealed by Ranger 7." *Transactions of the International Astronomical Union; Vol. XIIB, Proceedings of the Twelfth General Assembly, Hamburg, 1964,* London and New York: Academic Press, 1966, pp. 662–672.

Shoemaker, E. M., "Preliminary Analysis of the Fine Structure of the Lunar Surface in Mare Cognitum." *The Nature of the Lunar Surface: Proceedings of the 1965 IAU-NASA Symposium,* Hess, W. N., Menzel, D. H., and O'Keefe, J. A., Eds., Baltimore: The Johns Hopkins Press, 1966, pp. 23–77.

Shoemaker, E. M., *Progress in the Analysis of the Fine Structure and Geology of the Lunar Surface from the Ranger VIII and IX Photographs.* Technical Memorandum No. 33-266, Pasadena, California: Jet Propulsion Laboratory, June 15, 1966. *Proceedings of the Caltech–JPL Lunar and Planetary Conference,* September 13–18, 1965.

Urey, H. C., "Observations on the Ranger Photographs." *The Nature of the Lunar Surface: Proceedings of the 1965 IAU-NASA Symposium,* Hess, W. N., Menzel, D. H., and O'Keefe, J. A., Eds., Baltimore: The Johns Hopkins Press, 1966, pp. 3–21.

Urey, H. C., *Observations on the Ranger VIII and IX Pictures.* Technical Memorandum No. 32-266, Pasadena, California: Jet Propulsion Laboratory, June 15, 1966. *Proceedings of the Caltech–JPL Lunar and Planetary Conference,* September 13–18, 1965.

Urey, H. C., "Study of the Ranger Pictures of the Moon." *Proceedings of the Royal Society,* A, Vol. 296, 1967, pp. 418–431.

Urey, H. C., "Meteorites and the Moon." *Science,* Vol. 147, No. 3663, March 12, 1965, pp. 1262–1265.

Urey, H. C., "Dust on the Moon." *Science,* Vol. 153, No. 3742, September 16, 1966, pp. 1419–1420.

Whitaker, E. A., "Ranger Exploration of the Moon." *Bioastronautics and the Exploration of Space,* December 1965, pp. 39–58.

Whitaker, E. A., "The Surface of the Moon." *The Nature of the Lunar Surface: Proceedings of the 1965 IAU-NASA Symposium,* Hess, W. N., Menzel, D. H., and O'Keefe, J. A., Eds., Baltimore: The Johns Hopkins Press, 1966, pp. 79–98.

Whitaker, E. A., "Ranger Program and the Lunar Surface." *Gordon Conference,* Tilton, New Hampshire, June 28–July 2, 1965.

Whitaker, E. A., "Color Contrast on the Moon." *American Institute of Aeronautics and Astronautics Symposium* (Dr. Tiffany read paper), New York, January 1966.

Whitaker, E. A., "Mare Structure." *Symposium on the Lunar Surface,* Tucson, Arizona, January 27–28, 1967.

MISSION REPORTS AND PHOTOGRAPHIC ATLASES

Ranger VII Photographs of the Moon, Part I, Camera "A" Series. Lithograph Edition, NASA SP-61, Washington: National Aeronautics and Space Administration, Scientific and Technical Information Division, September 1964.

Ranger VII Photographs of the Moon, Part II, Camera "B" Series. Lithograph Edition, NASA SP-62, Washington: National Aeronautics and Space Administration, Scientific and Technical Information Division, February 1965.

Ranger VII Photographs of the Moon, Part III, Camera "P" Series. Lithograph Edition, NASA SP-63, Washington: National Aeronautics and Space Administration, Scientific and Technical Information Division, August 1965.

Ranger VIII Photographs of the Moon, Cameras "A," "B," and "P." Lithograph Edition, NASA SP-111, Washington: National Aeronautics and Space Administration, Scientific and Technical Information Division, 1966.

Ranger IX Photographs of the Moon, Cameras "A," "B," and "P." Lithograph Edition, NASA SP-112, Washington: National Aeronautics and Space Administration, Scientific and Technical Information Division, 1966.

Ranger VI Mission Description and Performance. Technical Report No. 32-699, Pasadena, California: Jet Propulsion Laboratory, December 15, 1966.

Ranger VII, Part 1, Mission Description and Performance. Technical Report No. 32-700, Pasadena, California: Jet Propulsion Laboratory, December 15, 1964.

Ranger VIII and IX, Part 1, Mission Description and Performance. Technical Report No. 32-800, Pasadena, California: Jet Propulsion Laboratory, January 31, 1966.

OTHERS

Alter, Dinsmore, *Lunar Atlas Supplement.* North American Aviation (545-Z-1), 1966.

Alter, Dinsmore, *Pictorial Guide to the Moon.* (Updated and Expanded Edition) New York: T. Y. Crowell Company, 1967.

Baldwin, R. B., "The Crater Diameter–Depth Relationship from Ranger VII Photographs." *Astronomical Journal,* Vol. 70, October 1965, pp. 545–547.

Barricelli, N. A., and Metcalfe, R., "Measurements of the Depth of Loose and Loosely Bonded Material on the Lunar Surface Based on Ranger VII, VIII, and IX Photographs." *Planetary Space Sciences,* Vol. 15, Northern Ireland: Pergamon Press Ltd., 1967, pp. 49–51.

Berg, Charles A., "Lunar Erosion and Brownian Motion." *Nature,* Vol. 204, October 31, 1964, p. 461.

Billingsley, F. C., "Processing the Ranger and Mariner Photographs and Other Digital Video Data." *SPIE 10th Technical Symposium,* San Francisco, August 1965: Published in *SPIE Journal,* Vol. 4, April–May, 1966, pp. 147–155.

Bouska, B., "Crater Diameter–Depth Relationship from Ranger Lunar Photographs." *Nature,* Vol. 213, January 14, 1967, p. 166.

Brinkmann, R. T., "Lunar Crater Distribution from the Ranger 7 Photographs." *Journal of Geophysical Research,* Vol. 71, January 1, 1966, pp. 340–342.

Choate, R., "Lunar Slope Angles and Surface Roughness from Ranger Photographs." *Fourth Symposium on Remote Sensing of Environment,* University of Michigan, 1966.

Conel, J. E., "What the Rangers Revealed About Lunar Geology." *Astronautics and Aeronautics,* January 1969, pp. 64–69.

Cross, C. A., "The Size Distribution of Lunar Craters." *Monthly Notices of the Royal Astronomical Society,* Vol. 134, No. 3, 1966, pp. 245–252.

DeWeiss, F. A., "Structure of the Lunar Lithosphere." *New York Academy of Sciences, Conference on Planetology and Space Mission Planning,* New York, November 3 and 4, 1965, paper. *New York Academy of Sciences, Annals,* Vol. 140 December 16, 1966, pp. 114–128.

Fielder, G., "Crater Alignments on the Moon." Submitted to *Planetary and Space Sciences* for publication.

Fielder, G., and Marcus, A., "Further Tests for Randomness of Lunar Craters." *Monthly Notices of the Royal Astronomical Society,* Vol. 136, 1967, pp. 1–10.

Fielder, G., "Vulcanism and Faulting on the Moon." *Mantles of the Earth and Planets,* Chapter 5, Runcorn, S. K., Ed., London: J. Wiley, 1967.

Firsoff, V. A., "A Preliminary Selenological Analysis of the Ranger Photographs." *British Astronomical Association, Journal,* Vol. 77, February 1967, pp. 106–111.

Gault, E. E., Quaide, W. L., and Oberbeck, V. R., "Interpreting Ranger Photographs from Impact Cratering Studies." *The Nature of the Lunar Surface: Proceedings of the 1965 (IAU-NASA) Symposium,* Hess, W. N., Menzel, D. H., and O'Keefe, J. A., Eds., Baltimore: The Johns Hopkins Press, 1966, pp. 125–140.

Gold, T., "Ranger Moon Pictures—Implications." *Science,* Vol. 145, September 4, 1964, pp. 1046–1048.

Green, J., "Interpretation of Ranger VII Photographs." *New York Academy of Sciences, Conference on Geological Problems in Lunar Research,* New York, May 16–19, 1964, paper. New York Academy of Sciences, Annals, Vol. 123, July 15, 1965, pp. 999–1002.

Green, J., "The Lunar Prospect." *Apollo—A Program Review. Society of Automotive Engineers, National Aeronautic and Space Engineering and Manufacturing Meeting,* Los Angeles, October 5–9, 1964; Proceedings, SP-257. New York: Society of Automotive Engineers, Inc., 1964, pp. 113–124.

Jaffe, L. D., "Lunar Surface Strength." *Icarus,* Vol. 6, No. 1, pp. 75–91, January, 1967.

Jaffe, L. D., "Strength of the Lunar Dust." *Journal of Geophysical Research,* Vol. 70, 1965a, pp. 6139–6146.

Jaffe, L. D., "Depth of the Lunar Dust." *Journal of Geophysical Research,* Vol. 70, 1965b, pp. 6129–6139.

Jaffe, L. D., "Lunar Dust Depth in Mare Cognitum." *Journal of Geophysical Research,* Vol. 70, 1966a, pp. 1095–1103.

Jaffe, L. D., "Lunar Overlay Depth in Mare Tranquillitatis and Alphonsus." *Icarus* Vol. 5, 1966b, pp. 545–550.

Kirhofer, W. E., and Willingham, D. E., *Ranger IX Photographic Parameters.* Technical Report No. 32-966, Pasadena, California: Jet Propulsion Laboratory, September 15, 1966.

Kirhofer, W. E., and Willingham, D. E., *Ranger VIII Photographic Parameters.* Technical Report No. 32-965, Pasadena, California: Jet Propulsion Laboratory, November 1, 1966.

Kirhofer, W. E., and Willingham, D. E., *Ranger VII Photographic Parameters.* Technical Report No. 32-964, Pasadena, California: Jet Propulsion Laboratory, November 1, 1966.

Kopal, Zdenek, "The Nature of Secondary Craters Photographed by Ranger VII." *Icarus,* Vol. 5, March, 1966, pp. 201–213.

MacCauley, John F. "Geologic Results from the Lunar Precursor Probes." *Anaheim 4th Annual Meeting of the AIAA,* October 23–27, AIAA Paper No. 67-862.

Larink, Johannes, "New Data on the Structure of the Lunar Surface." *Symposium über die Erforschung des Mondes und des interplanetären Raumes,* Munich, West Germany, April 22, 1966, paper, p. 3. In German.

Mesner, M. H., and Gravel, A. J., "The TV-Camera System for Ranger," *IEEE Transactions on Broadcasting,* Vol. BC-11, July, 1965, pp. 1–5.

Miller, B. P., "Distribution of Small Lunar Craters Based on Ranger VII Photographs." *Journal of Geophysical Research,* Vol. 70, May 1, 1965, pp. 2265–2266.

Miller, B. P., "Ranger TV Subsystem." *Electro Optics,* Camden, New Jersey: Radio Corporation of America, 1964, pp. 42–45.

Miyamoto, S., "Maria Surface of the Moon." *International Lunar Society Journal,* Vol. 2, December 1964, pp. 132–136.

Newell, H. E., *Ranger VII Preliminary Results Conference, Jet Propulsion Laboratory, Pasadena, California.* Scientific and Technical Information Division, July 31, 1964, p. 26.

Nicks, O. W., "The Ranger Lunar Missions." *COSPAR, International Space Science Symposium, 6th,* Buenos Aires, Argentina, May 13–19, 1965, paper, p. 54.

O'Keefe, J. A., "Interpretation of Ranger Photographs." *Science,* Vol. 146, October 23, 1964, pp. 514–515.

Parker, P. J., "The Ranger Programme." *Spaceflight,* Vol. II, No. 1, January 1969, pp. 24–26.

Pickering, W. H., "Ranger—On the Upward Trail." *Astronautics and Aeronautics,* Vol. 3, January 1965, pp. 18–21.

Porter, R. W., "Space Research in the United States—1964." *COSPAR, Planetary Meeting, 8th,* Buenos Aires, Argentina, May 10–12, 19–21, 1965, paper, p. 12.

Quaide, W. L., and Gault, D., "Interpretation of Ranger 7 Photographs in the Light of Impact Experiments." *Transaction of American Geophysical Union,* Vol. 45, 1964, p. 628.

Rackham, T. W., "A Comparison of Lunar Photography from Space Probes and Ground-Based Observatories." *Icarus,* Vol. 6, 1967, pp. 440–444.

Rackham, T. W., "A Note on Lunar Ray Systems." *Icarus,* Vol. 4, December 1965, pp. 544–546.

Rifaat, A., "Isodensitometric Measurements of Lunar Slopes from the Ranger Photographs." *Measure of the Moon: Proceedings of the Second International Conference on Selenodesy and Lunar Topography held in the University of Manchester, England, May 30–June 4, 1966,* Kopal, Z., and Goudas, C. L., Eds., New York: Gordon and Breach, Science Publishers, Inc., 1967, pp. 455–462.

Rindfleisch, T. C., *Lunar Topography from Photometry Considerations.* Technical Report No. 32-786, Pasadena, California: Jet Propulsion Laboratory, September 15, 1965.

Rindfleisch, T. C., "Getting More Out of Ranger Pictures By Computer." *Astronautics and Aeronautics,* January 1969, pp. 70–74.

Schurmeier, H. M., "Lunar Exploration." *Lunar Missions and Exploration,* Leondes, C. T., and Vance, R. W., Eds., New York: John Wiley and Sons, Inc., 1964, pp. 498–531.

Schurmeier, H. M., Heacock, R. L., Miller, B. P., "The Ranger VII Mission." *Transactions of the International Astronomical Union; Vol. XIIB, Proceedings of the Twelfth General Assembly, Hamburg, 1964,* London and New York: Academic Press, 1966, pp. 656–657.

Schurmeier, H. M., Heacock, R. L., Wolf, A. E., "The Ranger Missions to the Moon." *Scientific American,* January 1966.

Sjogren, W. L., "Estimate of Four Topocentric Lunar Radii." *Measure of the Moon: Proceedings of the Second International Conference on Selenodesy and Lunar Topography held in the University of Manchester, England, May 30– June 4, 1966,* Kopal, Z., and Goudas, C. L., Eds., New York: Gordon and Breach, Science Publishers, Inc., 1967, pp. 341–343.

Sjogren, W. L., Trask, D. W., Vegos, C. J., and Wollenhaupt, W. R., "Physical Constants as Determined from Radio Tracking of the Ranger Lunar Probes." *American Astronautical Society, Space Flight Mechanics Specialist Conference,* University of Denver, Denver, Colorado, July 6–8, 1966, Paper 66-105.

Smith, G. M., Willingham, D. E., Kirhofer, W. E., "Ranger VII Camera Calibration and Performance." *Transactions of the International Astronomical Union; Vol. XIIB, Proceedings of the Twelfth General Assembly, Hamburg, 1964,* London and New York: Academic Press, 1966, p. 657.

Smith, G. M., Vrebalovich, T., and Willingham, D. E., "Eyes on the Moon." *Astronautics and Aeronautics,* March 1966.

Trask, D. W., "Distribution of Lunar Craters According to Morphology from Ranger VIII and IX Photographs." *Icarus,* Vol. 6, No. 2, March 1967, pp. 270–276. Communicated by E. M. Shoemaker.

Walker, E. H., "Comments on the Photographs Obtained by Ranger 7," *Astronautyka,* Vol. 8, No. 1, 1965, p. 306.

Willingham, D. E., *The Lunar Reflectivity Model for Ranger Block III Analysis.* Technical Report No. 32-664, Pasadena, California: Jet Propulsion Laboratory, November 2, 1964.

Notes*

Chapter One

1. *Public Papers of the Presidents of the United States: Dwight D. Eisenhower, 1957* (Washington: Government Printing Office, 1958 [210], p. 724.

2. Two International Polar Years (IPY 1 and 2) for the study of geophysical phenomena had been conducted previously; IPY 1 in 1882–83, when meteorological, magnetic, and auroral stations were first established in the arctic regions, and IPY 2 fifty years later in 1932–33. Both contributed significantly to man's knowledge of the earth's magnetism and of the ionosphere. IPY 2 and the IGY (briefly known as IPY 3 but altered to IGY with the expansion in the scope of inquiry) were sponsored by the various international scientific unions organized in 1919. The International Council of Scientific Unions (an administrative body comprising the chief officers of the unions) coordinated the efforts and established the Comité Special de l'Année Géophysique International (CSAGI) to oversee the project. The idea to advance the date for an IPY 3 by twenty-five years to take advantage of the advance in electronics and communications made during World War II originated at a dinner party in April 1950 at the home of James Van Allen in Washington, D.C. J. Tuzo Wilson, *IGY, The Year of the New Moons* (New York: Alfred A. Knopf, 1961), pp. 7–10; also Hugh L. Dryden, "IGY—Man's Most Ambitious Study of His Environment," *National Geographic,* February 1956, pp. 385–298.

3. Since nearly every scientist who participated in the Ranger Project was a Ph.D., the academic title is omitted.

4. Frank J. Malina, "Origins and First Decade of the Jet Propulsion Laboratory," *The History of Rocket Technology* (Detroit: Wayne State University Press, 1965), Eugene M. Emme, ed., pp. 46–66; Frank J. Malina, "The U.S. Army Air Corps Jet Propulsion Research Project, GALCIT.

---

* The hyphenated numbers in parentheses at the ends of individual citations are catalog numbers of documents on file in the history archives of the JPL library.

Project No. 1, 1936–1946: A Memoir," and "America's First Long Range Missile and Space Exploration Program: The ORDCIT Project at the Jet Propulsion Laboratory, 1943–1946: A Memoir," *Essays on the History of Rocketry and Astronautics: Proceedings of the Third Through Sixth History Symposia of the International Academy of Astronautics* (Washington: National Aeronautics and Space Administration, 1977), R. Cargill Hall, ed.

5.　William Pickering with James Wilson, "Countdown to Space Exploration: A Memoir of the Jet Propulsion Laboratory, 1944–1958," *Essays on the History of Rocketry and Astronautics,* Hall.

6.　*Project Red Socks* (Pasadena, California: Jet Propulsion Laboratory, California Institute of Technology, October 21, 1957), pp. 2, 3 (2-581b); also letter from William Pickering to Lee DuBridge, with attachments, October 25, 1957 (2-581a).

7.　Interview of William Pickering by Cargill Hall, August 20, 1968, p. 6 (2-753). Avid competition over final responsibility for long-range rocket and space flight projects shaded relations between the United States Army and the United States Air Force during the 1950s.

8.　Interview of Herbert York by Cargill Hall, December 17, 1971 (2-2235).

9.　Interview of John Clark by James Wilson, August 14, 1971, p. 6 (3-490).

10.　Presidential approval cited in Air Force News Service Release No. 1303, subject: "Space Programs Revealed, Air Force Given Moon Probe Role," March 28, 1958; McElroy announcement in DOD News Release No. 288-58, March 27, 1958; see also Advanced Research Projects Agency (ARPA) Orders No. 1-58 and 2-58 of March 27, 1958. The news release contains the only reference to ARPA expectations and goals. By the time these programs had concluded, project objectives had been substantially altered at the operating level. For the Air Force Thor-Able it was "to obtain scientific data in cislunar space . . . " *1958 NASA/USAF Space Probes (Able-1) Final Report,* (Space Technology Laboratories, Inc., February 18, 1959), Volume 1: Summary, p. 2. For the Army, it was "to establish a trajectory in the vicinity of the moon, to make a significant scientific measurement, and to advance space technology." Henry Curtis and Dan Schneiderman, *Pioneer III and IV Space Probes* (JPL TR 34-11. Pasadena, California: Jet Propulsion Laboratory, California Institute of Technology, January 29, 1960), p. 1.

11.　ARPA Order No. 2-58, March 27, 1958, p. 1.

12.　*1958 NASA/USAF Space Probes (Able-1) Final Report* (Space Technology Laboratories, Inc., February 18, 1959), Volume 2: Payload and Experiments, pp. 18ff.

13.　*1958 NASA/USAF Space Probes Final Report,* Vol. 1, p. 11.

14. Ibid., p. 72; also "First U.S. Lunar Probe Fails," *Aviation Week and Space Technology,* August 25, 1958, pp. 20–21.

15. *First Semiannual Report to the Congress of the National Aeronautics and Space Administration* (Washington: Government Printing Office, 1959), pp. 14–15. More than ever eager to claim American progress toward a lunar strike, *Life* magazine headlined: "The Moon Shot: Why a 'Failure' is a Great Success," *Life,* October 27, 1958.

16. White House Press Release, Message of the President, subject: "United States Aeronautics and Space Activities, January 1–December 31, 1958," scheduled for delivery to the Congress February 2, 1959, pp. 10–11; also *1958 NASA/USAF Space Probes Final Report,* Vol. 1, p. 84.

17. Curtis and Schneiderman, *Pioneer III and IV Space Probes,* p. 2.

18. J. L. Stuart, *A Miniature Photographic Camera for Space Probe Instrumentation* (JPL TR 34-137. Pasadena, California: Jet Propulsion Laboratory, California Institute of Technology, October 28, 1960); also, *Space Programs Summary No. 3* for the period March 15, 1959, to May 15, 1959 (Pasadena, California: Jet Propulsion Laboratory, California Institute of Technology, June 1, 1959), p. 84.

19. Thomas R. Atkinson, et al., *Vidicon Camera and Tape Recorder System Development* (JPL Memorandum 30-4. Pasadena, California: Jet Propulsion Laboratory, California Institute of Technology, June 8, 1959), p. 1; and John R. Scull, *A System for Lunar Photography and Data Transmission* (JPL TR 34-142. Pasadena, California: Jet Propulsion Laboratory, California Institute of Technology, May 28, 1960), pp. 2–3.

20. Albert R. Hibbs and C. W. Snyder, "Results of Pioneer III Flight," *Proceedings of the Lunar and Planetary Exploration Colloquium* (Downey, California: North American Aviation, Inc.), Vol. 1, No. 4, January 12, 1959, pp. 48–50; and James Van Allen and L. A. Frank, *Survey of Radiation Around the Earth to a Radial Distance of 107,400 Kilometers* (Iowa City: State University of Iowa, January 1959).

21. Interview of Clifford Cummings by Cargill Hall, August 19, 1971, p. 2 (2-2218).

22. Atkinson, *Vidicon Camera and Tape Recorder System Development.* There is no question that this system was delayed in development: however, no written documentation has been uncovered that completely accounts for NASA's decision to replace the vidicon camera with the radiation experiment.

23. *First Semiannual Report to Congress of NASA,* p. 17.

24. Albert R. Hibbs, "Results of Pioneer IV Flight," *Proceedings of the Lunar and Planetary Exploration Colloquium*, Vol. 1, No. 5, March 18, 1959, p. 12.

25. S. Fred Singer, "Research in the Upper Atmosphere with Sounding Rockets and Earth Satellite Vehicles," *Journal of the British Interplanetary Society*, Vol. II, No. 2, March 1952, pp. 62–63; also, cf. William R. Corliss, *NASA Sounding Rockets, 1958–1968: A Historical Summary* (NASA SP-4401. Washington: National Aeronautics and Space Administration, 1971); on heritage, see Patrick Hughes, *A Century of Weather Service: A History of the Birth and Growth of the National Weather Service, 1879–1970* (New York: Gordon and Breach, Science Publishers, Inc., 1970), Part II; also William R. Corliss, *Scientific Satellites* (NASA SP-133. Washington: National Aeronautics and Space Administration, 1967), Chapter 1.

26. Cf., contemporary proposals for experiments in *Proceedings of the Lunar and Planetary Exploration Colloquium*, Vol. 1, passim; also, United States Congress, House, Select Committee on Astronautics and Space Exploration, *Space Handbook: Astronautics and Its Applications*, Staff Report, 85th Congress, 2nd Session, 1958, p. 216. The quest to discover extraterrestrial life remained largely in the province of life science, which, in this venture, was closely allied with planetary science.

Knowledge of the origin and evolution of the moon's features, two British astronomers declared in 1955, "can never be established with certainty so long as men are confined to earth. Only when the first spaceships take off for the moon, and we are able to view the surface at close quarters and actually analyze the lunar crust, will this (particular) question be finally settled." H. Percy Wilkins and Patrick Moore, *The Moon* (London: Faber and Faber, 1955), p. 401. This general attitude was expressed most forcefully by Thornton Page at a later date: "I like to think of myself as a 'big-telescope man'... ground-based, of course ... but I have to admit that getting outside the atmosphere for a better look, and bringing home samples of astronomical bodies, beats anything the Palomar telescope can offer." Thornton Page, "A View from the Outside," *Bulletin of the Atomic Scientists*, Vol. 25, September 1969, p. 61. Arthur Clarke, on the other hand, after reviewing the conflicting theories on the origin of the moon and its surface features in 1951, already had mused: "one sometimes wonders if the matter will be settled even when we have reached the Moon." Arthur C. Clarke, *The Exploration of Space* (New York: Harper & Brothers, 1951), p. 107.

27. Stuart, *A Miniature Photographic Camera;* Atkinson, *Vidicon Camera and Tape Recorder System Development; Exploration of the Moon, the Planets, and Interplanetary Space* (JPL Report 30-1. Pasadena, California: Jet Propulsion Laboratory, California Institute of Technology, April 30, 1959), Albert R. Hibbs, ed., pp. 18–20; Carl Gazley, Jr., and David J. Masson,

"Recovery of a Circum-Lunar Instrument Carrier," *Proceedings of the VIIIth International Astronautical Congress, Barcelona, 1957* (Springer-Verlag, Vienna, 1958), pp. 137–146; Merton E. Davies, *A Photographic System for Close-up Lunar Exploration* (Rand Corporation Report RM-2183. May 23, 1958); Merton E. Davies, "Lunar Exploration by Photography from a Space Vehicle," *Proceedings of the Xth International Astronautical Congress, London, 1959* (Springer-Verlag, Vienna, 1960), pp. 268–278; and proposals and discussion in Lee Stephenson, "Making Measurements by Hardlanding Vehicles," *Proceedings of the Lunar and Planetary Exploration Colloquium*, Vol. 1, No. 2, pp. 19–20 and passim; see also, for example, B. B. Chew, "A Look at the Future Space Effort," *Proceedings of the Lunar and Planetary Exploration Colloquium*, Vol. 1, No. 2, p. 1; *A Statement by the President and Introduction to Outer Space* (President's Scientific Advisory Committee, March 26, 1958), pp. 4–5; William W. Kellogg, "Research in Outer Space," report of the Technical Panel on the Earth Satellite Program, U.S. National Committee for the IGY, as printed in *Science*, Vol. 127, No. 3302, April 11, 1958, p. 799; and *Proposal for a Lunar Probe* (LMSD-49800. Lockheed Aircraft Corporation Missiles and Space Division, June 12, 1959).

28. With increased federal expenditures for basic research in the years immediately following the war, physics and the biomedical sciences did splendidly. But, as Daniel Greenberg has noted, geologists and chemists among other planetary scientists, had been "largely left out of this affluence" and could claim no exclusive source of financial support among government agencies, military or otherwise. Daniel S. Greenberg, *The Politics of Pure Science* (New York: The New American Library, Inc., 1967), p. 173.

29. The first Lunar and Planetary Exploration Colloquium, sponsored by the Rand Corporation, North American Aviation, and the California Research Corporation, was held in Downey, California, on May 13, 1958. Participants in succeeding months included Carl Sagan (U.C. Berkeley), Harold Urey (U.C. San Diego), Dinsmore Alter (Griffith Park Observatory), Eugene Shoemaker (U.S. Geological Survey), Gerard Kuiper (University of Arizona), Albert Hibbs (JPL–Caltech), and Gerhardt Schilling (Rand Corporation). The three principal objectives of the Colloquium were: "(1) to bring together people of common interest for the exchange of scientific and engineering information; (2) to define the scientific and engineering aspects of lunar and planetary exploration and to provide a means for their long-term appraisal; (3) to make available, nationally, the collective opinion of a qualified group on this subject." The Colloquium continued to meet quarterly at different locations on the West Coast through May 1963. (*Proceedings of the Lunar and Planetary Exploration Colloquium*, Vol. 1, No. 1, May 13, 1958, "Introduction.")

30.   Creation of and rationale for the National Academy of Sciences Space
       Science Board noted in *Science in Space* (New York: McGraw-Hill Book
       Company, Inc., 1961), Lloyd V. Berkner and Hugh Odishaw, eds., pp. 429–
       433.

31.   See "Space Science Board, Report of Activities, June 1958–June 1959."
       NASA Document, no date or author indicated. Cf., also, reports and
       proposals appearing in United States Congress, Senate, Special Committee
       on Space and Astronautics, *Compilation of Materials on Space and
       Astronautics*, 85th Congress, 2nd Session, No. 2, April 14, 1958. SSB
       members included Lloyd Berkner (Chairman), Harrison Brown, Leo
       Goldberg, H. Keffer Hartline, Donald Hornig, William Kellogg, Christian
       Lambertsen, Joshua Lederberg, W. A. Noyes, Colin Pittendrigh, Richard
       Porter, Bruno Rossi, Alan Shapley, John Simpson, Harold Urey, James Van
       Allen, O. G. Villard, Jr., Harry Wexler, George Wollard, Hugh Odishaw
       (Executive Director), and R. C. Peavey (Secretary). Berkner and Odishaw,
       Science in Space, pp. 429–433; *A Statement by the President and
       Introduction to Outer Space*, p. 9.

32.   Harold Spencer Jones, "The Inception and Development of the Interna-
       tional Geophysical Year," *Annals of the International Geophysical Year*
       (London: Pergamon Press, 1959), Volume 1, p. 383.

33.   Among these bodies were the ad hoc Upper Atmosphere Rocket Research
       Panel (1946), with representation from United States agencies and
       academic groups engaged in upper air research, the NACA Special
       Subcommittee on the Upper Atmosphere (1946), an interagency group that
       sponsored research and published tables of standard properties of the
       atmosphere at high altitudes, and the Geophysical Sciences Committee
       (1948) of the Research and Development Board in the Department of
       Defense. History reviewed in John Townsend, "History of the Upper Air
       Rocket Research Program at the U.S. Naval Research Laboratory, 1946–
       1947," Appendix III, pp. 27–45, of Homer E. Newell, *The Challenge to
       United States Leadership in Rocket Sounding of the Upper Atmosphere*, U.S.
       Naval Research Lab unpublished document, August 28, 1957 (5-415);
       Milton W. Rosen, *The Viking Rocket Story* (New York: Harper & Brothers,
       1955), pp. 21–23; and Homer E. Newell, "Exploration of the Upper
       Atmosphere by Means of Rockets," *The Scientific Monthly*, Vol. 64, No. 6,
       June 1947, p. 454. For the NACA Committee and Geophysical Sciences
       Committee see Calvin N. Warfield, *Tentative Tables for the Properties of the
       Upper Atmosphere* (NACA Technical Note No. 1200. Washington: NACA,
       January 1947), pp. 2–3; and S. Fred Singer, "Research in the Upper
       Atmosphere," pp. 70–71; also David S. Akens, *Historical Origins of the
       George C. Marshall Space Flight Center* (MSFC Historical Monograph No.

1. Huntsville, Alabama: National Aeronautics and Space Administration, December 1960), p. 8.

34.   Joseph A. Shortal, *History of Wallops Station* (Comment Edition. Wallops Island, Virginia: National Aeronautics and Space Administration, 1968), "Part II: From Supersonic to Hypersonic Flight Research, 1950–1954," pp. IX–17 (5-243).

35.   See Constance McLaughlin Green and Milton Lomask, *Vanguard: A History* (Washington: Smithsonian Institution Press, 1971), pp. 21, 97–98. In space research, the Technical Panel observed when the IGY began, "a satellite can be used to observe only three kinds of things, namely, photons, particles, and fields." National Academy of Sciences Technical Panel on the Earth Satellite Program, "Comments on a Continuing Program of Scientific Research Using Earth Satellite Vehicles," January 7, 1957, p. 6 (5-1064b).

36.   NACA Special Committee on Space Technology, *Recommendations Regarding A National Civil Space Program,* October 28, 1958, p. 1. The proposed program to attain these objectives consisted of (1) geophysical observation—the mapping of gravitational and magnetic fields and their interactions with particles and radiations approaching earth from the sun and outer space—and experiments with man and living organisms; (2) continued upper atmosphere experiments with rocket sondes; and (3) supporting ground-based research on such questions as "radiation effects on materials, instruments, and living organisms, and means of radiation protection," see pp. 5–7. See also, for example, contemporary remarks of Hugh Dryden: "In my opinion the goal of the space program should be the development of manned satellites and the travel of man to the moon and nearby planets ... " Hugh Dryden, "Space Technology and the NACA," text of an address before the Institute of the Aeronautical Sciences, New York, N. Y., January 27, 1958, p. 2 (5-140).

37.   *Proposed Development Plan for Able 3-4 (Earth Satellite, Lunar Satellite, Deep Space Probe),* June 1, 1959, p. 2-1 (2-2232).

38.   Sky science could proceed without the large rockets, precise guidance, and spacecraft capability required for planetary research, and made good use of the technology available. For example, employing the smaller Thor-Able launch vehicle, on March 11, 1960, NASA placed the spheroidal, spin-stabilized Pioneer 5 machine on a space trajectory carrying it and a number of sky science experiments successfully into a solar orbit between the earth and Venus. See NASA News Release, subject: "Pioneer V Payload," March 8, 1960 (5-1035a); NASA News Release, subject: "Pioneer V Booster," March 8, 1960 (5-1035b); and Glenn A. Reiff, "The Pioneer Spacecraft Program," Joint National Meeting, American Astronautical Society, Operations Research Society, June 17–20, 1969 (5-1037).

39. Interview of George E. Mueller (then Vice President of STL) by E. M. Emme, NASA Historian, January 15, 1960. Also NASA News Release No. 60-265, September 1960, p. 2.

40. NASA memorandum from Robert Jastrow to Homer Newell, subject: "Report on December 1 Meeting of the Lunar Science Group," December 11, 1959 (2-1933a).

41. NASA's Space Sciences Division, responsible for experimental applications in space exploration, in 1958 was organized into six branches: Instrumentation, Meteorology, Fields and Particles, Planetary Atmospheres, Astronomy, and Solar Physics. Townsend noted that "at the present time the division consists of about 60 people, most of whom transferred from the NRL where they constituted a group which has been conducting upper air research with rockets for the past twelve years." NASA memorandum from John Townsend to Homer Newell, subject: "Staff–Space Sciences Division Relationships," February 6, 1959, p. 1 (2-1923a).

42. Robert Jastrow, *Red Giants and White Dwarfs* (New York: Harper & Row, 1967), p. 2.

43. NASA Memorandum for the File from Homer Newell, subject: "Telephone Conference Report with Robert Cowen, *Christian Science Monitor*," January 6, 1965 (2-730).

44. NASA Memorandum for the File from Homer Newell, subject: "Miscellaneous Notes," January 20, 1959, pp. 2–3 (2-1760); NASA approval noted in Homer Newell, "Staff Conference Report," January 27, 1959. p. 1; rationale for Working Groups in NASA Memorandum from Homer Newell to A. Silverstein, subject: "Recommendations for a Lunar Science Group," August 26, 1959 (2-1929). First meeting of the NASA Working Group on Lunar Exploration convened at JPL on February 5, 1959. Draft of a NASA Memorandum for the File from William Cunningham, subject: "Working Group on Lunar Explorations," January 15, 1965, p. 1 (2-651); see also, Homer E. Newell, "Harold Urey and the Moon," *The Moon,* Vol. 7, Nos. 1/2, March/April, 1973, pp. 1–5.

45. Cf. United States Congress, *Second Semiannual Report of the National Aeronautics and Space Administration,* 86th Congress, 2nd Session, House Document No. 361, 1960, p. 2; *NASA Long Range Plan,* December 16, 1959, p. 33; and National Aeronautics and Space Administration, *A National Space Vehicle Program: A Report to the President,* January 27, 1959, pp. 3, 10 (2-798). The NASA Research Steering Committee for Manned Space Flight, chaired by Harry Goett, Director of NASA's new Goddard Space Flight Center at Greenbelt, Maryland, had concluded its deliberations and released its planning recommendations. Reiterating a suggestion advanced one year earlier by the Vehicles Group in the NACA

Special Committee on Space Technology, the NASA committee urged a lunar exploration program ending in manned landings on the moon. See The Working Group on Vehicular Program of the Special Committee on Space Technology, *Interim Report to the National Advisory Committee for Aeronautics: A National Integrated Missile and Space Vehicle Development Program*, April 1, 1958. p. 6 (3-151c); and complementary JPL study, Allyn B. Hazard, *A Plan for Manned Lunar and Planetary Exploration*, November 1959 (3-323). Virtually all of these early program plans and recommendations, from new launch vehicles to lunar and planetary missions, were in large measure a response to the external challenge represented by achievements of the Soviet space program. As Abe Silverstein observed in testimony before Congress, large launch vehicles were required "so that we could do what we thought the Administration and Congress had set as its goal, namely, to make this country first in space." United States Congress, House, Committee on Science and Astronautics, *1960 NASA Authorization*, Hearings before the Committee and Subcommittees Nos. 1, 2, 3, and 4, 86th Congress, 1st Session on H. R. 6512, 1959, p. 410; cf., similar JPL sentiment in *Ten-Year Plan* (JPL Publication No. 31-2. Pasadena, California: Jet Propulsion Laboratory, California Institute of Technology, December 3, 1959), p. 1.

46. *Proposal for Space Flight Program Study* (Pasadena, California: Jet Propulsion Laboratory, California Institute of Technology, November 7, 1958) (2-620c); and interview of James Burke by Cargill Hall, January 27, 1969, p. 7 (2-1391).

47. *Exploration of the Moon, the Planets, and Interplanetary Space* (JPL Report 30-1. Pasadena, California: Jet Propulsion Laboratory, California Institute of Technology, April 30, 1959), Albert R. Hibbs, ed., p. 2.

48. This predominant concern with the moon at NASA Headquarters, after a planetary program had begun, caused JPL officials by the end of the year to request a charter and name change to: Working Group on Lunar *and* Planetary Exploration. Letter from William Pickering to Abe Silverstein, December 17, 1959 (2-803). The request was agreed to by NASA. Letter from Abe Silverstein to William Pickering, January 26, 1960 (2-318).

49. NASA Vega Flight Schedule of April 4, 1959, cited in United States Congress, House, Committee on Science and Astronautics, *Review of the Cancelled Atlas–Vega Launch Vehicle Development, December 1958– December 1959*, Report to the Committee, April 1960, p. 5.

50. Draft of NASA Memo for the File from Cunningham, "Working Group on Lunar Explorations," January 15, 1965 (2-651); letter from Pickering to Silverstein, December 17, 1959 (2-803).

51.   Letter from Val Larsen to Robert Jastrow, February 26, 1959 (2-829). J. Arnold, E. Anderson, and M. van Dilla responded with a proposal to obtain a gamma-ray spectrum of the moon's surface on March 17, 1959. JPL Interoffice Memo from Albert Metzger to Cargill Hall, subject: "Comments on Ranger Chronology," January 8, 1970. NASA contracted for design and development of the small, single-axis, seismometer with a Caltech–Columbia University team in July 1959. R. D. Gurney, et al., *Final Report, A Seismometer for Ranger Lunar Landing* (Pasadena, California: Seismological Laboratory, California Institute of Technology, May 15, 1962), pp. 1–2.

52.   United States Congress, House, Committee on Science and Astronautics, *The First Soviet Moon Rocket,* Report of the Committee, 86th Congress, 1st Session, on H. R. 1086, 1959, p. 6.

53.   See, for example, W. Sambrot, "Space Secret," *Saturday Evening Post,* February 21, 1959; and United States Congress, House, Committee on Appropriations, *National Aeronautics and Space Administration Appropriations,* Hearings before the Subcommittee, 86th Congress, 1st Session, 1959, pp. 196–197; also United States Congress, House, Select Committee on Astronautics and Space Exploration, *The Next Ten Years in Space, 1959–1969,* Staff Report of the Committee, 86th Congress, 1st Session, House Document No. 115, 1959, passim; and Evert Clark, "Vega Study Shows Early NASA Problems," *Aviation Week and Space Technology,* June 27, 1960, p. 62.

54.   NASA memorandum from Robert Jastrow to Homer Newell, subject: "Soviet Plans for Lunar Exploration," April 20, 1959 (2-1927).

55.   Testimony of Milton Rosen, Abe Silverstein, and J. Allen Crocker in United States Congress, Senate, Committee on Aeronautical and Space Sciences, *NASA Authorization for Fiscal Year 1960,* Hearings before the Committee and Subcommittee on NASA Authorization, on S. 1582, 86th Congress, 1st Session, 1959, Part I: Scientific and Technical Presentations, pp. 25–26, 121, 215. This NASA interest in lunar exploration also noted prominently during the first National Conference on Space Physics held at this time, April 1959. See the proceedings in *The Journal of Geophysical Research,* Volume 64, No. 2, November 1959.

56.   NASA Memorandum to the Files from William Cunningham, subject: "Lunar Explorations," June 16, 1959 (2-1928).

57.   Testimony of James D. Burke in the Report of the JPL Failure Investigation Board, "Ranger RA-5 Failure Investigation," November 13, 1962 (2-459); revised Vega objectives and launch dates that conformed to this directive are affirmed in a letter from William Pickering to Abe Silverstein, July 10, 1959 (2-835).

58. Interview of Edgar Cortright by Cargill Hall, March 4, 1968 (2-762).

59. NASA memo from Newell to Silverstein, "Recommendations for a Lunar Science Group," August 26, 1959 (2-1929). Membership of the Lunar Science Group in 1959 included:

> Robert Jastrow, Goddard Space Flight Center, NASA, Chairman
>
> Harrison Brown, California Institute of Technology
>
> Maurice Ewing, Columbia University
>
> Thomas Gold, Cornell University
>
> Albert Hibbs, Jet Propulsion Laboratory
>
> Joshua Lederberg, Stanford University, Department of Genetics
>
> Gordon MacDonald, University of California, Los Angeles, Institute of Geophysics
>
> Frank Press, California Institute of Technology
>
> Bruno Rossi, Massachusetts Institute of Technology
>
> Ernst Stuhlinger, Army Ballistic Missile Agency
>
> Harold Urey, University of California, San Diego

U.S. Congress, *Second Semiannual Report of NASA,* p. 123.

60. NASA Memorandum for the File from T. Keith Glennan, July 24, 1959, with attachments "Participants at 23 July Meeting," and "Excerpts from the Preliminary U.S. Policy on Outer Space for Use at the 23 July Meeting" (2-2608). Rationale advanced by Glennan in support of this position among his NASA associates at this time included (1) Vega launch delays that appeared to preclude planetary missions in 1960–1961; (2) proximity—the moon was much closer, the flight time was several days instead of several months, and spacecraft would not require as great a reliability; (3) a launch opportunity came up once each lunar month; and (4) communications at lunar ranges were less complicated. Interview of Homer Stewart by Cargill Hall, November 22, 1969. This NASA lunar decision reiterated in the letter from Richard Horner to William Pickering, December 16, 1959 (2-1935b); in the *NASA Long Range Plan,* December 16, 1959; and in the letter from William Pickering to Abe Silverstein, August 4, 1959 (2-825).

61. Details reviewed in NASA, OSFD, "Staff Conference Report, Monday, 7 November 1959," p. 1 (2-1760); letter from William Pickering to Abe Silverstein, November 9, 1959, (2-843); and in JPL Interoffice Memo from John Keyser to Distribution, subject: "Minutes of the Vega Staff Meeting November 25, 1969," November 30, 1959 (2-1016).

62.   JPL Interoffice Memo from Gumpel, Sloan, and Buwalda to James Burke, subject: "Lunar Rough Landing Scientific Experiment," December 10, 1959; see also G. F. Schilling, *Lunar and Deep Space Programs* (Washington: Office of Space Flight Development, National Aeronautics and Space Administration, September 14, 1959), p. 1.

63.   See NASA, OSFD, "Staff Conference Report, Friday, 4 December 1959" (2-1760). Cancellation order in TWX from Ralph Cushman to George Green, December 14, 1959 (3-804); JPL Announcement No. 14 from William Pickering to All Personnel, subject: "Cancellation of the Vega Program," December 11, 1959 (2-801). The decision hinged on reliability expected through extended use of Atlas-Agena B. See the testimony of T. Keith Glennan in United States Congress, House, Committee on Science and Astronautics, *Review of the Space Program*, Hearings before the Committee, 86th Congress, 2nd Session, 1960, No. 3, Part I, pp. 170–171. A most comprehensive survey of Vega prior to cancellation is contained in *The Vega Program* (JPL Report 30-6. Pasadena, California: Jet Propulsion Laboratory, California Institute of Technology, August 3, 1959), Jack N. James, ed.

64.   Letter from Abe Silverstein to William Pickering, December 21, 1959 (2-470). The NASA decision in late 1959 against attempting photography from low-altitude lunar orbiters was made in reviews during early December: "It is ... a project of greater technological difficulty than the rough landing projects, and ... the rough landing must be attempted before the low-altitude satellite if we are going to move ahead as quickly as possible on the lunar program." NASA memorandum from Robert Jastrow to Homer Newell, subject: "Report of December 1 Meeting of the Lunar Science Group," December 11, 1959 (2-1933a). NASA's two Atlas-Able lunar orbiters remaining at that time—and scheduled for launch in 1960—did not incorporate photography among the experiments in the spacecraft payload.

65.   Testimony of Homer Newell in United States Congress, House, Committee on Science and Astronautics, *Investigation of Project Ranger*, Hearings before the Subcommittee on NASA Oversight, 88th Congress, 2nd Session, No. 3, 1964, p. 45. Emphasis upon rapid development and flight schedules was prompted by the Soviet lunar program noted above, and by an optimism prevailing at JPL and NASA that Centaur-launched lunar soft-landing missions would be possible in 1961 or in 1962. "The type of mission described as a hard or rough lunar landing ... is dependent on the length of time required to develop rugged instruments. Unless this time is relatively short, vehicles with the capability of soft-landing a payload will be available before the ruggedized (seismometer) instrument package is ready." *Space Programs Summary No. 5* for the period July 15, 1959 to

September 15, 1959 (Pasadena, California: Jet Propulsion Laboratory, California Institute of Technology, October 1, 1959), p. 69.

66. William H. Pickering, "Space—The New Scientific Frontier," an address before the American Institute of Chemical Engineers Annual Meeting in St. Paul, Minnesota, September 29, 1959, p. 5 (2-922); also, Pickering's statements are cited in *Aviation Week and Space Technology,* November 23, 1959, p. 26; this position is restated in the JPL *Ten-Year Plan* (JPL Publication No. 31-2. Pasadena, California: Jet Propulsion Laboratory, California Institute of Technology, December 3, 1959), p. 2.

67. NASA Memorandum for the File from Homer Newell, subject: "Trip Report for the Visit to Jet Propulsion Laboratory on 28 December 1959 by Homer E. Newell, Jr., Newell Sanders, J. A. Crocker, Morton J. Stoller," December 30, 1959, p. 2 (2-1935a).

68. However, Ranger almost wasn't Ranger. As a youth, Silverstein had owned a dog by the same name. During an unmercifully long and active life, his canine companion had demonstrated a talent for avoiding all human direction and otherwise had proven himself intractable and cantankerous. With a recollection of the obstinate beast still vivid to mind, Silverstein strenuously opposed the choice. Interview of Clifford Cummings by Cargill Hall, August 19, 1971, p. 1 (2-2218). Nevertheless, the name stuck at JPL, and was soon picked up at other organizations drawn into the project. In the absence of a more appropriate designation and with wide reference to the Atlas-Agena lunar enterprise as Ranger, NASA Headquarters consented to the appellation. On May 4, Cliff Cummings publicly announced the name. Julian Hartt, "Ranger Spacecraft Details Revealed by JPL Official," Los Angeles *Herald-Examiner,* May 4, 1960, p. 23.

Chapter Two

1. Letter from Don Ostrander to Wernher von Braun, December 29, 1959 (2-582).

2. *Interim Report of the Survey Team Established to Investigate the Use of Agena for the National Aeronautics and Space Administration,* January 15, 1960, p. 6. (2-604).

3. Ibid., p. 5.

4. Ibid., pp. 8–9.

5. NASA memorandum from Richard Horner to Albert Siepert and Abe Silverstein, subject: "Relationships with California Institute of Technology and Jet Propulsion Laboratory," November 16, 1959 (3-314).

6. JPL Interoffice Memo 24A from William Pickering to JPL Senior Staff, Section Chiefs, Section Supervisors, and Group Supervisors, subject: "Program Director for Ranger," January 26, 1960 (2-233).

7. Clifford I. Cummings, Trip Report No. RPD-7, "Conference at Washington, D.C.," April 5, 1960 (2-1031); letter from William Pickering to Abe Silverstein, December 29, 1959 (2-469).

8. JPL Announcement No. 51 from William Pickering to JPL Senior Staff, Section Chiefs, Section Managers, and Supervisors, subject: "Appointment of Ranger Project Manager," October 14, 1960 (2-303a).

9. The technical fix, aerodynamic drag brakes that operated much like the blades of a camera shutter, proved as simple and efficient as the shutoff valves that performed the equivalent function on liquid propellant missiles. Cf., James D. Burke, et al., "Range Control for a Ballistic Missile," U.S. Patent 3,188,958, granted June 15, 1965.

10. Letter from Richard Horner to William Pickering, December 16, 1959 (2-471).

11. NASA memorandum from Richard Horner to Don Ostrander and Abe Silverstein, December 29, 1959, cited in Ivan D. Ertel and Mary Louise Morse, *The Apollo Spacecraft: A Chronology* (NASA SP-4009. Washington: National Aeronautics and Space Administration, 1969), Volume I: "Through November 7, 1962," p. 34.

12. NASA memorandum from W. Schubert to Abe Silverstein, subject: "Agena Program Management," January 15, 1960 (2-2256).

13. Minutes of the Space Exploration Council Meeting, February 10–11, 1960, p. 7 (2-1416).

14. Ibid., p. 8.

15. Friedrich Duerr, "Preparing Ranger for Operations," *Astronautics,* September, 1961, p. 28.

16. See the Reports of the Agena-B Lunar Committee to the Agena-B Coordination Board during 1960 (in 2-2512).

17. Project Paperclip brought German scientists and engineers to the United States at the end of World War II. See Clarence G. Lasby, *Project Paperclip: German Scientists and the Cold War* (New York: Atheneum Press, 1971).

18. Letter from Richard Horner to Bernard Schriever, March 18, 1960. Enclosure: "Results of Centaur-Agena Management Conference, 1 March 1960" (2-2264a&b).

19. Letter from Don Ostrander to Wernher von Braun, June 10, 1960 (2-1040).

20. TWX from ARDC (RDGN), Andrews AFB, to Don Ostrander, subject: "NASA Agena B Program," July 16, 1960 (2-2248).

21. Letter from Don Ostrander to Wernher von Braun, August 3, 1960 (2-2247a); letter from Don Ostrander to O. J. Ritland, August 3, 1960 (2-2260).

22. Air Force Ballistic Missile Division Headquarters Daily Bulletin No. 71, April 12, 1960 (2-2249).

23. Letter from John Albert to James Burke, July 14, 1961 (2-1134).

24. JPL Interoffice Memo from Clifford Cummings to all JPL Division Chiefs, August 30, 1960 (2-1074).

25. *JPL-Industry Conference Proceedings, conducted by the Jet Propulsion Laboratory and National Aeronautics and Space Administration October 26, 1960* (Pasadena, California: Jet Propulsion Laboratory, California Institute of Technology, November 18, 1960), p. 18.

26. Interview of Gordon Kautz by Cargill Hall, December 17, 1971, p. 1 (2-2246). JPL Interoffice Memo from Clifford Cummings to Senior Staff, et al., subject: "Appointment of Ranger Assistant Project Manager," November 10, 1960 (2-1087).

27. Clifford I. Cummings, Trip Report No. RPD-7 (2-1031); see also JPL Interoffice Memo from James Burke to Brian Sparks, subject: "Visit by Dr. Glennan," July 20, 1960 (2-519).

28. Letter from Abe Silverstein to William Pickering, July 5, 1960 (2-328).

29. The Cortright team was initially composed of Gerhardt Schilling and Newton W. Cunningham (Lunar and Planetary Sciences), Oran W. Nicks and Benjamin Milwitzky (Lunar Flight Systems), and Fred Kochendorfer (Planetary Flight Systems). Oran Nicks, a young aeronautical engineer from Chance-Vought Aircraft, and Ben Milwitzky, another aeronautical engineer previously with NACA's Langley Laboratory, would supervise developments in JPL's lunar program in the months that followed. Interview of Edgar Cortright by Cargill Hall, March 4, 1968, p. 1 (2-762).

30. JPL Memo for the Record from J. L. Stamy, et al., subject: "Trip to Lockheed and BMD on Agena Funding," February 11, 1960, p. 1 (2-488).

31. *NASA/AFBMD Management Procedures,* signed by Major General O. J. Ritland and Abe Silverstein, January 1960 (2-1396a&b).

32. For a description of the Discoverer model, see Harold T. Luskin, "The Ranger Booster," *Astronautics,* September 1961, p. 30.

33.   *Ranger Project Development Plan,* May 24, 1961 (Pasadena, California: Jet Propulsion Laboratory, California Institute of Technology, Revised July 5, 1961), p. 12 (2-621); also NASA, *Minutes of the Space Exploration Program Council Meeting,* April 25–26, 1960, p. 11 (2-1406).

34.   Interview of James Spaulding by Cargill Hall, April 3, 1972.

35.   Five for Project Ranger and four for use in Goddard satellite programs. NASA memorandum from William Fleming to Albert Kelley, subject: "Agena-B Launch Vehicle Requirements," August 3, 1960 (2-2303); NASA memorandum from Albert Kelley to William Fleming, subject: "Agena-B Launch Vehicle Requirements," August 5, 1960 (2-2302); NASA memorandum from Don Ostrander to Abe Silverstein, subject: "Reorientation of Agena-B Launch Vehicle Contract," August 9, 1960 (2-2301).

36.   *The Ranger Project: Annual Report for 1961 (U)* (JPL TR 32-241. Pasadena, California: Jet Propulsion Laboratory, California Institute of Technology, June 15, 1962), p. 87.

37.   James Q. Spaulding and Frank A. Goodwin, *Report to JPL Management on Ranger–Agena Interface,* December 20, 1960 (2-1063).

38.   Letter from Russell Herrington to Hans Hueter, November 10, 1960 (2-1086).

39.   Letter from Albert Kelley to Don Ostrander, subject: "NASA–USAF Relations on Agena," October 24, 1960, p. 2 (2-2252).

40.   Hans Hueter, *Summation of Problem Areas Encountered with LMSD,* enclosure to letter from Wernher von Braun to Herschel Brown, December 28, 1960 (2-1153).

41.   *Space Programs Summary No. 37-7* for the period November 15, 1960, to January 15, 1961 (Pasadena, California: Jet Propulsion Laboratory, California Institute of Technology, February 1, 1961), pp. 4–5.

42.   Interview of James Burke by Cargill Hall, January 27, 1969, p. 7 (2-1391).

43.   JPL Interoffice Memo from Clifford Cummings to William Pickering, subject: "Recommendations as Subjects for Space Exploration Council Meeting," July 5, 1960 (2-1045).

44.   Minutes of the Space Exploration Program Council Meeting, July 14–15 1960, p. 6 (2-1418b).

45.   Cf., for example, T. Keith Glennan, *Transition Memorandum,* January 1961 (2-1755).

46. NASA memorandum from T. Keith Glennan to Participants at the Williamsburg Conference, subject: "Staff Paper on Project Management," October 14, 1960 (2-1083a).

47. NASA, "Fourth Semi-Annual Staff Conference," Williamsburg, Virginia, October 16–19, 1960, pp. 56–59 (2-1428). This action coincided with a similar recommendation contained in the NASA report, "Report of the Advisory Committee on Organization," October 12, 1960 (2-1426).

48. JPL Interoffice Memo from Robert Parks to Leonard Frankenstein, subject: "Comments on 'Proposed Instruction on Planning and Managing OSS Program'," June 13, 1963 (2-1950).

49. Letter from Edgar Cortright to John Martin, March 27, 1961 (2-348); letter from Abe Silverstein to William Pickering, February 16, 1961 (2-344); NASA memorandum from Abe Silverstein to Don Ostrander, subject: "Implementation of NASA Management Instruction 4-1-1," March 29, 1961 (2-346).

Chapter Three

1. JPL Interoffice Memo from William Pickering to Daniel Schneiderman—memo is lost—recollection of recipient.

2. See the testimony of Daniel Schneiderman on November 6, 1962, in RA-5 Failure Investigation Board Interviews, October 31–November 6, 1962, p. 1 (2-460b).

3. James D. Burke's testimony on November 2, 1962, ibid.; also rationale as described by Clifford I. Cummings, "The Shape of Tomorrow," *Astronautics,* Vol. 5, July 1960, pp. 24–25.

4. U.S. Army Ordnance Missile Command, *Satellite and Space Program Progress Report for NASA,* November 18, 1958 (2-578).

5. See Burke's testimony, RA-5 Failure Investigation Board Interviews (2-460b); *The Ranger Project: Annual Report for 1961(U)* (JPL TR 32-241. Pasadena, California: Jet Propulsion Laboratory, California Institute of Technology, June 15, 1962), p. 2.

6. JPL Interoffice Memo from John Keyser to Distribution, subject: "Vega Program; Preliminary Design Phase Assignments," April 14, 1959 (2-1002).

7. Interview of James Burke by Cargill Hall, January 27, 1969, pp. 2–3 (2-1391).

8. *Space Programs Summary No. 4* for the period May 15, 1959, to July 15, 1959 (Pasadena, California: Jet Propulsion Laboratory, California Institute of Technology, August 1, 1959), p. 7; United States Congress, Senate, Committee on Aeronautical and Space Sciences, *NASA Authorization for Fiscal Year 1960,* Hearings before the NASA Authorization Subcommittee, 86th Congress, 1st Session, on S. 1582 and H.R. 7007, 1959, Part II: "Program Detail for Fiscal Year 1960," p. 750; for design evolution see James D. Burke, "The Ranger Spacecraft," *Astronautics* September 1961, pp. 23–24.

9. JPL Interoffice Memo from William Pickering to Division Chiefs, et al., subject: "Establishment of Space Sciences Division," July 14, 1959 (3-225); letter from William Pickering to R. L. Bell, Colonel P. H. Seordas, and G. P. Chase, November 20, 1959 (2-865).

10. Burke, "The Ranger Spacecraft," pp. 24–25; the spacecraft functional specifications released in August and September 1959 reflected the decision, see *Payload Functional Description, Vega V-6,* JPL Functional Design Group, August 21, 1959 (2-945); and James D. Burke, "Design Criteria for Vega Lunar Capsule," December 7, 1959 (2-1349).

11. Homer E. Newell, *High Altitude Rocket Research* (New York: Academic Press, Inc., 1953), p. v.

12. See the testimony of Homer Newell, United States Congress, House, Committee on Science and Astronautics, *1962 NASA Authorization,* Hearings before the Committee and Subcommittees 1, 3, and 4, 87th Congress, 1st Session, on H.R. 3238 and H.R. 6029, 1961, No. 7, Part I, p. 244.

13. Letter from Homer Newell to Bruno Rossi, December 18, 1959 (2-1937); Robert Jastrow, ed., *The Exploration of Space, A Symposium of Space Physics*, April 29-30, 1959, sponsored by the National Academy of Sciences, the National Aeronautics and Space Administration, and the American Physical Society (New York: The Macmillan Company, 1960).

14. NASA Space Sciences "Staff Conference Report," February 19, 1959, p. 2 (2-1760). See also "From Interest to Resolve" in Chapter One of this volume.

15. JPL Interoffice Memo from Space Sciences Division to Distribution, subject: "Vega Missions," November 30, 1959 (2-846).

16. NASA Memorandum for the File from Homer Newell, subject: "Trip Report for the Visit to Jet Propulsion Laboratory on 28 December 1959 by Homer Newell, Jr., Newell Sanders, J. A. Crocker, Morton J. Stoller," December 30, 1959 (2-1935a); letter from William Pickering to Abe

Silverstein, December 29, 1959 (2-2483); also William H. Pickering, "Do We Have a Space Program?" *Astronautics,* January 1960, pp. 83–84.

17. Based on NASA Memo for the File, from Newell, p. 2 (2-1935a). The craft was, nonetheless, represented publicly as a "lunar spacecraft," Fourth Semiannual Report of the *National Aeronautics and Space Administration* (Washington: Government Printing Office, 1961), p. 57.

18. Letter from William Pickering to Gerhardt Schilling, January 20, 1960 (2-668); letter from Abe Silverstein to William Pickering, January 26, 1960 (2-669); letter from William Pickering to Gerhardt Schilling, May 5, 1960 (2-671); letter from Abe Silverstein to William Pickering, May 23, 1960 (2-1411a).

19. Daniel Schneiderman, et al., *Spacecraft Design Criteria and Considerations; General Concepts, Spacecraft S-1* (JPL Section Report No. 29-1. Pasadena, California: Jet Propulsion Laboratory, California Institute of Technology, February 1, 1960).

20. JPL Interoffice Memo from Clifford Cummings to all Division Chiefs, subject: "Ranger A Program Guidelines," February 17, 1960 (2-1024).

21. Letter from Abe Silverstein to William Pickering, January 26, 1960 (2-669); also "Mission Objectives and Design Criteria," in *Ranger Spacecraft Design Specification Book,* Spec. No. RA12-2-110 (Pasadena, California: Jet Propulsion Laboratory, California Institute of Technology, April 19, 1960), p. 2 (2-1094a).

22. The objectives of the original Ranger flights one through five were "(a) to create and test a new spacecraft design whose features can be exploited in the performance of lunar and interplanetary flight missions; and (b) using this spacecraft to perform two classes of scientific experiments: first, a group of measurements dealing with particles, fields and the solar atmosphere within one million miles of the Earth; and second, a group of measurements of lunar characteristics close to and on the surface of the Moon." *Ranger Project Development Plan (Rangers 1–5)* (Revision 2. Pasadena, California: Jet Propulsion Laboratory, California Institute of Technology, August 1, 1962), p. I-2.

23. Letter from Silverstein to Pickering, May 23, 1960 (2-1411a).

24. JPL Spec. No. RA12-2-110, *Ranger Spacecraft Design Specification Book,* p. 1 (2-1094a).

25. Testimony of Burke, RA-5 Failure Investigation Board Interviews (2-460b); *Space Programs Summary No. 37-3* for the period March 15, 1960, to May 15,1960 (Pasadena, California: Jet Propulsion Laboratory, California Institute of Technology, 1960), p. 26.

26.   Spacecraft technical details can be found in Appendix B.

27.   NASA Statement from T. Keith Glennan, subject: "Decision to Negotiate on the Lunar Capsule," April 27, 1960 (2-320).

28.   Aeronutronic, a Division of Ford Motor Company, *Design Study of a Lunar Capsule Final Report* (Publication No. U-870, Under Contract M48024, April 15, 1960), Volume I: "Design Study" (2-703a).

Chapter Four

1.   JPL Interoffice Memo from Victor Clarke to James Burke, subject: "RA-3, 4, 5 Performance," July 14, 1960 (2-562).

2.   JPL Interoffice Memo from James Burke to Harris Schurmeier, subject: "Ranger Weight Control," July 19, 1960 (2-2389); JPL Interoffice Memo from S. Rubinstein to Distribution, subject: "RA-3 Spacecraft Weight Reduction Program" August 1, 1960 (2-2530).

3.   NASA PMP Chart 11-0, dated August 10, 1960, citing Official NASA Flight Schedule of June 15, 1960 (2-968).

4.   In future, NASA Headquarters ordered a change in trajectory management procedure. See NASA Headquarters Memo from Robert Seamans to NASA Headquarters Staff and Field Center Directors, subject: "Launch Vehicle Characteristics," December 7, 1960 (2-2097).

5.   Minutes of the Sixth Meeting of the Agena-B Coordination Board, October 13, 1960, p. 4 (2-486); see also, JPL Interoffice Memo from C. G. Pfeiffer to James Burke, subject: "History of the Atlas/Agena Trajectory Management Problem," January 16, 1961 (2-2195).

6.   JPL Interoffice Memo from C. G. Pfeiffer to All Concerned, subject: "Ranger Guidance and Trajectory Sub-contract to STL" December 14, 1960 (2-1060).

7.   JPL Interoffice Memo from M. R. Mesnard to Distribution, subject: "RA-345 Omni-Antenna," January 12, 1961 (2-1099).

8.   NASA memorandum from T. Keith Glennan to Operation Heads of Headquarters Offices and Directors, NASA Centers, subject: "Further Comments on Monterey Conference," March 20, 1960 (2-323); see also, letter from T. Keith Glennan to William Pickering, June 17, 1960 (2-1236). If the issue ever reached his desk, James E. Webb, the new Kennedy-appointed NASA Administrator who was pledged to excel in astronautics, undoubtedly concurred.

9.   JPL Interoffice Memo from James Burke to Harris Schurmeier, subject: "RA-3 Weight Situation," February 16, 1961, p. 2 (2-563).

10.   Minutes of the Ninth Meeting of the Agena-B Coordination Board, March 1, 1961, p. 7 (2-491).

11.   JPL Interoffice Memo from James Burke to Distribution, subject: "Ranger Project Review Meeting for May 3, 1961," April 28, 1961 (2-1114).

12.   JPL Interoffice Memo from James Burke and Gordon Kautz to Brian Sparks, subject: "Ranger Project Status Report No. 9," June 5, 1961, p. 1 (2-1314).

13.   See unsigned document, "Probability of Ranger Success vs. Spacecraft Weight," February 22, 1961 (2-2529).

14.   Flight tests would not be used again in NASA's unmanned space program, although they continued to be rigorously applied in NASA's manned space program—except that no science was included.

15.   Later in Project Ranger, other models were added in the test cycle, a dynamic test model and a design evaluation model among them, not considered here.

16.   JPL Interoffice Memo from James Burke to All Concerned, subject: "Ranger Test Philosophy," June 13, 1960 (2-2522).

17.   See James Burke's review in JPL Interoffice Memo from James Burke to Brian Sparks, subject: "Project Manager's Remarks for Senior Council Meeting," August 24, 1962, p. 2 (2-2540).

18.   See Oran Nicks' review in NASA Headquarters Memo from Oran Nicks to Abe Silverstein, subject: "Analysis of JPL Headquarters Relationships and Recommendations for Improvements," October 6, 1961 (2-332b).

19.   *A Review of Space Research* (Publication 1079. The Report of the Summer Study conducted under the auspices of the Space Science Board of the National Academy of Sciences at the State University of Iowa, June 17–August 10, 1962. Washington: National Academy of Sciences—National Research Council, 1962), pp. 10–11.

20.   Later, Professor Lederberg's cautions appeared in an article "Moondust," *Science,* Volume 127, 1958, p. 1473.

21.   *A Review of Space Research,* pp. 10–13.

22.   Cited in Charles M. Atkins, *NASA and the Space Science Board of the National Academy of Sciences* (Comment Draft, HHN-62. Washington: National Aeronautics and Space Administration, September, 1966), pp. 108–109 (5-53).

23. NASA memorandum from Abe Silverstein to Harry Goett, subject: "Sterilization of Payloads," October 15, 1959 (2-1930).

24. Cf. the representations of Carl Sagan in *Proceedings of the Lunar and Planetary Exploration Colloquium* (North American Aviation, Inc.), Volume II, No. 3, p. 46; and Harold Urey, Volume I, No. 3, p. 31.

25. James D. Burke, "The Ranger Spacecraft," *Astronautics,* September 1961, p. 25; and the testimony of Oran Nicks in United States Congress, House, Committee on Science and Astronautics, *Investigation of Project Ranger,* Hearings before the Subcommittee on NASA Oversight, 88th Congress, 2nd Session, 1964, No. 3, p. 65.

26. Richard W. Davies and Marcus G. Comuntzis, *The Sterilization of Space Vehicles to Prevent Extraterrestrial Biological Contamination* (JPL EP 698. Pasadena, California: Jet Propulsion Laboratory, California Institute of Technology, August 31, 1959), p. 13.

27. JPL Interoffice Memo from James Burke to All Concerned, subject: "Sterilization," March 8, 1960 (2-994); JPL Interoffice Memo from George L. Hobby to Casper Mohl, subject: "Sterilization Procedure," April 26, 1960 (2-997); JPL Interoffice Memo from George L. Hobby to Manfred Eimer, subject: "Spacecraft Sterilization," April 27, 1960 (2-998).

28. Letter from Clifford Cummings to Robert Seamans, subject: "Procedures for Sterilization of Ranger A3 Spacecraft," May 25, 1961 (2-1119); see also, JPL Interoffice Memo from Rolf Hastrup to Distribution, subject: "Ranger Sterilization Program," May 24, 1961 (2-1118).

29. Letter from Robert Seamans to William Pickering, June 26, 1961 (2-1705).

30. "Functional Specification Ranger RA-3, RA-4, and RA-5 Spacecraft Mission Objectives and Design Criteria," in *Ranger Spacecraft Design Specification Book,* Spec. No. RA 345-2-110D (Pasadena, California: Jet Propulsion Laboratory, California Institute of Technology, August 8, 1962), p. 3 (2-1095e).

31. NASA Management Instruction No. 37-1-1, subject: "Establishment and Conduct of Space Sciences Program—Selection of Scientific Instruments," effective April 15, 1960 (2-447); John Naugle's testimony, "Management of Space Science," *Program Review: Science and Applications Management* (Washington: National Aeronautics and Space Administration, June 22, 1967), pp. 48–50 (2-757); SSSC membership cited in NASA Circular 73, May 27, 1960.

32. Interview of Albert Hibbs by Cargill Hall, October 2, 1972, pp. 23–24 (3-595); see also, James D. Burke, "Engineering Aspects of the Ranger Project," July 9, 1962, p. 2 (2-1363).

33. V. L. Filch and L. M. Lederman, "Vela Hotel: Comments on Simplified X Ray Detectors for Interim Use," Institute for Defense Analyses, Jason Division, January 24, 1961 (2-490).

34. James Burke, conference report, "Visit by Los Alamos and Sandia People," April 14, 1960 (2-478).

35. Letter from A. W. Betts to John Clark, April 15, 1960 (2-481a).

36. "A LASL-Sandia Proposed Vela Hotel Experiment for the Ranger A-1 and A-2 Probes," Los Alamos Scientific Laboratory, Sandia Corporation, May 3, 1960 (2-479).

37. Letter cited in the letter from Abe Silverstein to William Pickering, subject: "Vela Hotel Experiments for Ranger 1 and 2 Spacecraft," June 29, 1960 (2-480).

38. JPL Interoffice Memo from James Burke to Distribution, subject: "Vela Hotel Experiment for RA-1 and RA-2," June 7, 1960, pp. 2–3 (2-1038).

39. Letter from Silverstein to Pickering, June 29, 1960 (2-480).

40. See Table II in "The Vega-Ranger: Where Planet and Sky Science Meet," Chapter Three of this volume.

41. *Space Programs Summary No. 37-3* for the period March 15, 1960, to May 15, 1960 (Pasadena, California: Jet Propulsion Laboratory, California Institute of Technology, June 1, 1960), p. 5.

42. "Mission Objectives and Design Criteria," in *Ranger Spacecraft Design Specification Book,* Spec. No. RA12-2-110A (Pasadena, California: Jet Propulsion Laboratory, California Institute of Technology, November 1, 1960), p. 4 (2-1094b).

43. The instruments are described in *Scientific Experiments for Ranger 1 and 2* (JPL TR 32-55. Pasadena, California: Jet Propulsion Laboratory, California Institute of Technology, January 3, 1961); see also, Albert R. Hibbs, Manfred Eimer, and Marcia Neugebauer, "Early Ranger Experiments," *Astronautics,* September 1961, pp. 26–27.

44. JPL Interoffice Memo from James Burke to Brian Sparks, subject: "Visit by Dr. Glennan," July 22, 1960, p. 1 (2-519).

45. "A LASL-Sandia Proposed Vela Hotel Experiment for RA-3, 4, 5," Los Alamos Scientific Laboratory, Sandia Corporation, August 19, 1960.

46. Letter from James Burke to Edgar Cortright, September 9, 1960 (2-1078).

47. Letter from Clifford Cummings to Edgar Cortright, September 16, 1960 (2-1424).

48. Letter from R. F. Taschek to Edgar Cortright, September 14, 1960 (2-1079). The Vela Hotel Program did continue using other space vehicles, eventually resulting in the very successful system that became operational in the mid-1960s.

49. F. E. Lehner, E. O. Witt, W. F. Miller, and R. D. Gurney, *Final Report, A Seismometer for Ranger Lunar Landing* (Seismological Laboratory, Pasadena, California: California Institute of Technology, May 15, 1962).

50. Frank Press, Phyllis Buwalda, and Marcia Neugebauer, "A Lunar Seismic Experiment," *Journal of Geophysical Research* Vol. 65, October 1960, pp. 3097f; also R. L. Kovach and Frank Press, *Lunar Seismology* (JPL TR 32-328. Pasadena, California: Jet Propulsion Laboratory, California Institute of Technology, August 10, 1962).

51. James R. Arnold, "Gamma Ray Spectroscope of the Moon's Surface," *Proceedings of the Lunar and Planetary Exploration Colloquium,* Vol. I, No. 31, October 29, 1958.

52. *Lunar Impact TV Camera (Ranger 3, 4, 5): Final Engineering Report* (AED R-2076. Princeton, New Jersey: RCA Astro-Electronics Division, November 15, 1963).

53. NASA Summary Minutes of the Meeting of the Space Sciences Steering Committee held on October 16, 1961 (2-1215).

54. JPL Interoffice Memo from Walter Brown to Distribution, subject: "Ranger 3, 4, 5 Radar Data," March 27, 1961, p. 1 (2-1111b).

55. JPL Interoffice Memo from M. R. Mesnard to James Burke, subject: "Method of Commutating Radio Altimeter Reflectivity Data for Telemetry on RA-345," April 28, 1961 (2-1113).

56. NASA Summary Minutes of the Meeting of the SSSC, October 16, 1961 (2-1215); see also, H. E. Wagner, *Scientific Subsystem Operation: Ranger 3, 4, 5* (JPL TM 33-80. Pasadena, California: Jet Propulsion Laboratory, California Institute of Technology, February 3, 1962); and *Scientific Experiments for Ranger 3, 4, 5* (JPL TR 32-199. Pasadena, California: Jet Propulsion Laboratory, California Institute of Technology, December 5, 1961).

Chapter Five

1. JPL Announcement No. 39 from William Pickering to Senior Staff, et al., subject: "Appointment of Program Director and Program Deputy Director of the Deep Space Instrumentation Facility (DSIF)," May 6, 1960 (2-242).

2.   Eberhardt Rechtin, "Communications to the Moon and Planets,"
      Proceedings of the Radio Club of America, Volume 42, No. 1, April 1966,
      pp. 1-8. Rechtin, as events turned out, would later (1967-1970) become
      Director of ARPA.

3.   *Soviet Space Programs 1966-70* (Science Policy Research Division and
      Foreign Affairs Division of the Congressional Research Service and the
      European Law Division of the Law Library, Library of Congress.
      Washington: Government Printing Office, 1971), p. 152. The second station
      had been completed by 1970.

4.   "DSN Facility Activation Dates," a JPL chart circa April 1967 (2-141).

5.   Cf. Edward M. Walters, *The Origins of the Australian Cooperation in Space*
      (Comment Edition, HHN-82. National Aeronautics and Space Administra-
      tion, May 1969) (5-195). Politics and economics continually impinge on
      the world of science and deep space radio tracking. In 1975 NASA
      prepared to abandon the Johannesburg site, partially in response to
      demands from opponents of apartheid.

6.   Rechtin "Communications to the Moon and Planets," p. 3.

7.   Eberhardt Rechtin, *Space Communications* (JPL Technical Release 34-68.
      Pasadena, California: Jet Propulsion Laboratory, California Institute of
      Technology, May 1, 1960).

8.   William D. Merrick, Eberhardt Rechtin, Robertson Stevens, and Walter K.
      Victor, *Deep Space Communications* (JPL Technical Release 34-10.
      Pasadena, California: Jet Propulsion Laboratory, California Institute of
      Technology, January 29, 1960); Nicholas A. Renzetti, "DSIF in the Ranger
      Project," *Astronautics,* September 1961, pp. 35-37, 70.

9.   Manfred Eimer, Albert R. Hibbs, and Robertson Stevens, *Tracking the
      Moon Probes* (JPL EP 701, Revised. Pasadena, California: Jet Propulsion
      Laboratory, California Institute of Technology, December 28, 1959), p. 9;
      Manfred Eimer and Y. Hiroshige, "Evaluation of Pioneer IV Orbit
      Determination Program," *Seminar Proceedings: Tracking Programs and
      Orbit Determination, February 23-26, 1960* (Pasadena, California: Jet
      Propulsion Laboratory, California Institute of Technology, 1960), p. 43.

10.   *The Ranger Project: Annual Report for 1961(U)* (JPL TR 32-241. Pasadena,
      California: Jet Propulsion Laboratory, California Institute of Technology,
      June 15, 1962), p. 383.

11.   Nicholas A. Renzetti, ed., *A History of the Deep Space Network from
      Inception to January 1, 1969* (JPL TR 32-1522. Pasadena, California: Jet
      Propulsion Laboratory, California Institute of Technology, September 1,
      1971), Volume I.

12. JPL Interoffice Memo No. 81 from William Pickering to Senior Staff, Section Chiefs, and Section Managers, subject: "Laboratory Policy and Procedure for Space Flight Operations," November 9, 1960 (2-305); see also, Marshall Johnson, "Space Flight Operations," *Astronautics Information: Systems Engineering in Space Exploration Seminar Proceedings, May 1–June 15, 1963* (Pasadena, California: Jet Propulsion Laboratory, California Institute of Technology, June 1, 1965), p. 39.

13. JPL Interoffice Memo No. 114 from William Pickering to Senior Staff, Section Chiefs, and Section Managers, subject: "Establishment of Space Flight Operations Section," October 5, 1961 (2-299).

14. JPL Report from the Data Handling Committee to Brian Sparks, subject: "Interim Report from Data Handling Committee," June 22, 1961 (3-666); letter from Val Larsen to Abe Silverstein, subject: "Data Operations Command Facility Summary and Index," October 17, 1961 (2-1460a).

15. NASA News Release, Homer E. Newell, "Remarks on the Dedication of the Space Flight Operations Facility, Jet Propulsion Laboratory," May 14, 1964, p. 9 (3-170c).

16. See United States Congress, House, Committee on Science and Astronautics, *Management and Operation of the Atlantic Missile Range,* 86th Congress, 2nd Session, 1960.

17. Interview of James Burke by Cargill Hall, January 27, 1969, p. 4 (2-1391).

18. Letter from Richard E. Horner to Bernard Schriever, March 17, 1960 (2-2263); see also, Francis E. Jarrett, Jr., and Robert A. Lindemann, *Historical Origins of NASA's Launch Operations Center to July 1, 1962* (Comment Edition, KHM-1. Cocoa Beach, Florida: John F. Kennedy Space Center, National Aeronautics and Space Administration, October 1964), pp. 66–67 (5-225); and paragraph 4.a of the LOD/AFMTC Agreement of July 17, 1961.

19. See Abe Silverstein's comments, "Minutes of the Space Exploration Council Meeting, April 25–26, 1960," p. 11 (2-1406).

20. JPL Interoffice Memo from Clarence Gates to All Concerned, subject: "Launch to Injection Instrumentation," June 14, 1960, p. 2 (2-1414).

21. JPL Interoffice Memo from Joseph Koukol to Clifford Cummings, subject: "Comments of the Agena Tracking Panel Meeting at AMR, June 8, 1960," June 10, 1960, p. 1 (2-1039).

22. TWX from Sam Snyder, NASA Headquarters, to Kurt Debus, LOD, subject: "AMR Atlas Agena-B Launch Management," July 19, 1961, in *NASA Agena Program Presentation,* October 1, 1962, p. 3-18 (2-2269).

23.   "Range Use and Support Agreement Between the Launch Operations Directorate, George C. Marshall Space Flight Center, NASA, and the Air Force Missile Test Center, Air Force Systems Command USAF, at Patrick Air Force Base, Florida," signed by Kurt H. Debus, Director LOD, and L. I. Davis, Major General, Commander AFMTC, July 17, 1961 (2-2374).

Chapter Six

1.   "Mission Objectives and Design Criteria," in *Ranger Spacecraft Design Specification Book,* Spec. No. RA12-2-110A (Pasadena, California: Jet Propulsion Laboratory, California Institute of Technology), November 1, 1960, p. 2 (2-1094b).

2.   See Harold T. Luskin, "The Ranger Booster," *Astronautics,* September 1961, pp. 30–31, 73–74.

3.   See "The Deep Space Network" in Chapter Five of this volume.

4.   *Space Programs Summary No. 37-11, Volume I* for the period July 1, 1961, to September 1, 1961 (Pasadena, California: Jet Propulsion Laboratory, California Institute of Technology, October 1, 1961), pp. 40–41. Calibration and system evaluation exercises already had been developed to maintain the accuracy of the Deep Space Instrumentation Facility. Star tracking by optically pointing the large antennas maintained their mechanical and readout capabilities. Radio beacons were flown on balloons, helicopters, and fixed wing aircraft to provide signal acquisition practice and to demonstrate the smooth operation of the deep space tracking system.

5.   JPL internal document, "Deep Space Instrumentation Facility" (2-929).

6.   *Space Programs Summary No. 37-11, Volume I,* pp. 3–4; *The Ranger Project: Annual Report for 1961(U)* (JPL TR 32-241. Pasadena, California: Jet Propulsion Laboratory, California Institute of Technology, June 15, 1962), pp. 361–362; TWX from William Pickering to Abe Silverstein, subject: "Summary of Difficulties Encountered with Spacecraft, Launch Vehicle, and Supporting Equipment at Cape," August 9, 1961 (2-1456); JPL Interoffice Memo from Leonard Bronstein and A.E. Dickinson to Distribution, subject: "RA-1 Spacecraft Failure Analysis," August 10, 1961 (2-2483).

7.   The friction experiment was designed to measure the coefficient of various metals rotated by a small electric motor in the space environment. In this case, its electronics package stood accused as the magnetic culprit. See "Space Science and the Original Ranger Missions" in Chapter Four of this volume.

8. Interview of James Burke by Cargill Hall, January 27, 1969, p. 8 (2-1391).

9. *Space Programs Summary No. 37-11, Volume I,* pp. 41–45.

10. Ibid., p. 4; JPL Ranger Technical Bulletins Nos. 1 through 5, August 24–September 5, 1961 (2-931); *Space Programs Summary No. 37-15, Volume VI* for the period March 1, 1962, to June 1, 1962 (Pasadena, California: Jet Propulsion Laboratory, California Institute of Technology, June 30, 1962), p. 3; Nicholas A. Renzetti, *Tracking and Data Acquisition for Ranger Missions 1-5* (JPL TM 33-174. Pasadena, California: Jet Propulsion Laboratory, California Institute of Technology, July 1, 1964), pp. 18–19.

11. A. E. Dickinson, ed., *Spacecraft Flight Performance: Final Report* (Pasadena, California: Jet Propulsion Laboratory, California Institute of Technology, November 2, 1961), pp. 10–13 (2-465).

12. NASA memorandum from Edgar Cortright to James Webb, subject: "Ranger I (P-32) Status Report No. 4," September 1, 1961 (2-679); *Sixth Semiannual Report to the Congress: July 1, 1961, through December 31, 1961* (Washington: National Aeronautics and Space Administration, 1962), p. 64.

13. Dickinson, *Spacecraft Flight Performance,* p. 1.

14. *The Ranger Project: Annual Report for 1961(U),* p. 365; *Space Programs Summary No. 37-12, Volume I* for the period September 1, 1961, to November 1, 1961 (Pasadena, California: Jet Propulsion Laboratory, California Institute of Technology, December 1, 1961), p. 2.

15. Interview of Burke by Hall, January 27, 1969, p. 9 (2-1391).

16. Interview of Allen Wolfe by Cargill Hall, November 5, 1969, p. 2 (2-1533).

17. *Space Programs Summary No. 37-12, Volume I,* p. 2.

18. *The Ranger Project: Annual Report for 1961(U),* p. 366.

19. *Space Programs Summary No. 37-13, Volume I* for the period November 1, 1961, to January 1, 1962 (Pasadena, California: Jet Propulsion Laboratory, California Institute of Technology, February 1, 1962), p. 4; *The Ranger Project: Annual Report for 1961(U),* pp. 366–367, 396–397; A. E. Dickinson, "RA-2 Preliminary Spacecraft Performance Report," November 29, 1961 (2-2133); Renzetti, *Tracking and Data Acquisition for Ranger Missions 1-5,* pp. 21, 24.

20. See "A Difference in Weights and Measures" in Chapter Four of this volume.

21. Letter from Osmond Ritland to Henry Eichel, November 21, 1961, in *Agena B Failure Investigation Board,* November 30, 1961, Appendix A (2-2136).

22. James Burke, JPL Trip Report, subject: "Meeting of the USAF Failure Investigation Board at LMSC, November 27 and November 28, 1961, " prepared November 30, 1961 (2-2134). Along with Eichel, board members included Air Force Ranger Chief Major Jack Albert, various Lockheed participants, NASA personnel from Hueter's Agena Office and Debus' Launch Operations Directorate, and JPL Ranger Project Manager James Burke, Reliability Chief Brooks T. Morris, and JPL Launch Vehicle Integration Chief Harry J. Margraf.

23. Committee report on the review of the Reliability and Quality Assurance Programs of Lockheed Missiles and Space Company, conducted January 24–25, 1962 (2-2270b).

24. James Burke, JPL Trip Report, subject: "Ranger II Failure Review (12-4) and Agena Management Meeting (12-5), Washington, D.C.," prepared December 7, 1961 (2-414).

25. See "A Learning Experience" in this chapter.

26. JPL Interoffice Memo from M. R. Mesnard to G. F. Baker, subject: "Ranger Follow-on, Scientific Power 'On' During Countdown," January 8, 1962, p. 2 (2-1224); see also, "Mission Objectives and Design Criteria Ranger P-53 through P-56 Spacecraft (RA-6 through RA-9)" in *Ranger Spacecraft Design Specification Book,* Spec. No. RL-2-110, November 8, 1961 (2-1124).

27. *Space Programs Summary No. 37-13, Volume I,* p. 7; J. B. Rittenhouse, et al., "Friction Measurements on a Low Earth Satellite," *ASLE Transactions,* Volume 6, 1963, pp. 161–177.

Chapter Seven

1. Text as printed in "Excerpts from Task Force's Report to Kennedy on U. S. Position in Space Race," *The New York Times,* January 12, 1961, p. 1; cf. President's Science Advisory Committee, "Report of Ad Hoc Panel on Man-in-Space," November 18, 1960 (2-2368b).

2. Loyd S. Swenson, Jr., James M. Grimwood, and Charles C. Alexander, *This New Ocean: A History of Project Mercury* (NASA SP-4201. Washington: National Aeronautics and Space Administration, 1966).

3. "Man's Role in the National Space Program," a policy paper developed by the Space Science Board, National Academy of Sciences, Washington, D.C., March 27, 1961, pp. 1–2 (2-932c).

4.   "Support of Basic Research for Space Science," a policy paper developed by the Space Science Board, National Academy of Sciences, Washington, D.C., March 27, 1961 (2-932b).

5.   United States Congress, House, Committee on Appropriations, *Independent Offices Appropriations for 1961,* Hearings before the Subcommittee, 86th Congress, 2nd Session, 1960, Part 3, p. 466.

6.   United States Congress, House, Committee on Science and Astronautics, *H.R. 6169—A Bill to Amend the National Aeronautics and Space Act of 1958,* Hearings before the Committee, 87th Congress, 1st Session, 1961, p. 5.

7.   Memorandum from John Kennedy to Lyndon Johnson, April 20, 1961, cited in John M. Logsdon, *The Decision To Go To the Moon: Project Apollo and the National Interest* (Cambridge: Massachusetts Institute of Technology Press, 1970), p. 109.

8.   Interview of Oran Nicks by Irving Ertel, William Putnam, and James Grimwood, January 23, 1967, p. 2 (5-724).

9.   NASA Master Plan for Space Launches as of May 25, 1961 (2-2243); also NASA memorandum from Oran Nicks to Edgar Cortright, subject: "Items Requiring Review and Decision by the Director of OSFP," June 1, 1961 (2-1487a&b).

10.   United States Congress, Senate, Committee on Science and Astronautics, *NASA Authorization for Fiscal Year 1962,* Hearings before the Committee, 87th Congress, 1st Session, on H.R. 6874, 1961, p. 56.

11.   Ibid., p. 65.

12.   Letter from Oran Nicks to William Pickering, June 9, 1961 (2-1129).

13.   JPL Interoffice Memo from Clifford Cummings to James Burke, subject: "Ranger Follow-on," July 5, 1961 (2-1157); JPL Interoffice Memo from Clifford Cummings to Ted Candee, subject: "Letter Contract No. 950137, Ranger Follow-on TV Mission," September 11, 1961 (2-2241); JPL Interoffice Memo from Charles Hemler to File 1647, subject: "High Resolution T.V. Camera Mission Contractor Selection," July 14, 1961 (2-1704); and "Ranger History," a working draft in preparation for the Congressional hearings held the week of April 27, 1964, p. 18 (2-458).

14.   NASA memorandum from Oran Nicks to Edgar Cortright, subject: "Review of JPL–RCA Management of Ranger Camera Capsule," January 15, 1962 (2-336); Minutes of the Ranger TV System Meeting, held on July 7, 1961 (2-1711); *Ranger TV Subsystem (Block III) Final Report* (Princeton, New Jersey: Astro-Electronics Division, Radio Corporation of America, July 22, 1965), Volume I: "Summary," p. 25 (2-960).

15. Gladwin Hill, "Ranger Project Adds 4 Rockets to Televise Pictures of Moon," *The New York Times,* August 30, 1961, p. 4; also JPL News Release of August 29, 1961 (2-2124); and "Unmanned Lunar Program to be Accelerated," *Aviation Week and Space Technology,* August 14, 1961, p. 28.

16. *Ranger Block 3 Project Policy and Requirements* (JPL EPD 65. Pasadena, California: Jet Propulsion Laboratory, California Institute of Technology, March 8, 1963); JPL Interoffice Memo from James Burke and Harris Schurmeier to Ranger Program Distribution Lists 6 and 7, subject: "Spacecraft System Manager for Ranger," August 31, 1961 (2-1525).

17. JPL Interoffice Memo from James Burke to Division Chiefs, Section Chiefs, and Group Supervisors, subject: "Ranger Follow-on, " September 1, 1961 (2-2125).

18. JPL Interoffice Memo 313-12 from Donald Kindt to Raymond Heacock, subject: "RCA TV Subsystem Design Modification," February 15, 1962 (2-1231).

19. *Space Programs Summary No. 37-15, Volume I* for the period March 1, 1962, to May 1, 1962 (Pasadena, California: Jet Propulsion Laboratory, California Institute of Technology, May 31, 1962), p. 29; *Space Programs Summary No. 37-13, Volume II,* for the period November 1, 1961, to January 1, 1962 (Pasadena, California: Jet Propulsion Laboratory, California Institute of Technology, February 1, 1962), pp. 23–26.

20. *Space Programs Summary No. 37-16, Volume VI* for the period May 1, 1962, to August 1, 1962 (Pasadena, California: Jet Propulsion Laboratory, California Institute of Technology, August 31, 1962), p. 13.

21. Official NASA Flight Schedule, September 19, 1961 (2-968).

22. Letter from Oran Nicks to Clifford Cummings, September 27, 1961 (2-1147).

23. NASA memorandum from Oran Nicks to Edgar Cortright, subject: "Lunar Program Support to Manned Lunar Landing," December 6, 1961 (2-433).

24. *Ranger Project Development Plan* (Revised. Pasadena, California: Jet Propulsion Laboratory, California Institute of Technology, June 5, 1961), p. 35 (2-621); JPL Interoffice Memo from Clifford Cummings to Harris Schurmeier, September 19, 1961 (2-1145); JPL Interoffice Memo from Homer Stewart to Gordon Kautz, subject: "Ranger Follow-On Program Expectations," June 11, 1962 (2-1255).

25. For unmanned exploration, "the principal purpose of [Rangers 6–9] is to support the accelerated program to land an American on the Moon by 1970." *Sixth Semiannual Report to the Congress, July 1 through December*

31, 1961 (Washington: National Aeronautics and Space Administration, 1962), p. 67.

26.   Clifford I. Cummings, "The Lunar Program," *Minutes of Briefing on the Occasion of the Visit of Lyndon B. Johnson, Vice President of the United States of America to the Jet Propulsion Laboratory, October 4, 1961* (Pasadena, California: Jet Propulsion Laboratory, California Institute of Technology, October 19, 1961), p. 1; see also *Minutes of Briefing on the Occasion of the Visit of U.S. Senator Robert S. Kerr, Chairman of the Senate Committee on Aeronautical and Space Science and James E. Webb, Administrator, NASA to the Jet Propulsion Laboratory, October 3, 1961* (Pasadena, California: Jet Propulsion Laboratory, California Institute of Technology, October 20, 1961).

27.   Robert L. Rosholt, *An Administrative History of NASA, 1958–1963* (NASA SP-4101. Washington: National Aeronautics and Space Administration, 1966); cf. also NASA News Release, subject: "Administration and the Conquest of Space" Banquet Address by James Webb, NASA Administrator, at the National Conference of the American Society for Public Administration, Detroit, Michigan, April 14, 1966 (2-462).

28.   Cf., for example, the letter from Homer Newell to Joseph Karth, April 10, 1962 (2-897); and NASA memorandum from Homer Newell to James Webb, April 2, 1963 (2-2171).

29.   As Burke saw and described the technological progression: "The Rangers are enormously more complex than Pioneer 4, yet they are but a small step toward the technology that will be required for the reliable transportation of men and equipment to and from the moon . . . " James D. Burke, "The Ranger Spacecraft," *Astronautics,* September 1961, p. 23.

Chapter Eight

1.   See "When and How to Sterilize Spacecraft" in Chapter Four of this volume.

2.   TWX from James Burke to Benjamin Milwitzky, subject: "Ranger Project P-32–P-36 Status as of 30 September 1961," October 19, 1961 (2-1459); see also, *Final Technical Report: Lunar Rough Landing Capsule Development Program* (Newport Beach, California: Aeronutronic, a Division of Ford Motor Company, February 20, 1963) (2-1758).

3.   Letter from Oran Nicks to Clifford Cummings, October 26, 1961 (2-256a).

4. Letter from James Burke to Oran Nicks, November 9, 1961 (2-339); *Final Technical Report: Lunar Rough Landing Capsule Development Program,* (Newport Beach, California: Aeronutronic, a Division of Ford Motor Company, February 20, 1963) (2-1758), p. 3-140.

5. Sterilization procedure is contained in JPL Interoffice Memo from Rolf Hastrup to Distribution, subject: "Procedures for Sterile Assembly of RA3 in SAF," July 10, 1961 (2-1133).

6. *Space Programs Summary No. 37-12, Volume II* for the period September 1, 1961, to November 1, 1961 (Pasadena, California: Jet Propulsion Laboratory, California Institute of Technology, December 1, 1961), pp. 14–15.

7. Ibid., p. 16.

8. Letter from Clifford Cummings to Robert Seamans, January 2, 1962, p. 3 (2-454b); history of JPL sterilization procedures in JPL Interoffice Memo ERG #228, from Ranger Project Office to Frank Goddard, subject: "Ranger Sterilization Experience," April 5, 1963 (2-659b); and in *A Review of Space Research* (Publication 1079, the Report of the Summer Study conducted under the auspices of the Space Science Board of the National Academy of Sciences at the State University of Iowa, June 17–August 10, 1962. Washington: National Academy of Sciences–National Research Council, 1962), pp. 10–25 and Appendix III.

9. Cf., JPL Interoffice Memo from James Burke to All Concerned, subject: "RA-3 Schedule," November 21, 1961 (2-2519); and JPL Interoffice Memo from Marshall Johnson to Allen Wolfe, subject: "On the Launching of Ranger 3," October 30, 1961 (2-1163).

10. Letter from William Pickering to Robert Seamans, November 15, 1961, p. 2 (2-1464).

11. NASA memorandum from Oran Nicks to Edgar Cortright, subject: "Plans for Ranger 3 Review Meeting on December 28 at JPL," November 21, 1961 (2-1483); telephone interview of James Burke by Cargill Hall, August 3, 1972.

12. See "Space Science and the Original Ranger Missions" in Chapter Four of this volume.

13. Letter from Homer Newell to William Pickering, December 8, 1961 (2-251).

14. Letter from Oran Nicks to Clifford Cummings, December 27, 1961 (2-1471). Rationale is discussed in "Briefing for Space Science Panel of the President's Science Advisory Committee," by Homer Newell, April 2, 1962, p. 11 (11-58).

15. Letter from Homer Newell to Donald Hornig, November 15, 1961 (2-2130).

16. See, for example, "JPL Speeds Lunar Work for Apollo," *Missiles and Rockets,* November 27, 1961, p. 100; also the statement by NASA Deputy Associate Administrator Thomas Dixon on February 21, 1962, as cited in United States Congress, House Committee on Science and Astronautics, *Astronautical and Aeronautical Events of 1962,* NASA Report to the Committee, 88th Congress, 1st Session, 1963, p. 19; and "Unmanned Lunar Program to Aid Apollo," *Aviation Week and Space Technology,* July 2, 1962, pp. 160, 163.

17. Newell's position is summarized most succinctly in his letter to Representative Joseph Karth, April 10, 1962 (2-897).

18. "Mission Objectives and Design Criteria Ranger P-53 Through P-56 Spacecraft (RA-6 through RA-9)," in *Ranger Spacecraft Design Specification Book,* Spec. No. RL-2-110 (Pasadena, California: Jet Propulsion Laboratory, California Institute of Technology, November 8, 1961), pp. 1, 5 (2-1124); cf., Hugh L. Dryden, "Future Exploration and Utilization of Outer Space," *Technology and Culture,* Volume 2, No. 2, Spring 1961, p. 124.

19. Letter cited in JPL Interoffice Memo from James Burke to Clifford Cummings, subject: "Ranger 6–9 Scientific Experiments," March 14, 1962 (2-1235); see also "P53 (RA-6), P54 (RA-7), P55 (RA-8), and P56 (RA-9) Spacecraft Design Characteristics and Restraints Design Characteristics," in *Ranger Spacecraft Design Specification Book,* Spec. No. RL-3-110A (Pasadena, California: Jet Propulsion Laboratory, California Institute of Technology, February 12, 1962) (3-1549).

20. Interview of Charles Sonett by Cargill Hall, May 1972; and interview of Walter Jakobowski by Cargill Hall, August 26, 1968, p. 1 (2-760).

21. Gerhardt Schilling, Sonett's predecessor, explained: We ran "a scientific program," and there was "pressure from the academic community to participate. Even if a flight was only partially successful, or even unsuccessful with regard to the mission objective, we might get some results from a scientific instrument and thus avoid complete failure (there were precedents in the early space program)." Letter from Gerhardt Schilling to Cargill Hall, June 23, 1970 (2-1979).

22. Interview of Oran Nicks by Cargill Hall, August 26, 1968, p. 5 (2-761).

23. *Prospectus: NASA Office of Space Sciences Long Range Plan* (Washington: National Aeronautics and Space Administration, November 1962), p. 1 (11-54).

24. TWX from William Cunningham to James Burke, March 1, 1962 (2-1233).

25. See "Space Science and the Original Ranger Missions" in Chapter Four of this volume; and James Burke, note on science in Ranger (2-2487).

26. JPL Interoffice Memo from James Burke to Brian Sparks, subject: "Ranger Project Status Report No. 48," March 2, 1962 (2-1314).

27. Letter from James Burke to William Cunningham, March 6, 1962 (2-1234).

28. NASA memorandum from William Cunningham to Charles Sonett, subject: "Scientific Experiments for Ranger Missions P-53, P-54, P-55, and P-56," April 18, 1962 (2-646).

29. Letter from Abe Silverstein to William Pickering, October 12, 1961 (3-363). More engineers than anticipated had been needed for Marshall Johnson's space flight operations, others had been assigned to a Surveyor lunar orbiter study, and more were to be detailed on loan to assist Joseph Shea in the Office of Manned Space Flight at Headquarters. JPL Interoffice Memo No. 83 from Brian Sparks to Senior Staff, subject: "Task Group for Dr. Shea," March 8, 1962 (3-440).

30. JPL Memo from James Burke to Clifford Cummings, March 14, 1962 (2-1235).

31. Letter from Oran Nicks to James Burke, March 20, 1962 (2-683).

32. JPL Interoffice Memo from Charles Cole to Clifford Cummings and James Burke, subject: "Science Experiments, Ranger Follow-On," March 21, 1962 (2-1238).

33. NASA memo from William Cunningham to Charles Sonett, April 18, 1962 (2-646).

34. See JPL Interoffice Memo from Raymond Heacock to Gordon Kautz, subject: "Ranger Follow-On Scientific Experiments," April 13, 1962 (2-2067).

35. See the letter from James Burke to William Cunningham, April 2, 1962 (2-1241).

36. NASA memo from William Cunningham to Charles Sonett, April 18, 1962, p. 5 (2-646).

37. Confirmed in a letter from James Burke to William Cunningham, April 13, 1962 (2-684).

38. Ibid.

39. At the outset of Project Ranger, the JPL project office and Space Sciences Division had established priorities among the scientific instruments; see "The Vega-Ranger: Where Planet and Sky Science Meet" in Chapter Three, and "Space Science and the Original Ranger Missions" in Chapter Four of this volume.

40. JPL Interoffice Memo from James Burke to William Pickering, Brian Sparks, and Clifford Cummings, subject: "Ranger 6–9 Bus Experiments," April 18, 1962 (2-1244). Determining the relative significance among differentiated fundamental research competing for government support has always been a source of contention, and as a rule establishing priorities has been consistently refused by the spokesmen of science. See Daniel S. Greenberg, *The Politics of Pure Science* (New York: the New American Library, Inc., 1967), pp. 231–241. As Homer Newell explained NASA's policy to an approving Space Science Panel of the President's Science Advisory Committee: "Having chosen the various scientific areas of interest and importance, we do not attempt to establish relative importances among the areas unless forced by limitations of resources to do so. Such comparative evaluations are extremely difficult and highly subjective, being in effect choices between apples and pears." Homer E. Newell, "Briefing for Space Science Panel of the President's Science Advisory Committee," April 2, 1962 (11-58).

41. NASA Summary Minutes of the Meeting of the Space Sciences Steering Committee held on April 23, 1962 (2-647).

42. Letter from Homer Newell to William Pickering, May 4, 1962, p. 2 (2-260).

43. *Space Programs Summary No. 37-16, Volume VI* for the period May 1, 1962, to August 1, 1962 (Pasadena, California: Jet Propulsion Laboratory, California Institute of Technology, August 31, 1962), p. 13; *Space Programs Summary No. 37-17, Volume I* for the period July 1, 1962, to September 1, 1962 (Pasadena, California: Jet Propulsion Laboratory, California Institute of Technology, September 30, 1962), p. 28.

44. As indicated earlier in this chapter, hardware delivery for these spacecraft had been firmly fixed by NASA.

45. JPL Interoffice Memo from S. Rothman to Distribution, subject: "Status Report Ranger Follow On (7–9)," June 22, 1962 (2-1262).

46. JPL Interoffice Memo from Allen Wolfe to James Burke, subject: "RA 7–9 Experiment Status," June 26, 1962 (2-1263).

47. Interview of James Burke by Cargill Hall, January 27, 1969, p. 14 (2-1391).

48. JPL Interoffice Memo from Allen Wolfe to James Burke, subject: "Review of Power Deficiency for Rangers 8 and 9," July 13, 1962 (2-1278).

49. JPL Interoffice Memo from James Burke to Brian Sparks, subject: "Ranger Project Status Report No. 69," July 27, 1962 (2-1314).

50. JPL Interoffice Memo from Allen Wolfe to James Burke, subject: "Recommendation on RA 7–14 Power," August 2, 1962 (2-1268); NASA Memorandum to the File from Walter Jakobowski, subject: "A. Ranger Power Budget; B. RCA Funding Status," August 2, 1962 (2-643).

51. NASA memorandum from William Cunningham to Oran Nicks, subject: "Trip Report of August 7–8, Jet Propulsion Laboratory,"August 16, 1962 (2-1175); see also the letter from James Burke to William Cunningham, August 10, 1962 (2-1272).

52. Clark replaced Sonett as Chief Scientist in the spring of 1962, after Sonett moved from Headquarters to become Chief of the newly formed Space Science Division at NASA's Ames Research Center in California. NASA Announcement No. 5-5, "Appointment of Associate Director and Chief Scientist, Office of Space Sciences," May 23, 1962.

53. NASA memorandum from William Cunningham to John Clark, subject: "Non-Visual Bus Experiments for Ranger Follow-On Missions," with attachment "Ranger Follow-On Scientific Experiments," August 23, 1962 (2-688).

54. "It was not only the time," Kautz recalled, "but he [Burke] knew the tradeoff that was being made in dedicating himself to get those [experiments] on in accordance with the directive that had been made. I think it was a known sacrifice, and I'm not at all sure that Headquarters ever really knew of the effort that was made to get them on in good faith." Interview of Gordon Kautz by Cargill Hall, December 17, 1971, p. 3 (2-2246).

Chapter Nine

1. *Space Programs Summary No. 37-13, Volume I* for the period November 1, 1961, to January 1, 1962 (Pasadena, California: Jet Propulsion Laboratory, California Institute of Technology, February 1, 1962), p. 3.

2. John Troan, "U.S. Plans Three Robot Moon Probes in '62," *Washington Daily News,* December 27, 1961, p. 3; also last paragraph of "Masterpiece for the Moon," an editorial in *The New York Times,* January 19, 1962, p. 30.

3. Cf. Marvin Miles, "U.S. Moon Shot May Top Russia's," *Los Angeles Times,* January 14, 1962, p. F-1; prior U.S. lunar flight missions are discussed in Chapter One and Appendix B of this volume.

4. William E. Kirhofer, *Post-Injection Standard Trajectory, Ranger P-34 (RA-3)* (JPL Engineering Planning Document 55. Pasadena, California: Jet Propulsion Laboratory, California Institute of Technology, January 5, 1962).

5. *Space Programs Summary No. 37-14, Volume II (U)* for the period January 1, 1962, to March 1, 1962 (Pasadena, California: Jet Propulsion Laboratory, California Institute of Technology, April 1, 1962), p. 5.

6. JPL Interoffice Memo from Albert Hibbs to Experimenters for the Ranger TV Mission, subject: "Dissemination of Ranger-3 Photographs," January 4, 1962 (2-1221b).

7. "Functional Specification Ranger RA-3, RA-4, and RA-5 Spacecraft Mission Objectives and Design Criteria," in *Ranger Spacecraft Design Specification Book*, Spec. No. RA345-2-110 D (Pasadena, California: Jet Propulsion Laboratory, California Institute of Technology, December 9, 1960) (2-1095e).

8. Nicholas A. Renzetti, *Tracking and Data Acquisition for Ranger Missions 1-5* (JPL TM 33-174. Pasadena, California: Jet Propulsion Laboratory, California Institute of Technology, July 1, 1964), p. 31.

9. It proved a valuable exercise because less than a month later the same field repair was made on the Mercury-Atlas that carried Lt. Colonel John Glenn into earth orbit aboard Friendship 7.

10. *Space Programs Summary No. 37-14, Volume II (U)* for the period January 1, 1962 to March 1, 1962 (Pasadena, California: Jet Propulsion Laboratory, California Institute of Technology, April 1962), p. 17.

11. *Space Programs Summary No. 37-14, Volume I* for the period January 1, 1962, to March 1, 1962 (Pasadena, California: Jet Propulsion Laboratory, California Institute of Technology, April 1, 1962), p. 4; Renzetti, *Tracking and Data Acquisition for Ranger Missions 1-5*, p. 31.

12. *Space Programs Summary No. 37-14, Volume I*, p. 44.

13. Ibid., pp. 3-9; *Space Programs Summary No. 37-14, Volume II (U)*, pp. 3-4; *Space Programs Summary No. 37-15, Volume VI* for the period March 1, 1962, to June 1, 1962 (Pasadena, California: Jet Propulsion Laboratory, California Institute of Technology, June 30, 1962), pp. 7-9; N.A. Renzetti, *Tracking and Data Acquisition for Ranger Missions 1-5*, pp. 31-38; Ranger 3 Flight Final Statement, January 28, 1962 (2-2500); JPL Ranger 3 Technical Bulletins 1 through 4, January 29, 1962, through February 9, 1962.

14. Letter from James Burke to Oran Nicks, February 8, 1962 (2-258).

15. JPL Interoffice Memo from James Burke to Brian Sparks, subject: "Ranger Project Status Report No. 44," February 2, 1962 (2-1314); Joseph A. Beacon, *Flight Operations Management,* Master's Thesis researched at the University of Southern California, January 1969, pp. 38–39 (3-633).

16. Letter from James Arnold to Cargill Hall, August 9, 1973 (2-2391).

17. Gladwin Hill, "Ranger Crosses Path of the Moon; TV Attempt Fails," *The New York Times,* January 29, 1962, p. 1.

18. "The Disobedient Rocket," *Time,* February 9, 1962, pp. 67–68.

19. Evert Clark, "Ranger 9 Flight Stirs Reliability Question," *Aviation Week and Space Technology,* Volume 76, February 5, 1962, p. 30.

20. Quoted by Marvin Miles in "Three Causes Found for Ranger's Failure," *Los Angeles Times,* February 8, 1962, Part II, p. 2.

21. William E. Kirhofer, *Post-Injection Standard Trajectory, Ranger P-35, (RA-4)* (JPL Engineering Planning Document 56. Pasadena, California: Jet Propulsion Laboratory, California Institute of Technology, April 9, 1962).

22. *Space Programs Summary No. 37-15, Volume I,* pp. 4–8.

23. Tom Allen, "Race for the Moon," *New York Sunday News,* April 8, 1962; see also, for example, Dick West, "Lunar Close-ups Expected Soon," *Los Angeles Times,* April 22, 1962, Part H, p. 5.

24. N.A. Renzetti, *Tracking and Data Acquisition for Ranger Missions 1–5,* pp. 44–50; *Space Programs Summary No. 37-15, Volume I,* p. 3; *Space Programs Summary No. 37-15, Volume VI,* pp. 9–10; *Space Flight Operations Memorandum Ranger IV* (Engineering Planning Document 91. Pasadena, California: Jet Propulsion Laboratory, California Institute of Technology, July 5, 1962); James D. Burke, *Preliminary Operations Letter, Ranger 4* (JPL Reorder 62-144. Pasadena, California: Jet Propulsion Laboratory, California Institute of Technology); JPL Ranger 4 Technical Bulletins 1 through 3, April 24, 1962, April 25, 1962, and April 26, 1962.

25. Quoted in "Leap Toward the Moon," *Time,* May 4, 1962, p. 40.

26. Bill Becker, "Ranger 4 Crashes on the Moon," *The New York Times,* April 27, 1962, p. 1; Lu Spehr, "Ranger 4 Strikes on Dark Side," *Pasadena Star-News,* April 26, 1962, p. 1.

27. "Man on Moon by 1968, Astronaut Chief Declares," *Los Angeles Herald-Examiner,* April 26, 1962.

28. This headline ran in the *Los Angeles Times,* April 27, 1962, Part I, p. 2.

29. "The Ranger Hits the Moon," an editorial in *The New York Times,* April 27, 1962, p. 34.

30. "A Week of Space Triumph," *Los Angeles Times,* April 29, 1962, Section C, p. 6.

31. Quoted in "Khrushchev Claims U.S. Missed Moon, but Scientists Refute Him," *Los Angeles Times,* May 11, 1962, Part I, p. 1.

32. Statement by William Pickering confirming Ranger 4 impact on the Moon (2-2412); see also, *Astronautical and Aeronautical Events of 1962,* Report of the National Aeronautics and Space Administration to the Committee on Science and Astronautics, U. S. House of Representatives, 88th Congress, 1st Session (Washington: Government Printing Office, June 12, 1963), p. 75.

Chapter Ten

1. Letter from Homer Newell to Joseph Karth, April 10, 1962 (2-897).

2. Attending with Homer Newell were his deputy, Edgar Cortright, Chief Scientist John Clark, Lunar and Planetary Program Director Oran Nicks, and John Naugle, Alan Crocker, and John Nicolaides.

3. NASA Minutes of the First Senior Council for the Office of Space Sciences, June 7, 1962, pp. 4, 8–9, 13 (2-1051).

4. For example, see Richard Witkin, "Project Apollo: Man's Race For The Moon," *The New York Times,* July 30, 1962, p. 1. The United States, Witkin asserted, was expected to beat the Soviets to a manned landing on the moon. But "whoever wins this race, it will be an event with no parallel in history. Not even Columbus' opening of the New World or the Wright Brothers' first flight had consequences as profound as may emerge from the first lunar voyage." The comment is illustrative of the profound interest and anticipation shared by many Americans.

5. John M. Logsdon, *NASA's Implementation of the Lunar Landing Decision* (HHN 81, Comment Edition. Washington: National Aeronautics and Space Administration, August 1969), pp. 63–65 (5-259).

6. Office of Manned Space Flight, "Requirements for Data in Support of Project Apollo" (Issue No. 1. Washington: National Aeronautics and Space Administration, June 15, 1962), pp. 1, 17 (2-2064).

7. Holmes' methods in managing Apollo would soon lead him into a direct confrontation with an equally strong-willed James Webb, who acted as NASA Administrator in fact as well as in name. The showdown in time was settled by President Kennedy—in favor of Webb.

8.   The JPL position is formally delineated in Clifford I. Cummings, *Jet Propulsion Laboratory Lunar Program Guidelines* (Engineering Planning Document No. 22. Pasadena, California: Jet Propulsion Laboratory, California Institute of Technology, August 15, 1962).

9.   See "After the Apollo Decision: What Science and Where?" and "Science Reasserted in Project Ranger" in Chapter Eight of this volume.

10.   "I propose," Burke had advised JPL leaders, "that we attack the third problem [manned vs. unmanned lunar exploration] by first trying to find out whether Newell and Holmes are as far apart as their subordinates seem to be. If they are, I suppose we will have to try to convince Dr. Seamans and Mr. Webb that there is a problem. If Newell and Holmes are not so far apart, we should sit down with Cortright and Shea to arrive at some written ground rules for the continued conduct of the Ranger Project." JPL Interoffice Memo from James Burke to William Pickering, Brian Sparks, and Clifford Cummings, subject: "Ranger 6–9 Bus Experiments," April 18, 1962 (2-2539).

11.   Cited in Lu Spehr, "Caltech Defends JPL Job," *Star News* [Pasadena], July 22, 1962, p. 1.

12.   Cited in "Congressman Criticizes Canaveral Space Failures," *The Tampa Tribune,* July 25, 1962.

13.   Mission description in *Mariner-Venus 1962, Final Project Report* (NASA SP-59. Washington: National Aeronautics and Space Administration, 1965); H. J. Wheelock, *Mariner Mission to Venus* (New York: McGraw-Hill Book Company, 1963); and John Lear, "The Voyage of Mariner 2 to Planet Venus," *Saturday Review,* January 5, 1963, pp. 87–91. The Soviet Venera 1 Venus spacecraft had failed to transmit radio signals a few days after its launch on February 12, 1961. A similar fate awaited the Soviet Mars 1 spacecraft launched shortly after Mariner 2, on November 1, 1962.

14.   *A Review of Space Research* (Publication 1079, the Report of the Summer Study conducted under the auspices of the Space Science Board of the National Academy of Sciences at the State University of Iowa, June 17–August 10, 1962. Washington: National Academy of Science–National Research Council, 1962), pp. 1–22.

15.   TWX from Oran Nicks to Brian Sparks, July 20, 1962, p. 4 (2-1174); see also NASA Minutes of Seamans' OSS Review of 19 July 1962, prepared by John Nicolaides, July 20, 1962, p. 2 (2-1759). Although new buildings were under construction by 1962, many JPL employees remained in rented trailers. There was simply no place to put contractor personnel who would have to be detailed to JPL to learn about Ranger.

16. Letter from William Pickering to Robert Seamans, subject: "Ranger Project activities in support of manned lunar flight programs," August 15, 1962 (2-368).

17. While Pickering awaited a reply, on September 6, Cunningham issued the ground rules for the extension of Project Ranger: Ranger flights 10 through 14 would carry an improved set of RCA television cameras in 1964; four additional seismometer capsule missions (flights 15 through 18) were also tentatively scheduled for launch in 1965. The schedule assigned the objectives of the Office of Space Sciences first priority, support of Project Apollo second. NASA memorandum from William Cunningham to Oran Nicks, subject: "Ranger Mission Plans Following RA-14," September 6, 1962 (2-689). The Office of Space Sciences did not appear to contemplate any changes whatever.

18. Letter from Robert Seamans to William Pickering, September 24, 1962 (2-367).

19. NASA Memorandum for the Record from Oran Nicks, subject: "Ranger Project Activities Discussion on 11 October 1962," November 15, 1962 (2-2331); see also JPL Interoffice Memo from Clifford Cummings and James Burke to William Pickering, subject: "Notes for October 11 Meeting of WHP, Newell, Holmes and Seamans," October 9, 1962 (2-2585).

20. NASA memorandum from Joseph Shea to Oran Nicks, subject: "Plans for Unmanned Lunar Orbiter Development," October 23, 1962 (2-431).

21. NASA memorandum from Homer Newell and D. Brainard Holmes to Robert Seamans, subject: "Establishment of a Joint OSS/OMSF Working Group," October 22, 1962 (2-351); also the letter from D. Brainard Holmes to William Pickering, November 20, 1962 (11-95b); and NASA News Release No. 62-251, subject: "Unit to Coordinate Manned and Unmanned Space Flight," for release November 27, 1962 (11-97). For their part, Holmes, Shea, and Gilruth issued the *Manned Spacecraft Center General Management Instruction 2-3-1: Inflight Scientific Experiments Coordination Panel* (October 15, 1962). Created as a counterpart to the Space Sciences Steering Committee in the unmanned program, the new panel acted to coordinate all scientific experiments proposed for manned missions, and was composed of members drawn from the Manned Spacecraft Center in Houston, Texas, with an ex officio member from the Office of Space Sciences.

22. Official NASA Flight Schedule, September 19, 1961, as adjusted in Official NASA Flight Schedule, March 20, 1962 (2-968).

23. NASA Memorandum for the Files from Oran Nicks, subject: "Trip to JPL April 24–26 (1962)," May 2, 1962, p. 3 (2-335).

24. JPL Interoffice Memo from A. E. Dickinson to Ranger-4 Distribution, subject: "Status of Ranger-4 Flight Analysis," May 8, 1962, in United States Congress, House, Committee on Science and Astronautics, *Investigation of Project Ranger,* Hearings before the Subcommittee on NASA Oversight, 88th Congress, 2nd Session, 1964, No. 3, pp. 489–492.

25. *Space Programs Summary No. 37-16, Volume I* for the period May 1, 1962, to July 1, 1962 (Pasadena, California: Jet Propulsion Laboratory, California Institute of Technology, July 31, 1962), p. 34.

26. *Space Programs Summary No. 37-17, Volume VI* for the period July 1, 1962, to October 1, 1962 (Pasadena, California: Jet Propulsion Laboratory, California Institute of Technology, October 31, 1962), p. 8.

27. JPL Interoffice Memo from Rolf Hastrup to George Hobby, subject: "Consideration in the Establishment of Lunar Spacecraft Sterilization Requirements," July 25, 1962 (2-1735).

28. TWX from William Pickering to Homer Newell, August 31, 1962 (2-1280). The waiver was approved in TWX from William Cunningham to Clifford Cummings, September 5, 1962 (2-1281).

29. JPL Interoffice Memo ERG #146 from Rolf Hastrup to Clifford Cummings, subject: "Recommended Changes in Lunar Sterilization Policy," August 30, 1962, pp. 1, 6 (2-1277).

30. Cf., letter from Clifford Cummings to Homer Newell, September 28, 1962 (2-1692).

31. *Space Programs Summary No. 37-15, Volume I* for the period March 1, 1962, to May 1, 1962 (Pasadena, California: Jet Propulsion Laboratory, California Institute of Technology, May 31, 1962), p. 8; *Space Programs Summary No. 37-18, Volume I* for the period September 1, 1962, to November 1, 1962 (Pasadena, California: Jet Propulsion Laboratory, California Institute of Technology, November 30, 1962), p. 4.

32. *NASA Agena Program Presentation,* October 1, 1962, p. 1-6 (2-2269).

33. Harold Washburn, "Summary of Experimenters Meeting on Ranger 5," held September 25, 1962 (2-1284).

34. Nicholas A. Renzetti, *Tracking and Data Acquisition for Ranger Missions 1–5* (TM 33-174. Pasadena, California: Jet Propulsion Laboratory, California Institute of Technology, July 1, 1964), p. 57.

35. JPL Interoffice Memo No. 103 from Brian Sparks to Senior Staff, et al., subject: "Support for the RA 5 Mission," October 11, 1962 (2-266).

36. "Ranger 5 Lofted on Lunar Course," *The New York Times,* October 19, 1962, p. 1-2.

37.   *Ranger 5 Flight Report* (Engineering Planning Document 147. Pasadena, California: Jet Propulsion Laboratory, California Institute of Technology, January 4, 1963); JPL Ranger 5 Technical Bulletin No. 1, October 19, 1962; James D. Burke, *Preliminary Spacecraft Operations Letter, Ranger 5* (JPL Reorder 62-405. Pasadena, California: Jet Propulsion Laboratory, December 6, 1962); NASA memorandum from Homer Newell to James Webb, subject: "Ranger 5 Post-Launch Report No. 2," October 30, 1962, with attachment: NASA Memorandum for the Files from Walter Jakobowski, subject: "Ranger 5 Post Launch Report No. 2," October 26, 1962 (2-692); *Space Programs Summary No. 37-18, Volume I*, pp. 2–8.

Chapter Eleven

1.   "An analysis of scientific measurements made by instruments on the Ranger 3 lunar spacecraft shows that the intensity of gamma rays in interplanetary space probably is as much as ten times higher than anticipated . . . " NASA News Release 62-212, October 15, 1962. Published results in "Letters to the Editor," *Journal of Geophysical Research,* Volume 67, No. 12, November 1962, pp. 4878–4880; and "Detection of an Interstellar Flux of Gamma Rays," *Nature,* Volume 204, No. 4960, November 21, 1964, pp. 766–767.

2.   Heat sterilization, in fact, was not even mentioned in the text of the JPL report, though it appears in passing in an attachment, under questions asked of those interviewed: "B. Does sterilization cause problems? Some problems had occurred, but most interviewees were not very positive about the connection." JPL document, "Ranger RA-5 Failure Investigation, Report of JPL Failure Investigation Board," November 13, 1962, Attachment, "Secretary's Summary of the RA-5 Failure Investigation Board Activities," p. 2 (2-459). The outspoken Hastrup, however, does not appear among those JPL personnel listed as interviewed. See "One More Time" in Chapter 10 of this volume.

3.   "Ranger RA-5 Failure Investigation, Report of JPL Failure Investigation Board," pp. 2, 6.

4.   Cited in United States Congress, House, Committee on Science and Astronautics, *1964 NASA Authorization,* Hearings before the Subcommittee on Space Sciences and Advanced Research and Technology, 88th Congress, 1st Session on H.R. 5466, 1964, No. 3, Part 3a, p. 1606.

5.   Letter from Homer Newell to William Pickering, October 20, 1962 (2-264).

6.   NASA memorandum from Homer Newell to Albert Kelley, subject: "Review of Ranger Spacecraft," October 29, 1962 (2-2470f).

7. NASA memorandum from Oran Nicks to Albert Kelley, subject: "Review of Ranger Spacecraft," October 29, 1962 (2-2470e).

8. Considering this aspect of the JPL equation, James Van Allen, the renowned sky scientist and long-time friend of William Pickering, observed: "They have tremendous esprit at JPL; it's almost offensive. It's like the Marines." Cited in "Space Exploration: Voyage to the Morning Star," *Time,* March 8, 1963, p. 79; see also the letter from William Pickering to T. Keith Glennan, November 20, 1958 (2-407a); and William H. Pickering in *Space Science Seminar* (JPL Publication 30-10. Pasadena, California: Jet Propulsion Laboratory, California Institute of Technology, August 28, 1959), Part I: "General Introduction," p. 16.

9. NASA memorandum from Albert Kelley to Homer Newell, subject: "Ranger Board of Inquiry, Final Report," December 5, 1962.

10. National Aeronautics and Space Administration, "Final Report of the Ranger Board of Inquiry," November 30, 1962, p. 5. (2-2463).

11. Ibid., p. 4.

12. Condensed from ibid., pp. 6–11.

13. Condensed from ibid., pp. 12–15.

14. See "A Planetary Machine for Space Science" in Chapter Three of this volume. For the most part, Pickering and his JPL associates concurred with the NASA report, but they took strong exception to items that were overlooked, and to those findings that seemed to hold the Laboratory accountable for events it did not control. The delay in creating a Ranger Program Office at NASA Headquarters, for example, was not mentioned, nor was the fact that JPL had urged NASA to place further Ranger spacecraft with an industrial contractor only to have that recommendation turned down. Objections were also registered over some of the proposed details for testing future spacecraft. See JPL "Comments on the Ranger Investigation," February 7, 1963 (2-460a); also JPL Interoffice Memo from Eberhardt Rechtin to Brian Sparks, subject: "Discussion of Rebuttal to Kelley Report," February 20, 1964 (2-415).

15. JPL Engineering Change Order No. 3703 initiated by S. Rubinstein, October 25, 1962 (2-1325).

16. NASA Memorandum for the Files from William Cunningham, subject: "Ranger Program Objectives," December 21, 1963 (2-375); TWX from Homer Newell to Robert Parks, December 21, 1962 (2-1336); letter from Homer Newell to William Pickering, December 31, 1962 (2-316); letter from William Pickering to Edgar Cortright, December 28, 1962 (2-329).

17. JPL Announcement No. 85 from William Pickering to All Personnel, subject: "Establishment of Lunar and Planetary Projects Office," December 18, 1962 (2-268); JPL Announcement No. 86 from William Pickering to All Personnel, subject: "Ranger Project," December 18, 1962 (2-269).

18. See "Organizing for Ranger: JPL and Headquarters" in Chapter Two of this volume; also, "Ranger's Man on the Ground: Harris McIntosh Schurmeier," *The New York Times,* July 31, 1964, p. 6.

19. Harris M. Schurmeier and Allen E. Wolfe, "Ranger Design Program," December 17, 1962, p. 1 (2-2073).

20. Harris M. Schurmeier, untitled JPL document regarding review of Ranger activities, April 22, 1964 (2-1850).

21. William H. Pickering, "Objectives of the Jet Propulsion Laboratory," February 15, 1963 (2-965); JPL Interoffice Memo No. 110 from Brian Sparks to Technical Division Chiefs, subject: "Division Project Organization," April 4, 1963 (2-1591); JPL Interoffice Memo No. 177 from William Pickering to Senior Staff, Section Chiefs, and Section Managers, subject: "Charter for Quality Assurance and Reliability Office," April 9, 1963 (2-441); also James D. Burke, "Recommendations on Ranger," December 18, 1962 (2-1700).

22. However, they made no mention of the personnel turnover at JPL. NASA News Release No. 62-268, subject: "Ranger Improvement Program," December 19, 1962 (2-934). That fact was reported quickly enough by the aerospace trade journals. See, for example, "Successive Failures Precipitate JPL Shakeup," *Aviation Week and Space Technology,* Volume 77, December 24, 1962, p. 17.

23. Letter from Edgar Cortright to Joseph Karth, December 28, 1962, as reprinted in U.S. Congress, *1964 NASA Authorization,* pp. 1596–1598; and letter from Homer Newell to William Pickering, December 31, 1962 (2-316).

24. Letter from Harold Urey to Homer Newell, October 24, 1962 (2-2300). A few days later geologist Frank Press, the Ranger seismometer experimenter and Director of Caltech's Seismological Laboratory, submitted a statement to Newell and to Donald Hornig on the President's Scientific Advisory Committee, which he said represented "the view held by many planetary scientists." It recommended that NASA at least acknowledge scientific parity for the nonvisual lunar experiments by immediately including three more scientific capsule packages in Project Ranger:

     We are aware of the competition between capsule experiments and TV experiments. We are also informed about demands of the Apollo program on the Ranger series. We have heard the argument that the Ranger capsule faces a lower probability of success because of the complexity of the landing maneuver. We feel, however, that information concerning the physical state and constitution of the moon is of tremendous scientific importance ... and also of high pertinence to the Apollo program.

If NASA considered the Ranger capsule missions vital when the unmanned lunar exploration program began, Press and his cohorts thought it senseless to discontinue them now. "Frankly speaking we cannot justify in our minds the almost exclusive emphasis on TV in the remaining Ranger series in view of the uncertain quality of the pictures and the limited scientific and engineering results that Ranger TV will provide." Letter from Frank Press to Homer Newell, October 22, 1962 (2-691). Taken together, all of the unhappy planetary scientists who wrote and called and cabled NASA Headquarters urged a rethinking of the priorities and objectives of the unmanned lunar program. Cf., letter from G. H. Sutton, for M. Ewing and others, to Homer Newell, October 23, 1962; letter from Harold Urey to James Webb, October 29, 1962, cited in a letter from Hugh Dryden to Harold Urey, November 14, 1962 (2-2330); letter from James Arnold to Homer Newell, October 30, 1962 (2-1527a); cable from James Arnold to Homer Newell, October 31, 1962 (2-1527b).

This sentiment was by no means universal, however. Other planetary scientists preferred visual imaging over nonvisual experiments, and were pleased with the television emphasis placed upon the future Ranger missions. Of these, Ranger experimenters Eugene M. Shoemaker and Gerard P. Kuiper were the most influential spokesmen. See the letter from Gerard Kuiper to James Burke, December 4, 1962 (2-2166).

25.   Letter from Homer Newell to Harold Urey, November 5, 1962, pp. 1–2 (2-2299); see also United States Congress, House, Committee on Appropriations, *Independent Offices Appropriations for 1962,* Hearings before the Subcommittee, 87th Congress, 1st Session, 1961, Part 2.

In May 1961 Lloyd V. Berkner, Chairman of the Space Science Board of the National Academy of Sciences, had proudly lauded the high-water mark attained by space science before the first National Conference on the Peaceful Uses of Outer Space. Virtually every American space mission to date had been directed to scientific pursuits, and the results were impressive. Harold M. Schmeck, Jr., "U.S. Held Leader in Space Science," *The New York Times,* May 28, 1961, p. 15. But the intervening approval of Project Apollo and the first manned flights of the Mercury earth-orbiting spacecraft had long since captured popular attention and fired hopes for surpassing the Russians. Taken together with the unmanned lunar flight failures, and despite Newell's general optimism, in little more than one year's time the fortunes of space science had unquestionably nosedived.

26.   Explorer 35 (Interplanetary Monitoring Platform E), though placed in lunar orbit on July 22, 1967, supported eight *sky* science experiments to "obtain scientific data at lunar distances on the characteristics of the interplanetary plasma and the interplanetary magnetic field," and was not directed to explore the moon for either planetary science or Project Apollo as a primary goal. NASA Mission Objectives for the IMP-E Mission, NASA Archives S-861-57-06. Because of its selenocentric orbit, the flight is sometimes referred to as a "lunar mission." See also "Sky and Planetary Science" in Chapter One of this volume.

Chapter Twelve

1.   JPL Interoffice Memo from Allen Wolfe and J. B. Berger to Ranger Green Day Distribution, subject: "Minutes, Ranger Spacecraft Systems Review (Green Day) Held February 8, 1963," February 13, 1963, pp. 1–2 (2-1313); *Space Programs Summary No. 37-20, Volume VI* for the period January 1, 1963, to March 31, 1963 (Pasadena, California: Jet Propulsion Laboratory, California Institute of Technology, April 30, 1963), pp. 6–9.

2.   *Ranger TV Subsystem* (Astro-Electronics Division of the Radio Corporation of America. Princeton, New Jersey: January 17, 1963) (2-1741).

3.   Pertinent details of this "split" RCA television subsystem as finally produced are contained in K. J. Stein, "Ranger TV Performed Near Design Point," *Aviation Week and Space Technology,* Volume 81, August 17, 1964, p. 100f; and Donald H. Kindt and J. R. Staniszewski, *The Design of the Ranger Television System to Obtain High-Resolution Photographs of the Lunar Surface* (JPL Technical Report 32-717. Pasadena, California: Jet Propulsion Laboratory, California Institute of Technology, March 1, 1965).

4.   JPL Interoffice Memo from Harris Schurmeier to William Pickering, subject: "Summary of Problem Areas Presented in Senior Staff Meeting 2/16/63," February 22, 1963 (2-1564); JPL Interoffice Memo from Harris Schurmeier to Distribution, subject: "Ranger Project Guidelines," February 15, 1963 (2-1360).

5.   Notes of an interview of Homer Newell by Cargill Hall and Eugene Emme, March 5, 1968 (2-385).

6.   See "Organizing for Ranger: Huntsville and the Air Force" in Chapter Two, and "Launch Operations" in Chapter Five of this volume; also the letter from Clifford Cummings to Edgar Cortright, December 22, 1960 (2-1437); letter from William Pickering to Robert Seamans, June 9, 1961 (2-1451); letter from Abe Silverstein to William Pickering, October 3, 1961 (2-1457); TWX from William Pickering to Don Ostrander, June 13, 1961

(2-2373); "Statement of LOD Position Concerning Management of Launch Operations for NASA Agena-B Program at AMR" (2-1256a); Minutes of NASA AMR Agena Management Meeting of June 25, 1962, prepared by C. M. Cope (2-2069); JPL Report prepared by James D. Burke, subject: "LOD/AMR Meeting of June 25, 1962," June 25, 1962 (2-1261); and JPL Interoffice Memo from Harris Schurmeier to Distribution, subject: "Headquarters Meeting on LOC Operations," August 8, 1962 (2-2405).

7. See *NASA Agena Program Presentation,* October 1, 1962 (2-2269).

8. *Astronautical and Aeronautical Events of 1962,* Report of the National Aeronautics and Space Administration to the Committee on Science and Astronautics (Washington: Government Printing Office, June 12, 1963), p. 267; JPL Report by James Burke, subject: "Agena Management Meeting at LMSC, December 18, 1962 (2-1359a)"; and NASA memorandum from Homer Newell to the Directors of NASA's Field Centers, subject: "NASA Agena Program Management," January 24, 1963 (2-1359).

9. Letter from Harris Schurmeier to Seymore Himmel, subject: "Ranger Project Recommendations for Improvement of Reliability of the Atlas-Agena Launch Vehicle," March 11, 1963 (2-2275).

10. "RA [Block] III Launch Vehicle Review," Report of the JPL Review Board, August 1, 1963, pp. 15, 16 (2-2281); see also JPL Interoffice Memo from Brooks Morris to Harris Schurmeier, subject: "Risk to RA-6 Flight from Atlas G Guidance," September 12, 1963 (2-2443).

11. "Agreement between the Department of Defense and National Aeronautics and Space Administration regarding management of the Atlantic Missile Range of DOD and the Merritt Island Launch Area of NASA," January 17, 1963, cited in Barton C. Hacker and James M. Grimwood, "Planning the Experiments," chapter in *History of Gemini* (Comment Edition. Washington: National Aeronautics and Space Administration, July 31, 1968), p. 31 (2-1476).

12. *USAF-NASA Memorandum of Agreement, NASA Office of Space Sciences, Agena Launch Vehicle Program,* approved by Howell M. Estes, Jr., Lieutenant General, USAF, Vice Commander, Air Force Systems Command, and Robert C. Seamans, Associate Administrator of NASA, August 9, 1963 (2-2424).

13. See "Rangers for Apollo" in Chapter Seven of this volume.

14. Cited in "Old Devil Moon," *Newsweek,* April 15, 1963, p. 62; see also, "Station on the Moon? Russia Blasts Off New Space Shot," *Los Angeles Times,* April 3, 1963, Part I, p. 3; and Marvin Miles, "Moon Spacecraft Two Years Ahead of U.S. Vehicles, Russians Boast," *Los Angeles Times,* April 4, 1963, Part I, p. 1.

15.   Unites States Congress, Senate, Committee on Aeronautical and Space Sciences, *NASA Authorization for Fiscal Year 1964,* Hearings before the Committee, 88th Congress, 1st Session on S. 1245, 1963, Part 2: "Program Detail," pp. 1031–1032. Extensive earth-based research on lunar surface conditions was, of course, conducted at available facilities. Cf., Jerry Bishop, "Moon Mystery: Riddle of Lunar Surface Unsolved as Developers Rush Space Ship Work," *The Wall Street Journal,* June 13, 1963, p. 1, and "63 Moon Probe Held Vital for U.S. Aims," *Los Angeles Times,* January 1, 1963, Part I, p. 16.

16.   JPL Interoffice Memo from Allen Wolfe to Ranger Personnel, subject: "Ranger 6 Schedule," June 18, 1963 (2-1615).

17.   Patrick Rygh, "Block III SFO/DSIF Design Review and Procedure Revisions," a draft of a JPL Report, April 22, 1964 (2-1849).

18.   See, for example, JPL Interoffice Memo from Robert Parks to Division Chiefs, Division Mariner C and Ranger Project Representatives, subject: "Change Control Procedure," with enclosed document, "Method for Design and Engineering Documentation Change Control," April 5, 1963 (2-1372).

19.   Interview of Brooks Morris by Cargill Hall, September 20, 1969, p. 7 (2-1479); and interview of Harris Schurmeier by Cargill Hall, September 25, 1970, p. 12 (2-2209).

20.   *Space Programs Summary No. 37-22, Volume VI* for the period May 1, 1963, to July 31, 1963 (Pasadena, California: Jet Propulsion Laboratory, California Institute of Technology, August 31, 1963), pp. 8–9.

21.   TWX from Harris Schurmeier to William Cunningham, subject: "Answering Request for Status of Vidicon Situation," August 21, 1963 (2-644); NASA memorandum from William Cunningham to Oran Nicks, subject: "RCA Vidicon Status Report for Use at the AA Project Status Review," September 9, 1963 (2-645).

22.   *Space Programs Summary No. 37-20, Volume VI,* pp. 12–13; also, Minutes of the Ranger "Tuesday-Thursday Meeting," held April 11, 1963 (2-1584); JPL Interoffice Memo No. 3151-1 from Allen Wolfe and J. B. Berger to Ranger S/C Review Distribution, subject: "Minutes, Ranger Spacecraft Systems Review—1 May 1963," May 9, 1963 (2-1313).

23.   JPL Interoffice Memo from J. J. Nielsen to Harris Schurmeier, subject: "Minutes of Ranger TV Sub-System Quarterly Review, Princeton, New Jersey, May 2, 1963" (2-1607); also TWX from Donald Kindt and C. J. Bennett to J. G. Davison, May 9, 1963 (2-1551).

24. Letter from Bernard Miller to Donald Kindt, subject: "Mission Operations—Ranger TV Subsystem," July 24, 1963, cited in JPL Interoffice Memo from Ralph Moyer and Leonard Bronstein to Patrick Rygh, subject: "Mission Analysis – 3. Ranger Block III TV System Control," October 11, 1963 (2-1663).

25. JPL Memo from Moyer and Bronstein, October 11, 1963 (2-1663); JPL Interoffice Memo from Ralph Moyer and Leonard Bronstein to Patrick Rygh, Addendum No. 1 to the Memo of October 11, 1963, subject: "Mission Analysis – 3. Ranger Block III TV System Control," December 11, 1963 (2-1664); JPL Interoffice Memo from Leonard Bronstein to Patrick Rygh, subject: "Ranger 6 Mission Analysis – 4. TV Turn-off if DSIF-SFOF Communications Fail," December 6, 1964 (2-1683).

26. JPL Interoffice Memo from Donald Kindt to Allen Wolfe, subject: "Request for Waiver on TV Subsystem Design Review," August 20, 1963 (2-1553); JPL Interoffice Memo from Allen Wolfe to Walter Downhower and Donald Kindt, subject: "TV Subsystem Design Review," August 23, 1963 (2-1554).

27. NASA memo from Cunningham to Nicks, September 9, 1963 (2-645).

28. Official NASA Flight Schedule, September 10, 1964 (2-968).

29. JPL document, "Ranger 6 Diode Failure Report," n.d. (2-1528).

30. JPL Interoffice Memo from Harris Schurmeier to Distribution, subject: "Minutes of Ranger Project Meeting held October 21, 1963," October 29, 1963 (2-1573); letter from William Pickering to Homer Newell, October 29, 1963 (2-287); JPL Interoffice Memo No. 204 from William Pickering to All Personnel, subject: "Ranger Block III Flights," October 30, 1963 (2-2285).

Chapter Thirteen

1. See "Family Relations" in Chapter Ten of this volume.

2. NASA Notes on the meeting of the Office of Space Sciences Senior Council, January 3–4, 1963, p. 13 (2-404b).

3. NASA memorandum from William Cunningham to Homer Newell, subject: "Non-visual Bus Experiments on Rangers 6–9," February 15, 1963 (2-695). This action is confirmed in NASA Summary Minutes of the Meeting of the Space Sciences Steering Committee, February 18, 1963 (2-1763); and in the letter from Homer Newell to William Pickering, February 28, 1963 (2-184a).

4.   Summary of the Ranger Experimenters Meeting held January 28, 1963, prepared by Harold Washburn, February 6, 1963 (2-661a). This is confirmed in NASA Summary Minutes of Meeting No. 3-63 of the Planetology Subcommittee of the Space Sciences Steering Committee held on March 5–6, 1963, prepared by Verne Frykland, pp. 9–11 (2-1771).

5.   See "Family Relations" in Chapter Ten of this volume.

6.   NASA Office of Manned Space Flight, "Requirements for Data from Unmanned Spacecraft in Support of Project Apollo," March 1, 1963, pp. 1–2 and passim (2-1362).

7.   NASA Memorandum from Homer Newell to James Webb, March 26, 1963 (2-376).

8.   United States Congress, House, Committee on Science and Astronautics, *Panel on Science and Technology Fifth Meeting,* Proceedings before the Committee, 88th Congress, 1st Session, 1963, No. 1, pp. 50–51.

9.   Philip H. Abelson, "Manned Lunar Landing," an editorial in *Science,* April 19, 1963, Vol. 140, p. 259. Maryland speech in "Moon Project Attacked As Costly 'Leaf-raking'," *Washington Evening Star,* April 27, 1963, p. 1.

10.   Walter Sullivan, "Manned Moon Flight Supported In 8 Scientists' Retort to Critics," *The New York Times,* May 27, 1963, pp. 1–2.

11.   For his part, in the absence of activity in Houston, on June 5 Newell established a Manned Space Science Group as a new division within the Office of Space Sciences. Headed by geologist and Ranger experimenter Eugene Shoemaker, the Group reported to Newell institutionally, and to Brainerd Holmes and the Office of Manned Space Flight functionally. It would be responsible "for planning scientific training and the selection of astronauts, and recommending experiments for manned science exploration" of the moon. NASA Summary of the Office of Space Sciences Senior Council Meeting held June 5, 1964, prepared June 27, 1963, p. 14 (2-1051). Because Newell's office did not control Project Apollo, however, the move produced few noticeable results. See also, Newell's testimony in United States Congress, House, Committee on Science and Astronautics, *1966 NASA Authorization,* Hearings before the Committee, 89th Congress, 1st Session, 1965, No. 2, Part 1, p. 144; and the appeal of Newell and Jastrow to critics in Robert J. Jastrow and Homer E. Newell, "Why Land on the Moon?" *Atlantic Monthly,* Vol. 212, August 1963, pp. 41–45.

12.   "3.1 Mission Objectives," *Apollo System Specification,* Office of Manned Space Flight Program Directive (M-D M8000.001), May 2, 1963, p. 3-1.

13.   NASA News Release, text of the address by James E. Webb, Administrator of the National Aeronautics and Space Administration, at the Second National Flight Forum Symposium, Hartford, Connecticut, May 15, 1963 (2-2077). Weaver had authored the critical article, "What a Moon Ticket Will Buy," *Saturday Review,* Vol. 45, August 4, 1962, p. 38. Webb earlier had asked DuBridge not to commit himself to an adverse position before considering all aspects of Project Apollo. See the letter from James Webb to William Pickering, with a copy of a letter from Webb to Lee DuBridge, June 29, 1961 (2-1452a&b).

14.   TWX from Edgar Cortright to Robert Rodney, Western Operations Office of NASA, Santa Monica, October 3, 1962 (2-1180).

15.   See "NASA's Lunar Objectives Reconsidered" in Chapter Ten of this volume.

16.   NASA News Release No. 62-220, subject: "Five New Rangers Added to Space Sciences Program," October 15, 1962 (2-2410).

17.   Letter from James Burke to William Cunningham, subject: "Ranger Lunar Facsimile Capsule Project," November 15, 1962 (2-265).

18.   Allen E. Wolfe, Notes on the Ranger Quarterly Review held February 19, 1963 (2-1546); Official NASA Flight Schedule of February 28, 1963 (2-968); JPL Interoffice Memo No. 170 from William Pickering to Senior Staff, Section Chiefs, and Section Managers, subject: "JPL Lunar and Planetary Launch Schedule," March 1, 1963 (2-430).

19.   Letter from William Pickering to Edgar Cortright, February 20, 1963 (2-185a).

20.   NASA Memorandum for the Record by Robert Seamans, subject: "Important Factors Relating to the Selection of Northrop Corporation as the Ranger Contractor," March 7, 1963 (2-2334); interview of Oran Nicks by Cargill Hall, August 26, 1968, p. 2 (2-761); Edgar Cortright's testimony in United States Congress, House, Committee on Science and Astronautics, *1964 NASA Authorization,* Hearings before the Subcommittee on Space Sciences and Advanced Research and Technology, 88th Congress, 1st Session on H.R. 5466, 1963, No. 3, Part 3a, pp. 1599–1600; "Northrop Ranger Award," *Aviation Week and Space Technology,* March 18, 1963, p. 26.

21.   Unsent letter from William Pickering to Robert Seamans, March 7, 1963 (2-308a).

22.   Letter from Homer Newell to William Pickering, March 8, 1963 (2-183).

23.   NASA News Release No. 63-50, subject: "NASA Selects Ranger Industrial Firm," March 8, 1963 (2-1365).

24.   U.S. Congress, *1964 NASA Authorization,* p. 1599.

25.   Ibid., pp. 1600–1601.

26.   A. P. Alibrando, "Space Agency Faces Possibility of $500-Million Slash in Budget," *Aviation Week and Space Technology,* Vol. 78, March 18, 1963, p. 30.

27.   JPL Interoffice Memo 313-738 from D. Alcorn to Allen Wolfe, subject: "Ranger Block IV Mission Objectives," March 7, 1963 (2-1578); see also similar sentiments expressed in JPL Interoffice Memo from Raymond Heacock to Charles Campen, subject: "Ranger Series Reprogramming," June 27, 1963 (2-1383).

28.   Letter from William Cunningham to Harris Schurmeier, March 14, 1963 (2-18a).

29.   See "New Management and New Objectives" in Chapter Eleven, and "Planning in the Face of Change" in Chapter Thirteen of this volume.

30.   TWX from Harris Schurmeier to William Cunningham, April 1, 1963 (2-1579).

31.   Letter from William Cunningham to Harris Schurmeier, April 8, 1963 (2-1580).

32.   NASA Summary Minutes of the Meeting of the Space Sciences Steering Committee on April 15, 1963, prepared by Jean LeCompte, pp. 1–3 (2-1768); also letter from Homer Newell to William Pickering, April 26, 1963 (2-699); and letter from Homer Newell to William Pickering, June 19, 1963 (2-291).

33.   Homer E. Newell, address to the American Society of Newspaper Editors, April 20, 1963, as cited in a preparation of the NASA Special Communications Staff, Office of Technology Utilization and Policy Planning, *Comment on the National Space Program,* July 1, 1963, p. B-121 (5-218).

34.   Letter from Homer Newell to Harold Urey, April 25, 1963 (2-2337).

35.   Letter from Homer Newell to William Pickering, April 26, 1963 (2-180).

36.   TWX from Harris Schurmeier to Walter Jakobowski and William Cunningham, April 25, 1963 (2-1593).

37.   NASA memorandum from Benjamin Milwitzky to Oran Nicks, subject: "Surveyor Program Discussions, 17 December 1962," January 8, 1963 (2-374).

38. At Headquarters, Joseph Shea in particular, a deputy to Office of Manned Space Flight Director Brainerd Holmes, was skeptical whether Ranger and Surveyor would be able to provide Apollo support data in time, and he proposed substituting a manned Apollo lunar orbiter project in their place.

39. NASA memorandum from Edgar Cortright to Homer Newell, subject: "Recommended Reprogramming within the Office of Space Sciences," April 25, 1963 (2-1774).

40. JPL Interoffice Memo #3151-20 from Allen Wolfe and J. B. Berger, subject: "Minutes, Ranger Spacecraft Review Held 12 June 1963," June 18, 1963 (2-1313); see also James D. Burke, "Ranger Contingency Plans," June 12, 1963 (2-1612).

41. JPL Interoffice Memo from Harris Schurmeier to Distribution, subject: "Ranger Block IV Schedule," June 24, 1963 (2-1374).

42. Cited in Bill Sumner, "Ranger May Be Unhorsed: House Unit Puts Squeeze on JPL Expansion Plans," *The Independent* [Pasadena], June 27, 1963, p. C-1.

43. NASA Memorandum to the File from William Cunningham, subject: "Ranger Review with Associate Administrator," July 12, 1963 (2-718).

44. Letter from Homer Newell to William Pickering, July 12, 1963 (2-190); see also, interview of William Cunningham by Cargill Hall, March 6, 1968, p. 2 (2-474).

45. NASA Memorandum for the Record from Edgar Cortright, subject: "Some Comments on the NASA reorientation of the Ranger Program," July 15, 1963 (2-179); and Edgar Cortright's testimony in United States Congress, House, Committee on Science and Astronautics, *1965 NASA Authorization,* Hearings before the Subcommittee on Space Sciences and Applications, 88th Congress, 2nd Session, on H.R. 9641, 1964, No. 1, Part 3, p. 1575; see also, Edward H. Kolcum, "Three Ranger Hard-Landing Flights Eliminated; Four Others Delayed," *Aviation Week and Space Technology,* Vol. 79, July 29, 1963, pp. 17–18.

46. John Walsh, "NASA Ranger Misfortunes Attract Attention of Congress to Problems of Spacecraft Sterilization," *Science,* Vol. 141, July 12, 1963, p. 140.

Chapter Fourteen

1. William H. Pickering's testimony in United States Congress, House, Committee on Science and Astronautics, *Investigation of Project Ranger,* Hearings before the Subcommittee on NASA Oversight, 88th Congress, 2nd Session, 1964, No. 3, p. 171.

2. Northrop Corporation, "Proposal, Ranger Spacecraft System, Phase II," Volume I: "Management Technical," and Volume II: "Cost Proposal," (NSL 63-89, June 1963) (2-624 and 2-625); and JPL document, "Ranger Block V Project," Technical Report No. 32, Preliminary Draft, January 1964, p. 37 (2-1969).

3. Letter from George Salem, Associate Director Corporate Communications, Northrop Corporation, to Cargill Hall, April 15, 1974 (2-2460).

4. *Space Programs Summary No. 37-25, Volume I* for the period November 1, 1963, to December 31, 1963 (Pasadena, California: Jet Propulsion Laboratory, California Institute of Technology, January 31, 1964), pp. 2–3.

5. James D. Burke, draft of the Ranger Block V Preliminary Project Development Plan, attached to a JPL Interoffice Memo from Harris Schurmeier to Distribution, subject: "Ranger Block V Preliminary PDP Draft," May 28, 1963, p. 1 (2-1606a&b).

6. A review of the JPL Space Sciences Division effort appears in George Baker, *The Ranger Block V Science System,* JPL document, September 21, 1964 (2-2063).

7. JPL Interoffice Memo from Robert Meghreblian to Harris Schurmeier, subject: "Review of Experiments Proposed for Ranger Block V," August 5, 1963 (2-1638a).

8. Letter from Harris Schurmeier to William Cunningham, subject: "Ranger Block 5 Experiment Recommendations," August 9, 1963 (2-1640); cf., additional letter from Harris Schurmeier to William Cunningham, August 9, 1963 (2-1641).

9. See JPL responsibilities in JPL Interoffice Memo from Robert Parks to William Pickering, subject: "Ranger Block V Management Policy," July 24, 1963 (2-1775).

10. JPL Announcement No. 101 from William Pickering to Senior Staff, Section Chiefs, and Section Managers, subject: "New Assignment for G. Robillard and Robert F. Rose," August 9, 1963 (2-272).

11. For the original development, see "Space Science and the Original Ranger Missions" in Chapter Four of this volume.

12. NASA Summary Minutes of the Space Sciences Steering Committee Meeting held September 3, 1963, prepared by Margaret B. Beach, pp. 3–4 (2-1769).

13. Ibid., pp. 5–6; also, NASA memorandum from William Cunningham to Robert Seamans, subject: "Status Report No. 18—Ranger Program," September 8, 1963, p. 3 (2-725).

14. JPL Minutes of the Ranger "Tuesday-Thursday" Meeting of September 17, 1963, prepared by Gordon P. Kautz, September 20, 1963 (2-1584); and NASA memorandum from Homer Newell to Oran Nicks, subject: "Payload Approval for Initial Ranger Block V Missions," September 18, 1964 (2-726).

15. Letter from Homer Newell to William Pickering, September 18, 1963, pp. 1–2 (2-192).

16. Cf., NASA memorandum from R. P. Young to James Webb, subject: "Memorandum from Jay Holmes [on Air Force Public Relations Strategy]," August 31, 1962 (5-990a); testimony of Air Force General Curtis E. LeMay before the House Armed Services Committee, February 21, 1963, text as reprinted in *Army-Navy-Air Force Journal and Register*, March 2, 1963, p. 16; Senator Barry Goldwater, "Ten Ways to Catch Up in the Space Race," *Los Angeles Times*, January 15, 1963, Part II, p. 4; "Barry Goldwater on Space: GOP Candidate Wants Military, Not Civilians, to Run Space Program," *Science*, Vol. 145, July 31, 1963, pp. 470–471; on Congressional scrutiny of the NASA budget, see Thomas P. Jahnige, "The Congressional Committee System and the Oversight Process: Congress and NASA," *The Western Political Quarterly*, June 1968, pp. 230–231; and also, Vernon Van Dyke, *Pride and Power: The Rationale of the Space Program* (Urbana, Illinois: University of Illinois Press, 1964).

17. Memorandum from President John Kennedy to Vice President Lyndon Johnson, April 9, 1963 (2-333); and NASA memorandum from Oran Nicks to William Lilly, subject: "Response to Questions for Letter to the President," April 17, 1963 (2-333a).

18. See James E. Webb's address before the Milwaukee Press Club Gridiron Dinner as quoted in *Astronautics and Aeronautics, 1963* (NASA SP-4004. Washington: National Aeronautics and Space Administration, 1964), p. 145; interview of James Webb by Cargill Hall, October 26, 1972, pp. 7, 9–10 (2-2308); interview of James Webb by George Frederickson, Henry Anna, and Barry Kelmachter, May 15, 1969, p. 19 (5-715); *Space Quotes*, Volume II, No. 5, prepared by the NASA Reports and Special Communications Division, June 1964, p. 2 (5-719); and NASA memorandum from James Webb to Colonel George, March 10, 1964 (5-991).

19.   Favored Apollo, two years old in June, had grown like a baby Paul Bunyan and already consumed more than 50 percent of the annual budget of the entire agency.

20.   Interview of Webb by Hall, October 26, 1972, p. 10 (2-2308). However, after President Kennedy's death, in early 1964 NASA officials did decide to make an exception to this rule, and returned to Congress for a supplemental appropriation for Apollo of $141 million. The reasons are described in a letter from James Webb to Cargill Hall, December 20, 1974 (5-587a).

21.   Official NASA Flight Schedule, October 8, 1963 (2-968).

22.   "Ranger Block V Project," p. 6 (2-1969); and "Ranger Program, Management and Financial Report, Ranger Spacecraft Systems for October 1963," Northrop Space Laboratories Report No. NSL 63-267, p. 3; see also, *Spacecraft Systems Requirements Specification Ranger Block V* (Pasadena, California: Jet Propulsion Laboratory, California Institute of Technology, October 28, 1963) (2-1738).

23.   See "A Difference in Weights and Measures" in Chapter Four of this volume.

24.   Interview of Geoffrey Robillard by Cargill Hall, February 27, 1974, p. 8 (2-2447).

25.   JPL, *Ranger Block V Biweekly Status Report,* October 31, 1963, p. 8 (2-1677).

26.   NASA memorandum from John Rosenberry to Distribution List, subject: "OSS Staff Meeting October 24, 1963," October 30, 1963 (2-1759); and NASA memorandum from Oran Nicks to Homer Newell and Edgar Cortright, subject: "Major Current Problems in SL, November 1963," November 5, 1963 (2-1675); see also, NASA, *OSSA Review, November 5, 1963,* p. 51 (2-1505).

27.   NASA, *Briefing for the Administrator on Possible Expansion of Lunar and Planetary Programs,* December 2, 1963, p. 2 (5-675).

28.   Ibid., pp. 75, 82.

29.   Ibid., pp. 83, 84, 108.

30.   "Ranger Block V Project," p. 43 (2-1969).

31.   Department of Defense News Release No. 1556-63; Jack Raymond, "Air Force to Launch Space Station in Place of Dyna-Soar Glider," *The New York Times,* December 11, 1963, pp. 1, 22. The manned space laboratory that the Air Force had pushed so vigorously was destined never to leave the ground. Six years later in 1969, without a foreign menace to combat in

space, well behind schedule and over cost, the project would be quietly canceled.

32. "Station in Space," an editorial in *The New York Times,* December 11, 1963, p. 46.

33. TWX from Homer Newell to William Pickering, December 13, 1963 (2-1685).

34. Interview of Edgar Cortright by Cargill Hall and Eugene Emme, March 4, 1968, p. 4 (2-762); Cortright's recollection is supported by the record. Details of the reasoning that led to the canceling of Block V are contained in a draft of a NASA memorandum from Oran Nicks, subject: "Ranger Block V Cancellation," January 21, 1964 (2-729). Furthermore, in view of Homer Newell's personal commitment and prior pledges to nonvisual scientists, planet and sky, the decision to cancel Block V could only have been a reluctant one at best. See also, John W. Finney, "Five Lunar Shots Canceled by NASA," *The New York Times,* December 14, 1963, p. 14; and NASA News Release No. 63-276, subject: "NASA Cancels Five Follow-on Rangers," December 13, 1963 (2-2415).

35. Italics added, cited in William Hines, "Urey Says Economies Hurt Space Science," *The Evening Star* [Washington, D.C.], December 31, 1963, p. A5; see also, "Lunar Economy," *Christian Science Monitor,* January 4, 1964, Editorial; and a letter from James Arnold to Cargill Hall, June 12, 1970 (2-2194).

Chapter Fifteen

1. JPL Interoffice Memo from Raymond Heacock to Distribution, subject: "Ranger TV System Review Meeting," February 26, 1963, p. 2 (2-1550).

2. Interview of William Cunningham by Cargill Hall, March 6, 1968, p. 4 (2-474).

3. Letter from Homer Newell to Harold Urey, April 25, 1963 (2-2290); see also, letter from Harold Urey to Homer Newell, April 5, 1963 (2-2335).

4. See "Space Science and the Original Ranger Missions" in Chapter Four of this volume.

5. Letter from Gerard Kuiper to William Cunningham, June 18, 1963 (2-418b).

6. See "Requalifying Ranger: Progress and Problems" in Chapter Twelve of this volume.

7. NASA memorandum from William Cunningham to Oran Nicks, subject: "Role of Ranger TV Principal Investigator," July 10, 1963 (2-2339).

8. JPL Interoffice Memo from Harris Schurmeier to William Pickering, subject: "TV Experimenter Situation for Ranger," July 11, 1963 (2-418a).

9. A biography of Gerard Kuiper appears in John Lear, "The Portable Astronomer," *Saturday Review,* Vol. 47, September 5, 1964, p. 36ff.

10. NASA memorandum from William Cunningham to Homer Newell, subject: "TV Experimenters for Ranger Block III," July 19, 1963 (2-2341a); letter from Homer Newell to Gerard Kuiper, July 19, 1963 (2-2341b).

11. NASA Management Instruction No. 37-1-1, subject: "Establishment and Conduct of Space Sciences Program—Selection of Scientific Instruments," effective April 15, 1960 (2-447), see "Space Science and the Original Ranger Missions" in Chapter Four of this volume; and Management Instruction No. 4-1-1, subject: "Planning and Implementation of NASA Projects," issued January 19, 1961, see "Reorganizing for Ranger" in Chapter Two of this volume.

12. Handwritten comment on the Memorandum of Agreement, draft, Harris M. Schurmeier, September 6, 1963 (2-1386).

13. Memorandum of Agreement of September 11, 1963, as reprinted in Appendix A of *Ranger Block III Project Development Plan* (Project Document 8. Pasadena, California: Jet Propulsion Laboratory, California Institute of Technology, October 31, 1963) (2-13); reproduced in Appendix E of this volume.

14. The final details were established in the correspondence between JPL and Headquarters; see the letter from Oran Nicks to Robert Parks, November 20, 1963 (2-286); letter from William Pickering to Homer Newell, November 27, 1963 (2-194); and letter from Homer Newell to William Pickering, January 8, 1964 (11-98). They were embodied in the revised NASA General Management Instruction 37-1-1A of April 29, 1964. The Office of Space Science and Applications reorganized the science function in its program offices in February 1964; see NASA memorandum from Oran Nicks to SL Staff, subject: "Management Reassignments," February 17, 1964 (2-1490).

15. Letter from William Pickering to Homer Newell, December 6, 1963 (2-195).

16. Letter from Harris Schurmeier to William Cunningham, December 19, 1963 (2-1687); JPL Interoffice Memo from Max Goble to Robert Crabtree and James McGee, subject: "Test Phase Report 6, October 7 through December 18, 1963," December 18, 1963 (2-1686); also JPL Interoffice Memo from Allen Wolfe to Harris Schurmeier, subject: "RA-6 Acceptance

Meeting," December 30, 1963; see also "Redesigning for Improved Reliability" in Chapter Twelve of this volume.

17. NASA, *Briefing for the Administrator on Possible Expansion of Lunar and Planetary Programs,* December 2, 1963, p. 84 (5-675).

18. On November 28, 1963, President Johnson announced the establishment of the Kennedy Space Center at Cape Canaveral, and that Cape Canaveral "shall be known hereafter as Cape Kennedy." *Astronautics and Aeronautics, 1963* (NASA SP-4004. Washington: National Aeronautics and Space Administration, 1964), p. 451.

19. Interview of Thomas Vrebalovich by Cargill Hall, June 11, 1974, pp. 4-5 (2-2465). This was confirmed by NASA Associate Administrator for Manned Space Flight George Mueller. The lander design, he informed the House Committee on Science and Astronautics, "is in fact frozen at the present time. We are going to go ahead and build it." United States Congress, Senate, Committee on Aeronautical and Space Sciences, *NASA Authorization for Fiscal Year 1965,* Hearings before the Committee, 88th Congress, 2nd Session, on S. 2446, 1964, Part 2: "Program Detail," p. 489.

20. See JPL Interoffice Memo from Donald Willingham to Distribution, subject: "RAFO Trajectory Constraints," October 4, 1963; and Thomas Rindfleisch and Donald Willingham, *A Figure of Merit Measuring Picture Resolution* (JPL TR 32-666. Pasadena, California: Jet Propulsion Laboratory, California Institute of Technology, September 1, 1965).

21. Interview of Vrebalovich by Hall, June 11, 1974, pp. 3-4 (2-2465); see "Making a Case for More Rangers" in Chapter Thirteen of this volume for the mission objective; the Block III objective is also reprinted in *Ranger Block III Development Plan,* p. I-2 (2-13).

22. JPL Interoffice Memo No. 200 from William Pickering to Senior Staff, Section Chiefs, and Section Managers, subject: "Role of the Assistant Laboratory Director for Tracking and Data Acquisition," October 2, 1963 (2-273); for early developments in this area, see Chapter Five.

23. JPL Interoffice Memo No. 218 from William Pickering to Senior Staff, Section Chiefs, and Section Managers, subject: "Establishment of the Deep Space Network," December 24, 1963 (2-227).

24. *Space Programs Summary No. 37-26, Volume III* for the period January 1, 1964, to February 29, 1964 (Pasadena, California: Jet Propulsion Laboratory, California Institute of Technology, March 31, 1964), p. 5.

25. *Space Programs Summary No. 37-26, Volume I* for the period January 1, 1964, to February 29, 1964 (Pasadena, California: Jet Propulsion Laboratory, California Institute of Technology, March 31, 1964), p. 11; see

"Launch Vehicles Revisited" in Chapter Twelve of this volume for the NASA changes in launch operations.

26. NASA, *OSSA Review, January 23, 1964,* pp. 41–42 (2-1505).

27. JPL Interoffice Memo from Harris Schurmeier and Robert Parks to All Laboratory Personnel, subject: "Ranger A Flight Activities (alias RA-6, shortly to be RA VI)," January 27, 1964 (2-1808).

28. Marvin Miles, "Ranger Heads Straight for Target on Moon," *Los Angeles Times,* February 1, 1964, Part I, p. 1; William H. Pickering as quoted by Earl Ubell, "To the Moon, Hoping for a Smash Success," *New York Herald Tribune,* January 31, 1964. "It was," the *Tribune* ventured, "about all he could say considering the history of the project."

29. NASA News Release, subject: "News Conference on Ranger VI Impact on Moon," February 2, 1964, p. 9 (2-2509).

30. Ranger 6 flight events described in ibid.; also, interview of Vrebalovich by Hall, June 11, 1974, p. 26 (2-2465); *Ranger 6 Log,* January 30–February 2, 1964 (2-924); *Space Programs Summary No. 37-26, Volume VI* for the period January 1, 1964, to February 29, 1964 (Pasadena, California: Jet Propulsion Laboratory, California Institute of Technology, March 31, 1964), pp. 1–2; *Space Programs Summary No. 37-26, Volume I,* p. 2; Nicholas A. Renzetti, *Tracking and Data Acquisition for Ranger Missions 6–9* (JPL TM 33-275. Pasadena, California: Jet Propulsion Laboratory, California Institute of Technology, September 15, 1966), p. 10; *Ranger 6 Failure Analysis and Supporting Investigations* (JPL Engineering Planning Document 205. Pasadena, California: Jet Propulsion Laboratory, California Institute of Technology, March 27, 1964); and Marvin Miles, "U.S. Photo Spacecraft Races Toward Moon" *Los Angeles Times,* January 31, 1964, Part I, p. 1.

31. *United States Aeronautics and Space Activities, 1963: Report to Congress from the President of the United States,* January 27, 1964.

Chapter Sixteen

1. The views and lifespan of this contemporary aerospace journal coincided neatly with the most competitive "cold war" period in astronautics. For Coughlin's views, see "The Rise and Fall," *Missiles and Rockets,* May 25, 1964, p. 70.

2. William J. Coughlin, "A $150-million Failure," *Missiles and Rockets,* February 10, 1964, p. 50. Coughlin's source for the performance rating was apparently drawn from NASA Associate Administrator Robert Seamans,

who had observed in testimony before Congress on February 4 that "the last Ranger flight was extremely successful from the standpoint of the guidance technology, but as we score this, it was a 100-percent failure. So we are," he assured the legislators, "scoring hard." United States Congress, House, Committee on Science and Astronautics, *1965 NASA Authorization,* Hearings before the Committee, 88th Congress, 2nd Session, on H.R. 9641, 1964, No. 1, Part 1, p. 62.

3. Cf., remarks of Apollo Spacecraft Program Director Joseph F. Shea quoted from a speech in Milwaukee on February 3, 1964, in Harry S. Pease, "Sending Men to Moon in 1968 Forecast," *Milwaukee Journal,* February 4, 1964; and of John W. Eggleston, an Apollo space environmental specialist, in "Extra Apollo Moonshot Proposed," Associated Press Article, Brooks Air Force Base, Texas, February 6, 1964, cited in *JPL News Clips,* prepared by the Office of Public Information, February 7, 1964, p. 5. Other NASA officials moved swiftly to deny the veracity of these reports, TWX from Julian Scheer to William Pickering, February 6, 1964 (2-150), and a letter from Robert Gilruth to William Pickering, February 13, 1964 (2-1820); but the statements and subsequent refutations bespoke the low circumstances into which Ranger had fallen.

4. JPL Interoffice Memo 313-1398 from Walter Downhower to Donald Kindt, subject: "Committee of Section Chiefs Reviewing RA-6 Flight," February 12, 1964 (2-2078).

5. Karth quoted in "Legislators Back NASA on Ranger," *Aviation Week and Space Technology,* Vol. 80, February 10, 1964, p. 24; and letter from George Miller to William Pickering, February 5, 1964 (2-1813).

6. TWX from William Pickering to Homer Newell, February 11, 1964 (2-151); TWX from William Cunningham to Harris Schurmeier, February 17, 1964 (2-1826).

7. RA-6 Investigation Committee Final Report (Engineering Planning Document No. 205. Pasadena, California: Jet Propulsion Laboratory, California Institute of Technology, February 14, 1964)(2-1303).

8. Ibid., p. v.

9. Condensed from ibid., pp. 1-2 and 1-3.

10. NASA Announcement No. 64-27 by Robert Seamans, subject: "Establishment of the Ranger VI Review Board," February 3, 1964 (2-1811).

11. NASA memorandum from Robert Seamans to Earl Hilburn, subject: "Clarification of Ranger VI Review Board Membership established February 3, 1964," n.d. (2-2468); and NASA memorandum from Robert Seamans to Earl Hilburn, subject: "Ranger VI Review Board," February 3, 1964 (2-2467).

12. NASA News Release No. 63-141, subject: "Three New Appointments to Headquarters Staff," June 27, 1963 (2-2504).

13. See "Preparing for the Test Flights" in Chapter Four of this volume.

14. Meeting events as described in NASA memorandum from Walter Jakobowski to Homer Newell and Edgar Cortright, subject: "Comments on Ranger Failure Review Board Presentation on February 14," n.d. (2-1823); and NASA memorandum from William Cunningham to Homer Newell and Edgar Cortright, subject: "Ranger VI Failure Report Presentation of February 14," February 18, 1964 (2-2469i); also TWX from Cunningham to Schurmeier, February 17, 1964 (2-1826).

15. NASA memo from Jakobowski to Newell and Cortright, n.d., pp. 1–2 (2-1823).

16. NASA memo from Cunningham to Newell and Cortright, February 18, 1964, p. 2 (2-2469i).

17. William Cunningham, NASA, *OSSA Review, February 24, 1964,* pp. 49–50 (2-1505).

18. TWX from Homer Newell to William Pickering, February 24, 1964 (2-1828).

19. *Final Report of the Ranger 6 Review Board* (Washington: National Aeronautics and Space Administration, March 17, 1964), pp. 6–8 (2-2472).

20. Ibid., p. 12.

21. Letter from Robert Seamans to William Pickering, March 26, 1964 (2-2346).

22. Letter from James Webb to George Miller and Clinton Anderson, March 31, 1964, as reprinted in United States Congress, House, Committee on Science and Astronautics, *Investigation of Project Ranger* Hearings before the Subcommittee on NASA Oversight, 88th Congress, 2nd Session, 1964, No. 3, pp. 12–15.

23. Interview of Oran Nicks by Cargill Hall, August 26, 1968, p. 7 (2-761).

24. Draft of a NASA memorandum from Oran Nicks to Robert Seamans, subject: "Comments on Mr. Webb's Letter to Congress Regarding the Ranger Program," p. 1 (2-1844).

25. Condensed from ibid., pp. 2–9 of Attachment A; cf., Hilburn Board findings, "A Public Accounting," in this chapter, and JPL rationale, "The Aftermath," in Chapter Six of this volume.

26. Interview of Edgar Cortright by Cargill Hall and Eugene Emme, March 4, 1968, p. 12 (2-762).

27. NASA memorandum from Oran Nicks to Homer Newell and Edgar Cortright, subject: "Ranger 6 Review Board," April 9, 1964, pp. 1–2 (2-395).

28. As Pickering later confided to Newell, "It would appear that the Board was all too eager to rationalize a reason for extending its efforts into areas that otherwise could not be related to the specific Ranger 6 failure, and thus impose its opinions and prejudices upon the overall conduct of the Project." Letter from William Pickering to Homer Newell, May 22, 1964, p. 3 (2-2469c).

29. "JPL's Last Chance to Hit the Moon," Los Angeles *Herald-Examiner,* April 4, 1964.

30. "NASA Renews the Attack on JPL," editorial in *Los Angeles Times,* April 8, 1964, Part II, p. 4.

31. NASA News Release, text of the conference by James E. Webb, April 15, 1964, p. 20 (2-1972).

32. Miller quoted in Bill Sumner, "House Will Probe JPL Controversy," *Star News* [Pasadena], April 9, 1964, p. 1; also Robert Toth, "House Unit to Probe NASA, Jet Lab Dispute," *Los Angeles Times,* April 10, 1964, Part I, p. 1.

33. See, for example, the Karth–Newell colloquy in United States Congress, House, Committee on Science and Astronautics, *1965 NASA Authorization,* Hearings before the Subcommittee on Space Sciences and Applications, 88th Congress, 2nd Session, on H.R. 9641, 1964, No. 1, Part 3, pp. 1965–1966; and United States Congress, Senate, Committee on Aeronautical and Space Sciences, *NASA Authorization for Fiscal Year 1965,* Hearings before the Committee, 88th Congress, 2nd Session, on S. 2446, 1964, Part 2: "Program Detail," p. 591. After the Ranger 6 failure, Newell's longtime deputy Cortright asserted, "there was Congressional pressure to get rid of JPL, to reorganize JPL, take JPL over, all the things that we were considering [earlier] were suggested by Congress, some of it informally. Congressman Karth made suggestions like this to us privately." Interview of Cortright by Hall and Emme, March 4, 1968, p. 16 (2-762). See Karth background, "Making a Case for More Rangers" and "Lunar Orbiter and Congress Intervene" in Chapter Thirteen of this volume.

34. Letter from James Webb to Joseph Karth, April 23, 1964, pp. 1–2 (2-2479).

35. *Investigation of Project Ranger,* pp. 1, 3.

36. Ibid., p. 126.

37. Ibid., pp. 159–161.

38. Ibid., p. 214.

39.   Ibid., pp. 215–226.

40.   United States Congress, House, Committee on Science and Astronautics, *Project Ranger,* Report of the Subcommittee on NASA Oversight, 88th Congress, 2nd Session, 1964, pp. 20–31.

41.   Eventually set to extend three years, from January 1, 1964, through December 31, 1966. The previous extension had been for two years: January 1, 1962, through December 31, 1963.

42.   JPL Press Release 294, June 29, 1964; "JPL Deputy Named," *Missiles and Rockets,* Vol. 15, July 6, 1964, p. 8; JPL Announcement 8-64, August 11, 1964.

43.   William J. Coughlin, "Praise for Caesar," editorial in *Missiles and Rockets,* May 11, 1964, p. 46.

44.   "Ranger: Oversight Subcommittee Asks Why NASA Doesn't Prevail on JPL to 'Rigidize' Projectwise," *Science,* Vol. 144, May 15, 1964, p. 824.

45.   Letter from James Webb to George Miller, May 4, 1964 (2-2349).

Chapter Seventeen

1.   Interview of Gordon Kautz by Cargill Hall, December 17, 1971, p. 6 (2-2246); and interview of Edgar Cortright by Cargill Hall and Eugene Emme, March 4, 1968, p. 10 (2-762).

2.   Interview of Kautz by Hall, December 17, 1971, p. 4 (2-2246).

3.   JPL Interoffice Memo from Harris Schurmeier to Distribution, subject: "Organization for Special Efforts on TV Subsystem," February 27, 1964 (2-1830).

4.   JPL Interoffice Memo from Donald Kindt to Distribution, subject: "Assignment of Ranger TV Subsystem Project Engineer Deputy," March 20, 1964 (2-1835).

5.   Interview of Kautz by Hall, December 17, 1971, P. 5 (2-2246); and JPL Interoffice Memo from Harris Schurmeier to Ranger Block III Project Distribution List, subject: "AED/JPL Failure Report System for the TV Subsystem," March 25, 1964 (2-1836).

6.   NASA memorandum from Oran Nicks to William Cunningham, subject: "Test and Operations Program for Ranger Spacecraft," March 19, 1964 (2-392).

7. TWX from Harris Schurmeier to Seymour Himmel, et al., March 11, 1964, p. 2 (2-1833); and Official NASA Flight Schedule of April 14, 1964, p. 1 (2-968).

8. Letter from R. E. Hogan to Harris Schurmeier, subject: "Ranger Camera Electronics Unit 038 Investigation," April 23, 1964, p. 1 (2-1851).

9. Interview of Kautz by Hall, December 17, 1971, p. 9 (2-2246).

10. Ibid.; and TWX from Harris Schurmeier to William Cunningham, subject: "Bag of Hardware Incident," June 11, 1964 (2-2606).

11. Chester Gould, *Washington Post,* May 21, 1964, p. C11; and Jim Hicks, "Many a Slip 'Twixt Earth and Moon—and Measles Too," *Life,* August 14, 1964, p. 36a.

12. See "A Public Accounting" in Chapter Sixteen of this volume.

13. "Ranger Launch Vehicle Integration Summary," rough draft of JPL ED-333, December 27, 1964 (2-2080).

14. See "Planning the Ascent" in Chapter Six of this volume.

15. JPL Interoffice Memo from Alexander Bratenahl to Harris Schurmeier, subject: "A Possible Cause of the Accidental Turn-on of the RA-6 TV," May 26, 1964 (2-923).

16. JPL Interoffice Memo from Maurice Piroumian to Distribution, subject: "Presentation on Ranger Investigations Concerning Launch to Injection Environment," June 25, 1964 (2-1865).

17. Again, as in the case of Ranger 4 (see "One More Time" in Chapter Ten of this volume), mechanical and electrical reactions in the strange environment of space had not been foreseen. 18. JPL Interoffice Memo from Alexander Bratenahl to Charles Campen, subject: "Current Thinking on Effects of Booster Separation as Cause of Accidental Turn-on of RA-6 TV," July 30, 1964 (2-2084).

19. As Bratenahl later explained: "Schurmeier already had fixed the problem with the positive lock-out of the television subsystem [during ascent through the atmosphere], so I was certain Ranger 7 would work. Besides, it seemed improper to bother him at such a busy time." Interview of Alexander Bratenahl by Cargill Hall, August 25, 1970 (2-2192); see also, JPL Interoffice Memo from Robert Mackin to Robert Meghreblian, subject: "Electrostatic Charging of Launch Vehicles During Ascent," February 2, 1965 (2-2085).

20. NASA memorandum from William Cunningham to Edgar Cortright, subject: "Ranger 7 Television Subsystem Redesign, Rework, and Test at RCA," May 6, 1964, p. 4 (2-1856).

21.   TWX from William Pickering to Homer Newell, subject: "Ranger 'B' Schedule," May 11, 1964 (2-1858a); TWX from Homer Newell to William Pickering, May 12, 1964 (2-1858b).

22.   TWX from Harris Schurmeier to William Cunningham, May 18, 1964 (2-1859); TWX from Harris Schurmeier to William Cunningham, May 20, 1964 (2-1860); and TWX from Homer Newell to William Pickering, May 21, 1964 (2-2452).

23.   NASA memorandum from Homer Newell to Distribution, subject: "NASA Buy-Off for Ranger 7," May 25, 1964 (2-1727).

24.   Oran W. Nicks, "Report of OSSA Buy-Off Committee on Ranger, B Spacecraft Preshipment Meeting of June 15–16, 1964," attachment to the NASA memorandum from Homer Newell to Robert Seamans, subject: "Actions Taken by OSSA in Response to the 'Final Report' of the Ranger VI Review Board," July 25, 1964 (2-1971b); and letter from Homer Newell to William Pickering, June 17, 1964 (2-157).

25.   NASA memorandum from Maxime Faget to Willis Foster, subject: "Estimated Size of Ranger 6 Impact Crater and Recommendations for Subsequent Ranger Flight Missions," March 13, 1964 (2-393).

26.   Interview of Thomas Vrebalovich by Cargill Hall, June 11, 1974, p. 6 (2-2465).

27.   NASA memorandum from Homer Newell to Robert Seamans, subject: "Ranger B Status Review," July 10, 1964 (2-2471c).

28.   See "Space Flight Operations" in Chapter Five of this volume.

29.   Nicholas A. Renzetti, *Tracking and Data Acquisition for Ranger Missions 6–9* (JPL TM 33-275. Pasadena, California: Jet Propulsion Laboratory, California Institute of Technology, September 15, 1966), p. 47.

30.   Interview of Kautz by Hall, December 17, 1971, p. 9 (2-2246).

31.   Richard Witkin, "Ranger 7 on Course for the Moon," *The New York Times,* July 29, 1964, p. 13; also, Dave Swaim, "Jet Lab Personnel on Watch: Tension Grows at Blastoff," *Star News* [Pasadena], July 28, 1964, p. 1; and Marvin Miles, "U.S. Spacecraft Heading for Impact with Moon," *Los Angeles Times,* July 29, 1964, Part I, p. 1.

32.   "Ranger 7 Glides on Target: JPL Believes Success Near," *Star News* [Pasadena], July 30, 1964, p. 1; Richard Witkin, "Ranger 7 Re-Aimed for Moon's Bull's-Eye," *The New York Times,* July 30, 1964, pp. 1, 4; and Marvin Miles, "Perfect Strike for Moon Shot Seen," *Los Angeles Times,* July 30, 1964, Part I, p. 1.

33. Nichols' recounting as cited in Marvin Miles, "Ranger Gets Spectacular Photos," *Los Angeles Times,* August 1, 1964, Part I, p. 3.

34. Telephone interview of Patrick Rygh by Cargill Hall, January 21, 1975.

35. Ranger 7 mission events and performance as described in *Space Programs Summary No. 37-29, Volume I* for the period July 1, 1964, to August 31, 1964 (Pasadena, California: Jet Propulsion Laboratory, California Institute of Technology, December 15, 1964), Part 1: "Mission Description and Performance"; *Ranger 7 Log,* compiled November 4, 1964 (2-2431); and Raymond L. Heacock, Bernard P. Miller, and Harris M. Schurmeier, *The Ranger VII Mission* (Pasadena, California: Jet Propulsion Laboratory, California Institute of Technology, n.d) (2-1798); Renzetti, *Tracking and Data Acquisition for Ranger Missions 6-9; Space Flight Operations Memorandum, Ranger 7* (Engineering Planning Document 242. Pasadena, California: Jet Propulsion Laboratory, California Institute of Technology, October 12, 1964).

Chapter Eighteen

1. "Impact!" editorial in *The New York Times,* August 2, 1964, p. E 1.

2. White House text as reprinted in "Johnson Thanks Jet Lab," *Christian Science Monitor,* August 3, 1964, p. 5.

3. Quotes cited in Marvin Miles, "Ranger 7 Gets Brilliant Moon Photos," *Los Angeles Times,* August 1, 1964, Part I, p. 1; Richard Witkin, "Craft Hits Target Area; 4,000 Pictures Sent Back," *The New York Times,* August 1, 1964, pp. 1, 8; Dave Swaim, "Pickering Jubilant Over Moon Photos," *Star News* [Pasadena], July 31, 1964, p. 1; also, JPL film, "Lunar Bridgehead," (JPL 571-2A), November 25, 1964.

4. Quotes cited in John Pomfret, "President Hails New Lunar Feat," *The New York Times,* August 1, 1964, pp. 1, 8.

5. Ranger VII Post-Impact Press Conference, held on July 31, 1964, as reprinted by the Office of Public Education and Information, California Institute of Technology, National Aeronautics and Space Administration, p. 1a (2-745).

6. Ibid., p. 8.

7. Ibid., p. 20, and pp. 22–23.

8. See *Current News,* prepared by the Office of Public Information, National Aeronautics and Space Administration, August 6, 1964 (2-889).

9.   *Foreign Media Reaction to Ranger 7* (R-107-64. United States Information Agency document, August 5, 1964) (2-895). "Internationally," Richard Witkin concluded, "the Ranger 7 flight was a propagandistic bonanza for the United States. In the space olympics it probably represented as impressive a 'first' as the orbiting of the first Sputnik and the first manned orbital flight by Yuri A. Gagarin." Richard Witkin, "Ranger Spurs Space Program," *The New York Times,* August 2, 1964, p. 10 E.

10.   When the euphoria had subsided somewhat, *Life* magazine dispassionately flipped the Ranger 7 coin to consider the other side: if another failure had been tallied by JPL, the journal suggested, instead of paeans, "vegetables would have rained down from the gallery as never before." Jim Hicks, "Many a Slip 'Twixt Earth and Moon—and Measles too," *Life,* August 14, 1964, p. 36A.

11.   Quotes cited in Tom Wicker, "Johnson Sees the Success as Justifying the Cost," *The New York Times,* August 2, 1964, p. 1; and Robert Thompson, "U.S. Must Push Space Race, Johnson Asserts," *Los Angeles Times,* August 2, 1964, Part I, p. 1.

12.   Text, as reprinted in "Johnson's Remarks to Space Scientists," *The New York Times,* August 2, 1964, p. 50.

13.   *Congressional Record* of the House of Representatives, August 11, 1964, pp. 18370–18373.

14.   NASA News Release, subject: "Presentation on Ranger VII to Members of Congress," August 5, 1964, pp. 2, 13, and passim (2-2428).

15.   Quoted in Robert C. Toth, "$5.3 Billion for Space Program OKd in Senate," *Los Angeles Times,* August 6, 1964, Part I, p. 5.

16.   Richard Witkin, "Apollo Officials to Study Pictures," *The New York Times,* August 2, 1964, pp. 2, 51.

17.   Quotes cited in Richard West, "Ranger Photos Show Hospitality of Moon," *Los Angeles Times,* August 5, 1964, Part II, p. 1; Harold D. Watkins, "Ranger Photos Boost Confidence in Apollo," *Aviation Week and Space Technology,* Vol. 81, August 10, 1964, pp. 19, 21.

18.   TWX from Harris Schurmeier to William Cunningham, August 18, 1964 (2-1877); TWX from Oran Nicks to Harris Schurmeier, September 4, 1964 (2-398).

19.   NASA memorandum from Samuel Phillips to Oran Nicks, subject: "Ranger 8 Aim Point," October 16, 1964 (2-2062).

20.   Letter from Joseph Shea to Harris Schurmeier, November 16, 1964 (2-2086).

21. Minutes of the Ranger 8 and 9 Target Selection Meeting, held at the Jet Propulsion Laboratory, November 19, 1964, prepared by Harris Schurmeier, November 25, 1964, pp. 1–5 (2-1894).

22. NASA memorandum from Homer Newell to George Mueller, subject: "Ranger 8 Lunar Target Selection," January 19, 1965 (2-1511).

23. NASA memorandum from George Mueller to Homer Newell, subject: "Ranger Lunar Target Selection," January 26, 1964 (2-2061).

24. Telegram from Homer Newell to Gerard Kuiper, February 1, 1965 (2-2355).

25. NASA memorandum from Oran Nicks to Homer Newell, subject: "Ranger 8 Target Selection Review and Final Recommendations" February 9, 1965 (2-1514b).

26. Letter from Harry Hess to George Mueller, February 2, 1965 (2-2356).

27. Letter from Donald Wise to Urner Liddel, February 24, 1965 (2-2028).

28. Quoted in "Back in Orbit," *Newsweek,* August 17, 1964, p. 51.

29. Quoted in "Photos of Moon Sent to World's Leaders," *Independent* [Pasadena], August 17, 1964, p. 1.

30. NASA News Release, transcript of the Interim Scientific Results Conference, Ranger VII, August 28, 1964 (2-173).

31. Ibid., p. 40.

32. Ibid., pp. 42–43.

33. Ibid., pp. 50–60.

34. Ibid., pp. 62, 64.

35. NASA memorandum from Oran Nicks to Homer Newell, subject: "Report on Ranger VII Presentation to International Astronomical Union (I.A.U.)," September 8, 1964 (2-399); see, for example, "Moon Photos Stir Varied Valuation," *The New York Times,* September 1, 1964, p. 15.

36. John Lear, "What the Moon Ranger Couldn't See," *Saturday Review,* Vol. 47, September 5, 1964, p. 38.

37. Gerard Kuiper, "Letters to the Editor," *Saturday Review,* October 10, 1964, p. 32.

38. Jerry Bishop, "New Look at Ranger 7," *The Wall Street Journal,* September 17, 1964, p. 16.

39. Robert Cowen, "Was Ranger Worth the Cost?" *Christian Science Monitor,* November 18, 1964, Editorial.

40. Walter Sullivan, "Surface of Moon Said to be Lava: Kuiper Says Ranger Photos Are Conclusive Evidence," *The New York Times,* October 1, 1964, p. 50.

41. University of Arizona memorandum from Gerard Kuiper to the Co-Experimenters, Ranger Program, December 1, 1964 (2-1897).

Chapter Nineteen

1. As reprinted in Robert C. Cowen, "Was Ranger Worth the Cost?—II," *Christian Science Monitor,* January 4, 1965, Editorial.

2. See "Hard Questions for Space Science" in Chapter Eighteen of this volume, and, for example, the letter from Harold Urey to Gerard Kuiper, January 6, 1965 (2-2018).

3. William Cunningham, NASA, *OSSA Review, January 12, 1965* (2-1505); JPL Interoffice Memo from Harris Schurmeier to Executive Council, subject: "Ranger Status Report for January 4, 1965, No. 195," January 7, 1965 (2-1315).

4. NASA memorandum from William Cunningham to Edgar Cortright, subject: "Ranger 8 Prelaunch Status Review at ETR," February 9, 1965 (2-1513).

5. Vrebalovich, who acted as official liaison between the experimenters and the project office, was most disappointed with the experimenters' decision in this instance. "I kept telling them—'look, that's not your prerogative, your expertise is: Should I do a maneuver to enhance science? If the project engineers say no, that's one thing, but for you guys to say no for them, I don't think that's right'." Interview of Thomas Vrebalovich by Cargill Hall, June 11, 1974, p. 27 (2-2465). The experimenters, for their part, contended that any pictures were better than the chance of none at all.

6. For Ranger 8 flight events, see JPL Ranger C Status Bulletins Nos. 1 through 5, February 17 through 20, 1965 (2-1992); *Ranger VII and IX: Part I. Mission Description and Performance* (JPL TR 32-800. Pasadena, California: Jet Propulsion Laboratory, California Institute of Technology, January 31, 1966); *Space Programs Summary No. 37-34, Volume VI* for the period May 1, 1965, to June 30, 1965 (Pasadena, California: Jet Propulsion Laboratory, California Institute of Technology, July 31, 1965), pp. 1–12; also Marvin Miles, "Probe Hits Moon: Ranger 8 Sends Back Photos," *Los Angeles Times,* February 20, 1965, Part I, p. 1.

7. Quoted in Dave Swaim, "Jet Lab Jubilant on Shot: Project Called Textbook Flight," *Star News* [Pasadena], February 20, 1965, p. 1.

8. Transcript of the Ranger VIII Post-Impact Press Conference, JPL Auditorium, February 20, 1965, prepared by the Office of Public Education and Information, Jet Propulsion Laboratory, California Institute of Technology, p. 6 (2-904).

9. Ibid., p. 7.

10. Ibid., pp. 8–10.

11. Ibid., pp. 14–15; see also news accounts; for example, Marvin Miles, "Ranger Photos Likely to Stir Up Controversy," *Los Angeles Times,* February 22, 1965, Part I, p. 1; and "Via Ranger, Rills, and Dimples: Ranger 8," *Newsweek,* Vol. 65, March 1, 1965, pp. 64–65.

12. Fredric C. Appel, "Ranger 8 Hits Target On Moon and Radios Back 7000 Pictures to Aid Search for Landing Sites," *The New York Times,* February 21, 1965, p. 1.

13. "The Ranger Success," editorial in *The New York Times,* February 22, 1965, p. 20. The new photos, *U.S. News and World Report* added a few weeks later, had clearly "raised more questions than they answered, and started a major controversy . . . over the safety of putting a man on the moon." "Still a Big Secret: Is the Moon a Mantrap?" *U.S. News and World Report,* Vol. 58, March 8, 1965, p. 33.

14. Los Angeles *Herald-Examiner,* February 21, 1965; see also, Walter Sullivan, "Science: The Lunar Surface," *The New York Times,* February 28, 1965, p. 6 E.

15. "Remarks Following a Briefing at the National Aeronautics and Space Administration, February 25, 1965," *Public Papers of the Presidents of the United States: Lyndon B. Johnson, 1965* [81] (Washington: Government Printing Office, 1966), Book I, pp. 214 and 216.

16. Harris M. Schurmeier, "Ranger D Target Selection," report to Homer Newell, presented March 10, 1965, pp. 1, 2 (2-856).

17. Ibid., pp. 3–5.

18. Letter from Raymond Heacock to William Cunningham, March 9, 1965 (2-2031).

19. NASA memorandum from Oran Nicks to Homer Newell, subject: "Ranger D Site Selection," March 10, 1965 (2-1517); and NASA memorandum from Homer Newell to Robert Seamans, subject: "Ranger 9 Lunar Aim Point Priorities," March 15, 1965 (2-1518).

20. JPL News Release No. 324, subject: "Ranger D Launch Scheduled March 21, 1965," March 15, 1965 (2-2418); and "Ranger Gets Priority by Day Over Gemini," *The Cocoa Tribune,* March 16, 1965, p. 1.

21. Walter Sullivan, "The Week in Science: A Russian Steps Into Space," *The New York Times,* March 21, 1965, p. E 3; also *Astronautics and Aeronautics, 1965* (NASA SP-4006. Washington: National Aeronautics and Space Administration, 1966), pp. 131–132.

22. Letter from Harold Urey to Homer Newell, June 19, 1961 (2-2291).

23. JPL News Release No. 327, March 23, 1965 (2-912); NASA News Release No. 65-96, subject: "Ranger IX To Send World's First Live Moon Photos," March 23, 1965 (2-2435).

24. At that moment in Washington, Newell and Cortright were presenting the fiscal year 1965 NASA budget before Karth's Committee. "We had set up a TV," Newell recollected, "and committee and witnesses paused to watch. It was a dramatic display to the committee of what JPL and NASA could do." Comments of Homer E. Newell written on draft manuscript of this volume, July 1, 1975.

25. For Mission details see *Ranger VIII and IX: Part I; Space Programs Summary No. 37-33, Volume I* for the period March 1, 1965, to April 20, 1965 (Pasadena, California: Jet Propulsion Laboratory, California Institute of Technology, May 31, 1965), pp. 2ff; Nicholas A. Renzetti, *Tracking and Data Acquisition for Ranger Missions 6–9* (JPL TM 33-275. Pasadena, California: Jet Propulsion Laboratory, California Institute of Technology, September 15, 1966), p. 101; *Space Programs Summary No. 37-33, Volume VI* for the period March 1, 1965, to April 30, 1965 (Pasadena, California: Jet Propulsion Laboratory, California Institute of Technology, May 31, 1965), pp. 1–2.

26. Transcript of the Ranger 9 Press Conference held March 24, 1965, pp. 2–3 (2-2034a).

27. Telegram From Earl Hilburn to William Pickering, March 24, 1965 (2-2056).

28. Transcript of the Ranger 9 Post-Impact Press Conference held March 24, 1965, pp. 16–17 (2-2034b).

29. Cf., for example, James P. Bennett, "It's Moon Bull's Eye: Millions Ride With Ranger Via TV," Los Angeles *Herald-Examiner,* March 24, 1965; and Gladwin Hill, "Ranger Hits Moon and Sends Photos Seen Live on TV," *The New York Times,* March 25, 1965, p. 1.

30. "The Moon 'Live'," editorial in *Christian Science Monitor,* March 26, 1965; and "Eyes on the Moon," editorial in *The New York Times,* March 25, 1965, p. 36.

31. Cynthia Lowry, "Television Glamorized Moon Face," *Independent* [Pasadena], March 26, 1965, p. A7.

32. Text as reprinted in "President Cheers Ranger's Success," *The New York Times,* March 25, 1965, p. 22.

33. Notes for James Webb's Cabinet Briefing on the Ranger Program (2-2436); and "Ranger Pinpointed Two Landing Areas on Moon," *Los Angeles Times,* March 26, 1965, Part I, p. 1.

34. See Chapters Two and Twelve of this volume.

35. See recommendations in *Final Report of the Ranger Board of Inquiry* (Washington: National Aeronautics and Space Administration, November 30, 1962) (2-2463); and NASA memorandum from Oran Nicks to Abe Silverstein, subject: "Analysis of JPL–Headquarters Relationships and Recommendations for Improvements," October 6, 1961 (2-332b).

36. NASA memorandum from Homer Newell to Earl Hilburn, subject: "Studies Relating to Management Effectiveness in Scheduling and Cost Estimating NASA Projects," November 6, 1964, p. 5 (2-1754); see also Vernon Van Dyke, *Pride and Power: The Rationale of the Space Program* (Urbana, Illinois: University of Illinois Press, 1964), p. 140.

37. See Appendix G of this volume.

38. Quoted in William S. Beller, "Lunar Surface Controversy Rekindled," *Missiles and Rockets,* Vol. 16, April 26, 1965, p. 16.

39. Most subsequent accounts overlook the original Ranger missions and relegate the entire project to a reconnaissance role supportive of Apollo. See, for example, Richard S. Lewis, *The Voyages of Apollo: The Exploration of the Moon* (New York: Quadrangle/The New York Times Book Co., 1975), pp. 46–48.

40. Cf., letter from Frederick Seitz to James Webb, October 30, 1964 (5-713a), and the attachment to this letter, "Statement of the Space Science Board of the National Academy of Sciences on National Goals in Space, 1971–1985," October 28, 1964 (5-713b); also, letter from James Webb to Lyndon Johnson, February 16, 1965, and "Summary Report Future Programs Task Group, January 1965," as reprinted in United States Congress, Senate, Committee on Aeronautical and Space Sciences, *NASA Authorization for Fiscal Year 1966,* Hearings before the Committee, 88th Congress, 1st Session on S. 927, 1965, pp. 1027–1028, and pp. 1029–1035. See also, remarks of Amrom Katz in Armin J. Deutsch and Wolfgang B. Klemperer (eds.), *Space Age Astronomy* (an International Symposium sponsored by Douglas Aircraft Corporation, Inc., August 7–9, 1961, at the California Institute of Technology, in conjunction with the 11th General Assembly of the International Astronomical Union. New York: Academic Press, 1963), p. 474; Homer Newell's remarks in United States Congress, House, Committee on Science and Astronautics, *Investigation of Project*

Ranger, Hearings before the Subcommittee on NASA Oversight, 88th Congress, 2nd Session, 1964, No. 3, p. 6; and Harold Urey, with a chemist's reservations, in United States Congress, House, Committee on Science and Astronautics, *1966 NASA Authorization*, Hearings before the Subcommittee on Space Science and Applications, 89th Congress, 1st Session, on H.R. 3730, 1965, No. 2, Part 3, p. 434; see also Albert R. Hibbs, "The Surface of the Moon," *Scientific American*, Vol. 216, No. 3, March 1967, p. 61. As Thornton Page summed the spaceborne potential: "I like to think of myself as a 'big telescope man'—ground based of course—but I have to admit that getting outside the atmosphere for a better look ... beats anything that the Palomar telescope now can offer." Thornton Page, "A View from the Outside," *Bulletin of the Atomic Scientists*, Vol. 25, No. 7, September 1969, p. 61.

41.   See *Ranger VII: Part II. Experimenters' Analyses and Interpretations* (JPL TR 32-700. Pasadena, California: Jet Propulsion Laboratory, California Institute of Technology, February 10, 1965); and *Ranger VIII and IX: Part II. Experimenters' Analyses and Interpretations* (JPL TR 32-800. Pasadena, California: Jet Propulsion Laboratory, California Institute of Technology, March 15, 1966); also, Raymond L. Heacock, "Ranger: Its Mission and Its Results," *Space Log* (TRW, Redondo Beach, California), Summer 1965, pp. 15–23 (2-121); letter from Gerard Kuiper to Cargill Hall, August 4, 1970 (2-2058); and letter from Eugene Shoemaker to Cargill Hall, July 22, 15 (2-2506); see Appendix A for the status of lunar theory before Ranger 7.

42.   See William L. Sjogren, et al., *Physical Constants as Determined from Radio Tracking of the Ranger Lunar Probes* (JPL TR 32-1057. Pasadena, California: Jet Propulsion Laboratory, California Institute of Technology, December 30, 1966).

43.   Cf., Robert J. Parks, "Flight Projects," *Space Research Activities of the Jet Propulsion Laboratory* (A summary of a presentation to members of the Subcommittee on Space Sciences and Applications of the Committee on Science and Astronautics, United States House of Representatives, at Pasadena, California, March 23, 24, 1967. Pasadena, California: Jet Propulsion Laboratory, California Institute of Technology, 1967), p. 14 (2-11); *Ranger TV Subsystem (Block III) Final Report* (Astro-Electronics Division, Radio Corporation of America: Princeton, New Jersey, July 22, 1965), Volume I: "Summary," p. 77 (2-960); William H. Pickering, "Ranger—On the Upward Trail," *Astronautics and Aeronautics*, Vol. 3, January 1965, p. 20; statement of Homer Newell in United States Congress, Senate, Committee on Aeronautical and Space Sciences, *NASA Authorization for Fiscal Year 1964*, Hearings before the Committee, 88th Congress, 1st Session, on S. 1245, 1963, Part I: "Scientific and Technical Programs,"

p. 261; U.S. Congress, *1966 NASA Authorization,* p. 209; "Contributions of the Ranger Program," draft report from Homer Newell's files (2-2325); letter from Harris Schurmeier to William Cunningham, with attachment describing Ranger "uniqueness" and "state of the art improvements," October 5, 1964 (2-1882a&b); and William H. Pickering, "Systems Engineering at the Jet Propulsion Laboratory," Caltech Lecture Series: Systems Concepts for the Private and Public Sectors, April 27, 1971 (3-468).

44. U.S. Congress, *1966 NASA Authorization,* p. 470.

45. Letter from Hugh Bradner to Cargill Hall, September 17, 1975 (2-2507). For details, see Frank Lehner, "Brief Description of the Ranger Lunar Seismograph," *Proceedings of the IRE,* Vol. 50, 1962, p. 2297; Hugh Bradner, "Instrumenting the Sea Floor," *IEEE Spectrum,* November 1964, pp. 108–114; and Hugh Bradner, "Seismic Measurements on the Ocean Bottom: New Instruments are Used to Study Earth's Crustal Structure and Seismic Background," *Science,* Vol. 146, 1964, pp. 308–316; also Bradner, Dejerphanion, and Langlois, "Ocean Microseism Measurements with a Neutral Bouyancy Free-Floating Midwater Seismometer," *Bulletin of the Seismological Society of America,* Vol. 60, 1970, pp. 1139–1150.

46. Perfected by Robert Nathan, a JPL scientist with a PhD from Caltech in crystallography, it is first described in *Space Programs Summary No. 37-12, Volume I* for the period September 1, 1961, to November 1, 1961 (Pasadena, California: Jet Propulsion Laboratory, California Institute of Technology, December 1, 1961), p. 3.

47. R. H. Selzer, *Digital Computer Processing of X-Ray Photographs* (JPL TR 32-1028. Pasadena, California: Jet Propulsion Laboratory, California Institute of Technology, November 15, 1966). Technical details in Robert Nathan, *Digital Video-Data Handling* (JPL TR 32-877. Pasadena, California: Jet Propulsion Laboratory, California Institute of Technology, January 5, 1966); *Space Research Activities of the Jet Propulsion Laboratory,* p. 65 (2-11); Thomas C. Rindfleisch, "Getting More Out of Ranger Pictures by Computer," *Astronautics and Aeronautics,* January 1969, p. 70; see also, Homer Newell's statement in United States Congress, House, Committee on Science and Astronautics, *1967 NASA Authorization,* Hearings before the Subcommittee on Space Science and Application, 89th Congress, 2nd Session, on H.R. 12718, 1966, No. 4, Part 3, pp. 828–833; W. J. Hardy, "Motion Picture Script for Analyses and Summary of the Ranger VII, VIII, and IX Television Pictures," June 21, 1967 (2-2371); background in Kimmis Hendrick, "Photo Technique: Moon 'Brighter'," *Christian Science Monitor,* March 24, 1965, p. 4; Azriel Rosenfeld, *Picture Processing by Computer,* (New York: Academic Press, 1969); and interview of Vrebalovich by Hall, June 11, 1974, pp. 16–24 (2-2465).

INDEX

Abbott, Ira, 121
Abelson, Philip, 201-202
Able Lunar Orbiter, 14 (photo)
Advanced Research Projects Agency, 5, 6, 10
Aeronutronic Div., Ford Motor Co., 61, 115, 124-125, 138, 159, 203, 218
Agena B, 96 (diagram)
Agena B Coordination Board, 32-33, 43, 45, 64, 66, 91-93
Agena Failure Investigating Board, 109
Agena Survey Team, 25-26, 30-31
Air Force Ballistic Missiles Div., 31-34, 63-64
Air Force Space Systems Div., 109, 190-191
Albert, John E., 34, 35 (photo), 99, 106, 110, 169 (photo)
Alcorn, Deloyce, 206
Alexander, W. M., 54, 134
Allenby, J. R., 220
Alphonsus crater, 298-300, 302, 303-304 (photos)
Alter, Dinsmore, 12, 299
Anderson, Clinton P., 249
Anderson, Ernest C., 78
Anderson, H. R., 54, 134
Apollo Project, 112, 114, 115, 118, 120, 122-123, 156-159, 162-163, 177, 182, 189, 191-192, 199-200, 201-203, 206-207, 210, 217, 232, 279-285, 288, 290, 296, 309
Arnold, James R., 11, 15, 54, 78-79, 146, 149, 210, 216
ARPA. See Advanced Research Projects Agency
Atlas, 94-95
Atlas-Able, 13
  mission summary, 325
Atlas-Agena B, 23, 25, 63-67, 69, 74, 79, 94-95, 106, 109-110, 127, 139, 142-143, 161 (photo), 187-191
Atlas-Vega, 15, 16 (photo), 18, 20, 23
Atomic Energy Commission, 74
Bader, M., 134
Bailey, Frederick J., 173
Beckman, Arnold O., 249
Berkner, Lloyd, 12, 49
Betts, A. W., 75
Bourdeau, R. E., 134
Boyle, Marjorie J., 192
Bratenahl, Alexander, 260-261
Bronk, Detlev, W., 71
Brown, Harrison, 11, 15
Brown, Herschel J., 39

Brown, Walter E., 79, 115, 210, 216
Buckley, Edmond C., 87, 121
Burke, James D., 28-9, 30 (photo), 35 (photo), 42, 45-46, 57, 63-67, 70, 73-76, 79, 99, 101, 103, 105-106, 108-110, 116, 119, 123, 125, 127, 129-133, 135, 145, 148-149, 152-153, 159, 167, 169 (photo), 172, 176, 178, 181, 214-215, 226, 307-308
California Institute of Technology, 3, 52, 160, 205, 240, 249, 251-254
Campen, Charles, 223, 225-226, 261
Casani, John, 46
Case, Clifford, 192
Chubb, T. A., 54, 105, 111
Clark, John E., 6
Clark, John F., 75, 136, 283
Clarke, Victor, 139
Cole, Charles, 58, 132
Collins, Foster, 20
Comuntzis, Marc, 46
Continental Devices, 197
Coon, Kenneth C., 215, 218
Corporal missile, 4
Cortright, Edgar, 38 (photo), 75-77, 110, 113, 115-116, 122, 127, 130, 159-160, 180, 187, 203-206, 209-210, 221, 230, 236, 244, 251, 253, 262, 267 (photo), 290, 292
Coughlin, William J., 240, 254
Cousins, Norman, 286
Cummings, Clifford I., 24, 28, 29 (photo), 36, 43, 53, 57, 64-65, 73, 76-77, 106, 115-116, 118-120, 126-127, 159, 176
Cunningham, N. William, 122, 123 (photo), 130-133, 135-136, 152, 169 (photo), 173, 187, 190, 196, 199, 206-207, 210, 220, 225, 230-231, 234, 237, 244, 246-247, 257, 290, 298
Dangle, Eugene, 244
Davis, Leighton I., 91
Debus, Kurt H., 91-92, 105-106, 169 (photo)
Deep Space Instrumentation Facility. See Deep Space Network
Deep Space Network, 82-88, 99, 103-105, 108, 142, 150, 153, 233, 309-310, 314-315
  Deep Space Instrumentation Facility, 82
  DSN-spacecraft link, 87 (diagram)
  Goldstone antenna, 86 (photo)
  Johannesburg antenna, 100 (photo)
  station locations, 85
Denison, Frank G., 61, 64, 124-125

Dickinson, Bert, 269
Discoverer, 25, 39
Dobies, Edwin, 128
Donlan, Charles, 157
Downhower, Walter, 46, 236–238, 241
Dryden, Hugh L., 20, 21 (photo), 93, 114, 189, 246
DuBridge, Lee A., 4, 5, 160, 201–202, 249
Duerr, Friedrich, 32, 33 (photo), 169 (photo), 188
Duncan, Donal B., 125
Eichel, Henry H., 109–110
Eimer, Manfred, 132–133
Eisenhower, Dwight D., 3
Ewing, Maurice, 12, 54, 78
Explorer satellites, 4, 8, 94
Faget, Maxime, 262
Farley, T. A., 134
Fleming, William A., 32
Foster, John, 173
Fullbright, J. William, 280
Fulton, James, 113, 160
Gagarin, Yuri, 112
Garbarini, Robert, 262
Gemini Project, 299, 305
General Dynamics-Astronautics, 33, 63, 94, 139, 143, 189, 191, 260
Giberson, Eugene, 57, 159
Gilruth, Robert R., 201, 281
Glenn, John, 154
Glennan, T. Keith, 20, 21 (photo), 23, 43–5, 49, 65, 72
Goddard Space Flight Center, 31, 76, 191
Goett, Harry, 31, 43
Gold, Thomas, 12, 285, 309
Goldwater, Barry, 217, 271
Gray, Gordon, 20
Grumman Aircraft Corp., 192
Guggenheim Aeronautical Laboratory, 3
Harr, Karl G., Jr., 20
Hastrup, Rolph, 164–166
Heacock, Raymond, 128, 223, 226, 230 (photo), 274 (photo), 284, 292 (photo), 298, 300
Heppner, J. P., 54, 101, 105–106
Hercules Powder Co., 62
Hess, Harry, 283
Heyser, Richard, 236–237
Hibbs, Albert R., 49, 52, 74, 76, 101, 106, 128–129
Hilburn, Earl D., 244, 245 (photo), 247, 301
Himmel, Seymour C., 189–191
Hobby, George L., 73, 164
Holmes, D. Brainerd, 121–122, 156, 158–159, 163, 191–192, 201, 203, 217
Hornbeck, John, 173
Horner, Richard E., 20, 21 (photo), 31, 44, 113, 204

Howard, V. William, 212, 213 (photo), 218, 221
Hueter, Hans, 32, 42, 99, 106, 110, 188
Humphrey, Hubert H., 297 (photo), 306 (photo)
Huntsville. See Marshall Space Flight Center
Hyatt, Abraham, 114
IGY. See International Geophysical Year
image enhancement technique, 310–311 (photo)
Internatinal Astronomical Union, 285
International Council of Scientific Unions, 72
International Geophysical Year, 6, 12
Jakobowski, Walter, 122, 135, 173, 230–231, 244, 246
James, Jack, 208
Jastrow, Robert, 15, 20, 52
Jet Propulsion Laboratory, 1–2
    five-year plan for space exploration, 17–18
    organization of, 28, 34–37, 176, 180
    Pioneer lunar probe, 9 (photo)
    relationship with Lockheed, 39–43, 64–66
    relationship with NASA, 31–32, 34–39, 43–45, 130–134, 160–163, 249–255
    space flight operations, 88–93
    Space Flight Operations Facility, 90–91, 233, 235, 262–263 (photo), 267 (photo)
    spirit of, 174
Johnson, Lyndon B., 113, 217, 221, 271, 279 (photo), 280, 284, 296, 297 (photo), 305, 306 (photo)
Johnson, Marshall S., 89–90, 103, 145, 149
Johnson, Roy, 6
JPL. See Jet Propulsion Laboratory
Juno IV, 46–47
Kaplan, Joseph, 12
Karth, Joseph E., 156, 180, 203, 205, 210, 242, 252–254, 280
Kautz, Gordon P., 36, 37 (photo), 116, 127, 145, 179, 185–186, 215, 257–258, 264, 268
Kelley, Albert J., 42, 173–174
Kellogg, William, 12
Kendall, James, Jr., 260
Kennedy, John F., 113, 217–218
Khrushchev, Nikita, 154
Kindt, Donald, 193, 195, 241, 256
Kirhofer, William, 139
Kistiakowsky, George B., 20
Koppenhaver, James, 173
Kozyrev, Nikolay, 299
Kuiper, Gerard P., 12, 79, 128, 200, 216, 219, 225–226, 227 (photo), 233, 237, 262, 266, 273, 274 (photo), 280–281, 284–285, 286 (photo), 287, 289–290, 292 (photo), 301, 302 (photo)
LaGow, Herman, 173, 243, 262
Langley Research Center, 157, 209–210
Larkin, Walter, 103, 142
launch operations, explanation of, 91–92

launch vehicles. See vehicle name
Lear, John, 285-287
Lederberg, Joshua, 72, 202
Lewis Research Center, 189-190
Libby, Willard, 202
Liepmann, Hans, 260
Lingle, Walter, 244
Lockheed Missile and Space Div., 31, 39-43, 63, 65, 76, 95, 109-110, 139, 189, 191
Los Alamos Scientific Laboratory, 74-77
Low, George, 158, 192, 201
Ludwig, George, 10
Luedecke, Alvin R., 254
Luna 1, 10, 13, 18-19 (photo)
Luna 2, 20
Luna 3, 10, 20, 22 (photo)
Luna 4, 191
lunar flights, summary, 325-326
Lunar Orbiter Project, 209-210, 220, 246. See also Able Lunar Orbiter
lunar theory, 317-322
Luskin, Harold T., 39, 41 (photo), 42, 106, 110
Mackin, Robert, 260
Manned Orbiting Laboratory, 221
Manned Spacecraft Center, 120, 201, 232, 280-282
Mare Cognitum, 285, 290
Mare Tranquillitatis. See Sea of Tranquility
Margraf, Harry, 190
Mariner space probes
    Mariner C, 257
    Mariner R, 122, 135-136, 160, 161 (photo), 171-172
    Mariner 2, 160, 167, 177, 300
    Mariner 4, 301
Marshall Space Flight Center, 30-32, 34, 41-42, 44-45, 91, 110, 188-189
Martin, Earl, 99
McDonald, W. S., 134
McElroy, Neil, 6
McNamara, Robert, 221
Medaris, John B., 46
Meghreblian, Robert V., 214, 223
Mercury Project, 112-113
Merrick, William, 84
Metzger, Albert E., 54, 78, 146
Miller, Bernard P., 180, 186, 195, 256
Miller, George P., 242, 249, 252, 273, 280
Millikan, Robert A., 4
Moon. See lunar theory
Morris, Brooks T., 180
Mueller, George E., 282
Mulladay, J., 169 (photo)
NACA. See National Advisory Committee for Aeronautics
NASA. See National Aeronautics and Space Administration
National Academy of Sciences, 71-72

Space Science Board, 12, 51, 113, 162, 207, 283
National Advisory Committee for Aeronautics, 12, 20
National Aeronautics and Space Administration
    launch operations, control of, 19-191
    Lunar and Planetary Programs Div., 38, 113, 122, 127-128, 131, 201
    lunar program objectives, 23, 25-26, 160-163
    lunar program origins, 15, 17, 20, 23
    Office of Launch Vehicle Programs, 29-31, 91, 121
    Office of Manned Space Flight, 121-122, 158-160, 162-163, 192, 200-201, 281-282
    Office of Space Flight Programs, 27, 30-31, 49, 70, 87, 92, 106, 121
    Office of Space Science and Applications, 218, 220-221, 230, 234, 240, 243, 249-251, 254, 281
    Office of Space Sciences, 121-123, 127-130, 132, 157-159, 162-163, 173-174, 198-204, 208-210, 217
    organization of, 14-15, 38, 91-92, 120-121, 187-189, 218
    project management, 26-28, 31-32, 442-45, 91-92, 122-123, 188-198, 306-307
    realignment for Apollo, 120-122
    relationship with Air Force, 32-34, 39, 91-93, 187-191
    relationship with JPL, 31-32, 34-39, 43-45, 130-134, 160-163, 249-255
    scientific experiment selection policy, 73-80
    Space Exploration Program Council, 31, 43
    Space Sciences Steering Committee, 73, 78-79, 128-133, 135-137, 177, 207, 215, 219, 282
    spacecraft sterilization policy, 72
    Working Group on Lunar Exploration, 15, 17-18, 20
National Science Foundation, 71
Naugle, John, 268
Naval Research Laboratory, 14
Neher, H. Victor, 4, 54, 134
Neugebauer, M. M., 54
Newell, Homer, E., 12, 14, 18, 20, 23, 49, 51 (photo), 52-53, 73-75, 90-91, 121-122, 128-130, 133, 156-158, 163, 173, 180-181, 198-199, 201-208, 210, 217-218, 220-221, 224, 226, 230, 236, 239 (photo), 244, 246-247, 249, 262, 267 (photo), 271, 272 (photo), 278, 279 (photo), 280, 283, 298, 308-309
Nichols, George, 264, 268-269, 290
Nicks, Oran W., 70, 71 (photo), 97, 105-106, 113, 115-116, 122, 125, 127-130, 132-133, 160, 162-163, 173, 187, 201, 203, 207, 210, 215, 217, 220, 225, 230, 244, 249, 251, 257, 262, 281, 284-285, 298, 300

Northrop Corp., 204–205, 209, 212–215, 218, 221
O'Day, Marcus, 12
Odishaw, Hugh, 12
Ostrander, Donald R., 29, 31, 34, 91, 121
Pace, Robert, 34
Parks, Robert, 177, 178 (photo), 197, 209–210, 235, 236
Pauling, Linus, 201
Phillips, Samuel C., 281
photograph enhancement by computers. See image enhancement
Pickering, William H., 3, 4, 5 (photo), 23, 28, 31, 34, 43, 46, 53, 65, 106, 127, 133, 150, 154, 157, 159, 162–163, 165, 171, 177, 180, 197, 203–205, 226, 229, 233–234, 237, 239 (photo), 241–242, 249, 251, 253, 266, 267 (photo), 271, 272 (photo), 278, 279 (photo), 290, 301
Pieper, G. F., 134
Pioneer lunar probes
  mission summary, 325
  Pioneer 0, 7
  Pioneer 1, 8
  Pioneer 2, 8
  Pioneer 3, 8, 10
  Pioneer 4, 10
  spacecraft, 7 (photo), 9 (photo)
Piroumian, Maurice, 259
planetary science, 11–12, 14–15, 78–79, 128, 181–182, 207 210, 216, 218, 284–287, 309
President's Science Advisory Committee, 12, 217
Press, Frank, 12, 15, 54, 78, 216
project management. See National Aeronautics and Space Administration
Project Paperclip, 32
Radio Corp. of America, 78–79, 115–116, 186, 193–195, 223, 256–259
Ranger Project
  analysis of, 306–310
  Block I (see also Rangers 1 and 2)
    mission objectives, 56, 94, 329
    proof test model, 68 (photo)
    scientific experiments in, 77 (photo), 335
    spacecraft design, 46–48, 53–60
    technical details of, 329–331
    test flight preparation, 67–71
    trajectory diagram, 95
  Block II (see also Rangers 3, 4, and 5)
    maneuver sequence, midcourse, 146 (diagram)
    maneuver sequence, terminal, 148 (diagram)
    mission objectives, 57, 138, 329
    scientific experiments in, 62, 80 (diagram), 125, 336
    superstructure design, 62 (diagram)

technical details of, 329–331
weight reduction in, 63–67
Block III (see also Rangers 6, 7, 8, and 9)
  hardware delivery schedule, 131
  initiation of, 114–116
  launch schedule, 118, 197–198
  mission objectives, 115–116, 165, 179, 206, 329
  Project Development Plan, 128–129
  scientific experiments in, 134, 136 (diagram), 337
    organization of, 223–228, 341–344
  spacecraft development
    diagrams of, 187–188
    preliminary design, 118–119 (photo)
    redesign, 185–187
  technical details, 329–331
  television camera sequencing, 117–118
  television subsystem development, 186, 193–196, 256–258
  terminal maneuver, 291 (diagram)
Block IV
  cancellation of, 210
  initiation of, 202–203
  mission objectives, 206–208
  scientific experiments in, 338
Block V, 212–221
  cancellation of, 221
  initiation of, 202
  mission objectives, 214, 219
  scientific experiments in, 216, 338
Design Review Board, 179, 185, 192
financial history, 351
JPL investigation, 171–176
mission objectives, all flights, 329
naming of 23–24
NASA board of inquiry, 173–176
organization chart, 40
performance history, 355
priorities, 157
Project Development Plan, 45, 128–128
reorganization, 43, 176–177
schedule history, 347
scientific experiments for, 49, 51–54, 73–80, 128–137, 216, 335–338
spacecraft
  design, 48 (diagram), 54–57, 58 (diagram), 59–62
  packaging, 55 (photo), 56 (diagram)
  redesign of, 185–187
  sterilization of, 71–73, 124–127, 164–166, 174–177
  subcontractor selection, 203–205
  technical details, 329–331
  test procedures, 67–69, 245, 248, 250–251, 253, 261–262
Ranger 1, 94–104
  flight, 101–103

launch of, 102 (photo)
postflight analysis, 104
scientific experiments on, 335
spacecraft design, 58 (diagram)
systems testing of, 69 (photo)
trajectory diagram, 95
Ranger 2, 105–111
countdown for, 107 (photo)
flight of, 107–109
postflight analysis, 109–111
scientific experiments on, 335
spacecraft design, 58 (diagram)
trajectory diagram, 95
Ranger 3, 138–150
flight of, 143–147
launch constraints, 140 (diagram)
launch of, 144 (photo)
preparation for flight, 141 (photo)
postflight analysis, 147–150
scientific experiments on, 336
sterilization of, 125–126
superstructure design, 62 (diagram)
Ranger 4, 150–155
flight of, 152–154
launch preparation, 151 (photo)
postflight analysis, 163–164
scientific experiments on, 336
superstructure design, 62 (diagram)
Ranger 5, 163–170
flight of, 167–170
launch of, 168 (photo)
postflight analysis, 171–176
scientific experiments on, 336
sterilization of, 164–165
superstructure design, 62 (diagram)
Ranger 6, 223–255
assembly, 194 (photo)
failure investigation
    Congressional investigation, 252–254
    JPL–RCA board of inquiry, 241–243,
        258–261
    NASA review board, 243–252
flight of, 235–237
launch of, 235
lunar target for, 232–233
postflight analysis, 258–261
postflight press conference, 239 (photo)
scientific experiments on, 134, 337
television subsystem, 193–195, 196
    (photo), 197
Ranger 7, 256–280
flight of, 264–270
launch of, 265 (photo)
lunar photographs, 275–277
scientific experiments on, 134, 337
Ranger 8, 282–296
flight of, 289–290
lunar photographs, 293–295

lunar targets for, 282–283
scientific experiments on, 134, 337
Ranger 9, 296–306
flight of, 299–300
lunar photographs, 303–304
lunar targets for, 297
scientific experiments on, 134, 337
RCA. See Radio Corp. of America
Rechtin, Eberhardt, 57, 82, 83 (photo), 142,
    233
Red Socks Project, 5
Redstone Arsenal, 4
research sources, 312–314
Richter, Henry, 82
Rindfleisch, Thomas, 233
Ritland, Osmond J., 34, 109
Rittenhouse, J. B., 101, 106
Roadman, Charles, 121
Robillard, Geoffrey, 57, 215, 216 (photo), 218–
    219
Rocketdyne Div., North American Aviation, 95
Rohr Aircraft, 62
Rudolph, Arthur H., 173
Russia. See Soviet Union
Ryan Aeronautical Co., 62
Rygh, Patrick, 164, 165 (photo), 169, 180, 193,
    235–237, 267, 269
Sampson, William, 82
Sanders, N., 134
Sandia Corp., 74–77
Schienle, Donald R., 192
Schilling, Gerhardt, 74–75
Schneiderman, Daniel, 46–47, 54
Schurmeier, Harris M., 36, 56, 66, 89, 99, 116,
    176–178, 179 (photo), 180, 185–186, 189–
    192, 197, 203, 206, 208–210, 214–215,
    225–226, 234, 236–237, 239 (photo), 256–
    257, 260–261, 264, 266–267, 272, 281,
    290, 306 (photo)
science. See space science
Sea of Clouds, 262, 266, 275–277 (photos)
Sea of Tranquility, 233, 290, 293–295 (photos)
Seamans, Robert C., Jr., 44, 73, 93, 114–115,
    120–121, 126, 163, 189, 203–205, 210,
    243, 249, 251, 299, 305
Serbu, G. P., 134
Sergeant missile, 4
Shea, Joseph, 158, 163, 281
Shoemaker, Eugene M., 12, 79, 128, 163, 200,
    226, 228 (photo), 274, 284–285, 286
    (photo), 292, 301 (photo), 302 (photo)
Siepert, Albert, 44
Silverstein, Abraham, 14, 18, 20, 23, 27
    (photo), 28, 31, 38, 49, 53, 57, 72–73, 75,
    92, 106, 114, 121, 189
Simpson, George E., 267, 273
Simpson, J. A., 54
Singer, S. Fred, 12

sky science, 11–12, 51–52, 76–77, 100–101, 108–109, 114, 129–135, 199–200, 207
Small, John, 46
Smith, E. J., 134
Smith, Francis B., 173, 243, 262
Snyder, C. W., 54
Sonett, Charles P., 12, 123, 130, 132-133
Soviet Union, 1
  lunar program of, 10, 13, 18, 20, 23, 191
  mission summary, 236
  space race with, 23–24, 65–66, 70, 112–114, 138, 154, 181, 299, 307–309
Space Exploration Program Council. See National Aeronautics and Space Administration
space flight operations, explanation of, 88–91
Space Flight Operations Facility. See Jet Propulsion Laboratory
space race. See Soviet Union
space science, 11, 15, 73, 112–113, 122–123, 128–133, 156–157, 165, 181–182, 199, 201–202, 208, 212, 221–222, 284–288, 302, 308–309. See also sky science and planetary science.
Space Science Board. See National Academy of Sciences
Space Sciences Steering Committee. See National Aeronautics and Space Administration
Space Technology Laboratories, 6, 7 (Pioneer photo), 13, 65–67
Sparks, Brian, 76, 131, 135, 167, 176, 203
Spilhaus, Athelstan, 12
Spitzer, Lyman, 12, 202
Sputnik 1, 3, 10
sterilization. See spacecraft sterilization under Ranger Project
Stevens, Robertson, 82, 84
STL. See Space Technology Laboratories
Sullivan, Charles, 20
Suomi, Verner, 12
Surveyor Project, 79, 114–115, 128, 220, 246, 282
Thomas, Albert, 113, 160

Thor-Able, 7
Tiros satellite, 78
Townsend, John, 12
Urey, Harold C., 11, 15, 52, 79, 128, 181, 200, 202, 207, 221, 224, 226, 229 (photo), 284, 301 (photo), 302 (photo)
U.S.S.R. See Soviet Union
Van Allen, James, 4, 10, 12, 49, 52, 54, 202
Van Allen radiation belts, 8–10
Van Dilla, Marvin A., 54
Vega Project, 23–24, 47, 49–53
  launch schedules, 18
  spacecraft model, 50 (photo)
Vela Hotel, 74–77, 111
Venera 1, 65, 66 (diagram)
Victor, Walter, 82
Von Braun, Wernher, 4, 31–32, 43, 188
Voskhod 2, 299
Vrebalovich, Thomas, 223, 224 (photo), 232, 290
WAC Corporal rocket, 4
Wagner, Harry, 79
Walker, John M., 173
Washburn, Harold W., 141, 149, 223
Weaver, Warren, 201–202
Webb, James E., 87, 93, 113–114, 120, 121 (photo), 152–154, 156, 189–190, 202, 217, 221, 249, 252–254, 297 (photo), 306 (photo)
Wexler, Harry, 12
Whipple, Fred, 12
Whitaker, Ewen A., 226, 231 (photo), 274 (photo), 292 (photo), 301 (photo)
Wiesner, Jerome, 112
Willingham, Donald E., 233
Wise, Donald, 283
Witkin, Richard, 266
Wolfe, Allen E., 116, 117 (photo), 125, 127, 135, 145, 176, 179, 185, 192, 206, 264
Working Group on Lunar Exploration. See National Aeronautics and Space Administration
Wyatt, D. D., 160
Young, Jack, 44

THE AUTHOR

R. Cargill Hall is the Historian for the Jet Propulsion Laboratory. He came there in 1967 from Lockheed Missiles & Space Company, where he was an operations research analyst and a historian. He has published *Project Ranger: A Chronology* and served as Coordinator-Editor of the "Annual Chronology of Astronautical Events" of the International Academy of Astronautics. He also edited the two-volume *Essays on the History of Rocketry and Astronautics: Proceedings of the Third Through the Sixth History Symposia of the International Academy of Astronautics.*

Mr. Hall received his B.A. degree in political science from Whitman College in 1959 and his M.A. in political science and international relations from San Jose College in 1966. He was awarded the Goddard Historical Essay Trophy by the National Rocket Club in 1962 and 1963 and was named one of the Outstanding Young Men of America in 1968. Articles of his have appeared in the *American Journal of International Law, Journal of Air Law and Commerce, The Airpower Historian,* and *Technology and Culture.*